Learning to Teach Art and Design in the Secondary School

Learning to Teach Art and Design in the Secondary School proposes that making art, craft and design is a useful, critical and potentially transforming process, and therefore fundamental to a plural society. It offers a conceptual and practical framework for understanding the diverse nature of art and design in education at KS3 and the 14–19 curriculum. It provides support and guidance for learning and teaching in art and design by examining and evaluating the theoretical basis for different approaches to learning and their application to classroom practice.

With reference to current debates, this fully updated second edition explores a range of approaches to teaching and learning; it raises issues, questions orthodoxies and identifies new directions. It also contains a number of practical tasks for student teachers to undertake on their own or in groups, and extensive lists of further reading. The chapters examine:

- ways of learning;
- planning and resourcing;
- attitudes to making;
- critical studies;
- values in a plural curriculum.

The book is designed to provide underpinning theory and address practical issues for student teachers on PGCE and initial teacher education courses in Art and Design. It will also be of relevance and value to teachers in school with designated responsibility for supervision of Art and Design student teachers, HE-based tutors and NQTs during their induction year.

Nicholas Addison and **Lesley Burgess** are Senior Lecturers in art, design and museology at the Institute of Education, University of London.

Related titles

Learning to Teach Subjects in the Secondary School Series

Series Editors
Susan Capel, Marilyn Leask and Tony Turner

Designed for all students learning to teach in secondary schools, and particularly those on school-based initial teacher training courses, the books in this series complement *Learning to Teach in the Secondary School* and its companion, *Starting to Teach in the Secondary School*. Each book in the series applies underpinning theory and addresses practical issues to support students in school and in the training institution in learning how to teach a particular subject.

Learning to Teach English in the Secondary School, 2nd Edition
Jon Davison and Jane Dowson

Learning to Teach Modern Foreign Languages in the Secondary School, 2nd Edition
Norbert Pachler and Kit Field

Learning to Teach History in the Secondary School, 2nd Edition
Terry Haydn, James Arthur and Martin Hunt

Learning to Teach Physical Education in the Secondary School, 2nd Edition
Edited by Susan Capel

Learning to Teach Science in the Secondary School, 2nd Edition
Jenny Frost and Tony Turner

Learning to Teach Mathematics in the Secondary School, 2nd Edition
Edited by Sue Johnston-Wilder, Peter Johnston-Wilder, David Pimm and John Westwell

Learning to Teach Religious Education in the Secondary School
Edited by Andrew Wright and Ann-Marie Brandom

Learning to Teach Geography in the Secondary School
David Lambert and David Balderstone

Learning to Teach Design and Technology in the Secondary School
Edited by Gwyneth Owen-Jackson

Learning to Teach Music in the Secondary School
Edited by Chris Philpott

Learning to Teach ICT in the Secondary School
Edited by Steve Kennewell, John Parkinson and Howard Tanner

Learning to Teach Citizenship in the Secondary School
Edited by Liam Gearon

Starting to Teach in the Secondary School, 2nd Edition
Edited by Susan Capel, Ruth Heilbronn, Marilyn Leask and Tony Turner

Learning to Teach in the Secondary School, 4th Edition
Edited by Susan Capel, Marilyn Leask and Tony Turner

Learning to Teach Using ICT in the Secondary School, 2nd Edition
Edited by Marilyn Leask and Norbert Pachler

Learning to Teach Art and Design in the Secondary School

Second edition

A companion to school experience

Edited by Nicholas Addison
and Lesley Burgess

 Routledge
Taylor & Francis Group

LONDON AND NEW YORK

First published 2000
by RoutledgeFalmer

This edition published 2007
by Routledge
2 Park Square, Milton Park, Abingdon, Oxon, OX14 4RN

Simultaneously published in the USA and Canada
by Routledge
270 Madison Avenue, New York, NY 10016

Routledge is an imprint of the Taylor and Francis Group, an informa business

Typeset in Bembo by RefineCatch Limited, Bungay, Suffolk
Printed and bound in Great Britain
by TJ International Ltd, Padstow, Cornwall

British Library Cataloguing in Publication Data
A catalogue record for this book is available from the British Library

Library of Congress Cataloging in Publication Data
A catalog record for this book has been requested

ISBN 978–0–415–37773–7 (pbk)
ISBN 978–0–203–96246–6 (ebk)

Contents

Illustrations

PLATES

All plates appear in the colour plate section between pp. 168 and 169

Tasks

The Task Number identifies the Chapter (first number) and Unit (second number)

Notes on Contributors

Nicholas Addison is a Senior Lecturer in Art, Design and Museology at the Institute of Education, University of London and is Course Leader for the MA Art and Design in Education. He supervises doctoral students, is a tutor on the PGCE Art and Design secondary course and co-ordinator for the Primary PGCE Art and Design specialism. For 16 years he taught art and design and art history at a comprehensive school and a sixth form college in London. He has lectured in art history on BA courses and was Chair of the Association of Art Historians Schools Group (1998–2003). He has co-edited with Burgess, L. *Issues in Art & Design Teaching* (2003) also published by Routledge Falmer. His research interests include: critical studies, intercultural education, sexualities and identity politics. He directed the AHRB-funded research project, 'Art Critics and Art Historians in Schools' (2003) and co-directed with Lesley Burgess 'En-quire: Critical Minds' (2005–6) evaluating the significance of learning in galleries on contemporary art.

Andy Ash is a Lecturer in Art, Design & Museology at the Institute of Education, University of London and between 2000 and 2007 was the Course Director of PGCE secondary. Before moving to the Institute of Education he taught on the PGCE courses at Reading University and Homerton College Cambridge. Other teaching commitments have included lecturing on Foundation, GAD, GNVQ, and AL courses at The Henley College, Henley-on-Thames. Prior to this he taught Art and Design, CDT and Photography in Berkshire secondary schools. His research interests are in the field of 3D and sculpture and his PhD research is entitled *An Investigation into the Teaching of Sculpture at KS3*.

Lesley Burgess is a Senior Lecturer in Art, Design and Museology at the Institute of Education, University of London. She is Course Leader for the PGCE course in Art and Design and teaches on both the MA Art and Design in Education and Museums and Galleries in Education courses: between 1992 and 2001 she was Co-Director of the Artists in Schools Training Programme. Before moving to the Institute in 1990 she taught for 15 years in London comprehensive schools. She was a trustee of Camden Arts Centre. She has co-edited with Addison, N. *Issues in Art & Design Teaching* (2003) also published by Routledge Falmer. Her main research interests are curriculum development, issues of gender and Contemporary Art and Artists in Education. She co-directed, with the V&A, a DfEE-funded research project 'Creative Connections' (2002) which investigated the use of galleries and museums as a learning resource and co-directed with Nicholas Addison 'En-quire: Critical Minds' (2005–6) evaluating the significance of learning in galleries on contemporary art.

Paul Dash is a Lecturer in education and MA module co-ordinator. He is also a member of the Bilingualism and Culture Research Group. He is an external examiner for Roehampton University, London. Paul has previously taught in London schools for 22 years and for 3 years at the Institute of Education, University of London. In 2002 Paul received the Peake Award for Innovation and Excellence in University Teaching. In 2003 he received the Windrush Award for Contributions to Education. He is a practising painter and has shown in the Summer Exhibition at the Royal Gallery, the Open Exhibition at the Whitechapel Gallery, the Commonwealth Institute and various commercial galleries in London.

David Gee was Head of the Art and Design Department of North Westminster Community School until its closure in 2006, where he led one of the largest Visual Arts departments in the country. His teaching career has spanned 31 years in London comprehensive schools, adult education and the Institute of Education. He studied ceramics at Camberwell going on to Goldsmiths to complete his teacher training. He continues to practise as a potter, building a collection of large earthenware pieces for a prestigious annual charity exhibition at The Mall Galleries. He is interested in the development of contemporary thought and especially in theories of creativity and how these can be applied to the classroom. His long experience of teaching children and adults from a wide range of backgrounds, skills and interests has made him keenly aware of equal opportunity issues and the development of SEN practice through effective differentiation.

James Hall is Principal Lecturer in Art Education and Assistant Dean (Quality) at the Roehampton University, London. He co-ordinates the Art Teaching Studies Area in the Faculty of Education and teaches on undergraduate and postgraduate programmes. His research interests include the education, training and professional development of teachers of art and design and artists' prints. He is currently researching the development of professional knowledge and expertise of teachers of art and design in secondary schools. As a practising printmaker, he is a regular contributor to artists' bookworks.

Alison Hermon is a Senior Lecturer in Art and Design education at the University of Brighton, where she is currently teaching on the PGCE and undergraduate programmes. She has an MA in Art and Design Education from the Institute of Education, London. Her previous experience includes working as an Art and Design Co-ordinator in a special school for 12 years, teaching in mainstream secondary and primary schools, as well as museum and gallery education. For the past five years she has been delivering lectures to PGCE and MA students at Brighton University and the Institute of Education, University of London. Her current research is focused on Art and Design education for pupils with special educational needs.

Pam Meecham is a Senior Lecturer in museums and gallery education and course leader for the MA Museums and Galleries in Education. She was route leader for the PGCE course in Art and Design at Liverpool John Moores University before joining the Institute of Education, University of London. She also lectures and writes widely on art history, museum and gallery education and art and design education. Her publications include *Modern Art: A Critical Introduction* (2004) (2nd Edition) with Julie Sheldon and *Investigating Modern Art* with the Open University (1996) as well as Teacher Guides for the Tate Gallery, Liverpool. Her research interests include public monuments and American art during the 1950s, the latter being the subject of her PhD. She has directed a funded research project in collaboration with the Tate 'Visual Paths: writing in the gallery' (2002), conducted an evaluation of the four London Hub Museums Schools provision for MLA (2004) and most recently led a research project, funded by Photoworks, that analysed the work of a Photographer in Residence at Dartford Grammar School.

Roy Prentice is a Visiting Fellow in Art, Design and Museology at the Institute of Education, University of London. Formerly he was Head of Art, Design and Museology at the Institute and gained wide experience over many years as an art adviser for East Sussex LEA and as head of an art and design department in a London comprehensive school. His main research interest is the professional development of teachers of art and design – particularly the role of studio-workshop based learning. He is a practising painter.

Claire Robins is a Lecturer in Art, Design and Museology, MA module leader for the 'Learning and Teaching in Art and Design' and 'Artefacts as an Educational Resource' modules and teaches on the PGCE Art and Design secondary course. She is currently working on MPhil/PhD research which examines the practice of 'artists' interventions', where galleries and museums commission contemporary artists to elucidate, reinterpret and even critique their collections. She is also a practising artist with experience teaching both studio practice and theory in Further and Higher Education including, most recently, seven years at Camberwell College of Arts. Her research interests are contemporary theory and practice and their assimilation into educational contexts. In addition she has worked as a researcher on the DfEE-funded research project 'Creative Connections' (2002) that investigated Art and Design teachers' use of museums and galleries as a learning resource.

Kate Schofield is a Visiting Lecturer in Art, Design and Museology at the Institute of Education, University of London where she works as a PGCE tutor. (Between 1993 and 1997 she was Course Leader for the MA Museums and Galleries in Education.) After training as a textile designer, she gained extensive teaching experience in further education in Nottinghamshire before studying for an MA degree at London University. In her capacity as Chief Examiner, Senior Examiner and Reviser for each of three examination groups, she gained additional experience of the examination system in art and design: she has also acted as a consultant for SCAA.

John Steers was appointed General Secretary of the National Society for Art Education (now the National Society for Education in Art and Design) in 1981 after 14 years teaching art and design in secondary schools in London and Bristol. He was the 1993–96 President of the International Society for Education through Art and has served on its executive committee in several capacities between 1983 and the present. He has also served on many national committees and as a consultant to government agencies. He has published widely on curriculum, assessment and policy issues. He is a trustee of the Higher Education in Art and Design Trust and the Chair of the Trustees of the National Arts Education Archive, Bretton Hall. He is also a visiting senior research fellow at Roehampton University, London.

Foreword

In good times a well-founded publication can underpin optimism and imagination. In difficult times it reminds us of why we are here and where we are going. Thirty years ago such a book emerged from London University's Institute of Education. Entitled *Change in Art Education* by Dick Field it has remained a comfort and support to me since my days as a student teacher.

This book comes from the same institution at a time of massive changes in art, cultural practices and schooling. As such, I hope it strikes the same chord with many prospective teachers of art and design that Field's book achieved for me. The title of the book suggests that it is aimed at those often – but not exclusively – young adults about to embark on a career in our schools and I hope that they will grasp it eagerly and find that it informs a lifetime in arts education. Nonetheless I will be sad if that is the extent of its audience for I would recommend it to all those who are involved in the teaching of art and design.

This is because the book is a very clear, practical guide to ways of planning, delivering and assessing art and design in secondary schools but *more importantly* because it will stimulate and extend thinking about their work. We can organise teachers as much as we like but unless we produce thinking teachers, we can have no optimism for the future and so this publication is to be commended for it demands that we teachers 'question orthodoxy' *and* it helps us to achieve that.

This is not a surprise. The Art, Design and Museology department of the School of Arts and Humanities at the Institute of Education has a long and honourable tradition of work in the pedagogical field. The editors who inherit that tradition are committed, critical and clear-thinking and they have provided us with a timely and stimulating publication. I hope that in 30 years time we will be able to look back and celebrate its influence on a bright, new phase in art and design education.

Dr Dave Allen, Course Leader
Entertainment Technology
University of Portsmouth

Acknowledgements

We would like to thank the student teachers and teachers in partnership schools with whom we have worked over recent years: their continuing collaboration has made it possible to experiment, research and reflect critically on learning and teaching in art and design. Our partners in London galleries and museums merit special mention for providing us with insights into ways of making accessible to young audiences the expanding field of visual and material culture. Contributing authors represent some of the major centres for initial teacher education across England and our thanks go to them for their varying perspectives which complement the traditions and innovations of the Institute of Education (IoE). Colleagues and friends have provided invaluable insights, support and encouragement: they are too numerous to name individually. However, special mention must be given to Jessica Barr, Josephine Borradaile and Peter Thomas in Art, Design and Museology, School of Arts and Humanities at the Institute of Education for their expertise and organisation.

We are particularly grateful to the following, artists, craftspeople and designers for their generous contributions, both in the form of statements and reproductions of their work: Michael Brennand-Wood, *Walking in Space*; Caroline Broadhead, *Steppenwolf*; Lucy Casson, *Playing with Fire*; Meera Chauda, *Untitled*; Thomas Heatherwick, *Gazebo*; Cathy de Monchaux, *'Evidently Not'*; Keith Piper, *Trade Winds (detail)* image kindly provided by inIVA; Helen Storey, *Heart Development Hat*, Photography, Justine; Model, Korinna @ Models 1. We also wish to thank all PGCE students from the Institute who have given permission to reproduce their course work.

Finally, we would like to thank Anna Clarkson and her colleagues at Routledge for their advice and encouragement.

Nicholas Addison and Lesley Burgess

Abbreviations

AAH	Association of Art Historians
AAIAD	Association of Advisers and Inspectors in Art and Design
AAVAA	African and Asian Visual Arts Archive
AS Level	Advanced Subsidiary Level
BECTa	British Educational Communications and Technology agency
CAD	Computer Assisted Design
CDT	Craft, Design and Technology
CEDP	Career Entry Development Profile
DES	Department for Education and Science
DFE	Department for Education
DfEE	Department for Education and Employment
DfES	Department for Education and Skills
DWEMs	Dead White European Males
EAL	English as an Additional Language
Engage	National Association for Gallery Education
EO	Equal Opportunities
ERA	Education Reform Act
GCE	General Certificate in Education
GCSE	General Certificate in Secondary Education
GNVQ	General National Vocational Qualification
HE	Higher Education
HMI	Her Majesty's Inspector
HMSO	Her Majesty's Stationery Office
ICT	Information Communications Technology
IEP	Individual Education Plan
iJADE	International Journal of Art and Design Education
inIVA	Institute of International Visual Arts
INSET	In-Service Training

ITE	Initial Teacher Education
KS	Key Stage
LEA	Local Education Authority
NC	National Curriculum
NCET	National Council for Educational Technology
NSEAD	National Society for Education in Art and Design
NVQ	National Vocational Qualification
OFSTED	Office for Standards in Education
PGCE	Postgraduate Certificate in Education
PoS	Programmes of Study
PTE	Practical Teaching Experience
QCA	Qualifications and Curriculum Authority
QTS	Qualified Teacher Status
RAB	Regional Arts Board
RBL	Resource Based Learning
RSA	Royal Society of Arts
SCAA	Schools Curriculum and Assessment Authority
SEC	Secondary Examinations Council
SEN	Special Educational Needs
SENCO	Special Educational Needs Co-ordinator
SoW	Scheme of Work
TDA	Training and Development Agency for Schools
WAL	Women's Arts Library
yBa	Young British Artists

1 **Introduction**

Nicholas Addison and Lesley Burgess

What is the purpose of art, craft and design in education?
What is the philosophical basis for the inclusion of art and design?
Why is art and design only a foundation subject at KS3?
How can the subject be developed to acknowledge changes in
contemporary practice and the potential role of art, craft and design in the
twenty-first century?

WHAT DOES ART AND DESIGN DO?

Art is fundamental to society, a network of useful, pleasurable, challenging and
potentially transforming practices; indeed, it is artists, craftspeople and designers who
produce the environment in which we live. By acknowledging this fundamental
position and by drawing on the richness and diversity of contemporary and historical
practices, art and design teachers in secondary schools provide opportunities for
pupils to make and investigate art, craft and design in both creative and discursive
ways. This understanding of art and design acknowledges broader social and cultural
values and enables you as a student teacher and contemporary practitioner to go
beyond the intuitive and tacit understanding typical of traditional school art (Hughes
1998a; Downing and Watson 2004). You can help pupils to develop this understanding by placing practice in context; in this way they will begin to recognise how art can
be more than self-expression, that it functions on different levels to support, critique
and, significantly, produce shared meanings and cultural values.

As a student teacher, it is important that you engage pupils in art and design as a
meaning-making practice, a subject that can accommodate the expanding field of
material and visual culture, anything from designing furniture using recycled
materials to the production of a performance or the construction of a virtual gallery.
The significance of such forms lies in the uses to which they are put: utilitarian and

symbolic, affective and discursive, physical and spiritual. But their significance also depends on context, when and where they were produced and by whom. It is therefore not surprising that the field has generated contested discourses, from the purposeful construction of hierarchies and histories to their no less exacting deconstruction. This book helps you to plot a route through the complexities of these discourses so that the learning experiences you design for pupils can be seen to move beyond recreational practices towards critical and contextual understanding.

The first three years of secondary education are crucial to the significance of the subject because it is only then that pupils are taught by a specialist art and design teacher. Increasingly, art and design in primary education is seen as a vehicle for supporting the core curriculum, an adjunct to Literacy, Numeracy, Science and ICT; learning through art. This instrumental role, the ability of the subject to contribute to the whole curriculum, can be beneficial, but you must not allow it, as a service subject, to deny the place of art and design as a different and fundamental part of knowledge and understanding (Eisner 1998).

The expanding field of art, craft and design offers challenges and possibilities beyond the technical and formalist orthodoxies of secondary art education. Makers and critics are questioning traditional boundaries and your own art practice may well question historical distinctions you consider no longer valid. However, it is vital that you reflect on your own practice and relate it to the histories of art education. In this way the relationship between the past, present and future can be understood as a form of dialogue in which the differences between traditions are negotiated rather than placed in opposition.

Contemporary practice in art, craft and design blurs the boundaries between art and other forms of cultural production: the art object is no longer exclusively to be found in the gallery; the practitioner is no longer bound to their studio. Just as the sites of practice can be anywhere, from the natural environment to cyberspace, its methods can be interdisciplinary, from the anthropological to the psychoanalytical. As a student teacher taking on board an interdisciplinary approach you are invited to consider the methodological resources of other subjects in the school curriculum; Media Studies and semiotics, Geography and ecology, Religious Studies and inter-culturalism. It is important that you are aware and open to such possibilities and recognise the potential reciprocity between art and design and other areas of the curriculum.

There is no doubt that learning to teach art and design is a complex and multi-faceted process. It is therefore essential that you take the opportunity afforded by the PGCE course to develop your understanding by using all available resources: the expertise of fellow students, tutors, teachers and pupils, the wide range of local, national and international facilities, libraries and new technologies, the natural and built environment, galleries and museums, studios, classrooms (see Chapter 5). The richness of these resources provides inexhaustible possibilities for engaging pupils in stimulating learning activities, activities that enable them to 'make informed value judgements and aesthetic and practical decisions [to become] actively involved in shaping environments' (DfEE 1999: 14). In this way art and design can be both a creative and a critical subject.

The book enables you to:

- acknowledge the diversity of art, craft and design and its implications for learning and teaching;
- develop your subject knowledge;
- consider how to translate your practice in art, craft and design into pedagogy;
- recognise that theory and practice are interdependent;
- understand and employ methods and strategies for effective learning;
- develop reflexivity: question, evaluate and revise your teaching;
- question existing orthodoxies, identify and develop new directions;
- consider the purpose of art education in a plural society;
- develop a philosophy for art education.

WORKING DEFINITIONS

The term 'art and design' indicates the subject in the school curriculum which includes art, craft and design.

A lower case 'art education' indicates the field including art, art history, craft, criticism, design, and across phases (primary to HE).

On occasions the words 'art' and 'artist' are used generically to indicate all practice and practitioners in the production of visual and material culture: context should make this apparent.

'Art' is used to suggest the idea of a European, hierarchical tradition in which the fine arts: painting, sculpture, architecture are theorised as intellectual processes with possibly transcendental outcomes, and the applied arts, or 'crafts', are theorised as technical processes, producing utilitarian objects and where 'design' is seen as a historically specific industrial mode of production.

'School art' has come to suggest an insular tradition in which pupils' work is predicated on formalist and/or expressive modes perpetuated through exemplars (Downing and Watson 2004). To the uninitiated it represents skill, verisimilitude or recreation and therapy. Its references are largely male and western with tokenistic gestures to the art of others.

A BRIEF HISTORY: A DEVELOPING TRADITION

The idea that art, craft and design can significantly contribute to the education of all is a surprisingly recent phenomenon. The visual arts were excluded from the classical Liberal Arts, a hierarchy of disciplines divided into two domains, the 'quadrivium' and 'trivium', which identified the necessary components of an academic education, respectively: arithmetic, geometry, astronomy and music; grammar, rhetoric and logic. Not until the modern period were the sciences introduced as a separate and expanded domain. The visual arts are notably absent from academe until the Italian Renaissance when they surfaced as a parallel but isolated programme. It was the industrial revolution that forced modern governments to consider the role of design in educating a workforce capable of competing in world markets. This process began in Britain from the beginning of the Victorian period (Thistlewood 1992; Dalton 2001).

Three Victorians in particular were responsible for conditioning the way art education was to develop in Britain. Henry Cole conceived it as having an instrumental role in the development of the industrial modern nation state, John Ruskin proposed the arts as an aesthetic and moral entity, William Morris as a social and political activity. In the twentieth century, such theorists and educators as the philosopher John Dewey and the critic Herbert Read, the educators Franck Ciseck and Marion Richardson, focused on the experiential and expressive potential of art, its role in the education of the whole child, a means towards self-actualisation. More recently, educators have been keen to establish art as a cognitive activity, a distinct and unique way of coming to know the world. There are many permeations and deviations from these philosophies and they have affected art and design in schools to varying degrees. But it is important to understand that what happens in schools is in part a reflection of, or response to, broader practices, usually at one or two paces behind.

The promotion of critical approaches to the art curriculum first advocated in schools during the 1970s (Field 1970; Eisner 1972) marked a notable shift in practice. Since then there have been many initiatives promoting curriculum development. The report, *The Arts in Schools* (Robinson 1982) provided a rationale for the central place of art in the curriculum. It advocates: 'The arts have an essential place in the balanced education of our children and young people . . . [it provides] for a broad-based curriculum rather than one that is too occupied with academic learning' (pp. 3–4). There was also an attempt to accommodate the principles of equal opportunities through strategies of visibility, both multicultural and gendered, and, more significantly, anti-racist, anti-sexist initiatives (1980s). Before this could be consolidated the requirement to address the critical and contextual components of the National Curriculum (NC) (DES 1992) diverted attention to the canonic in relation to 'our' cultural heritage. However, the NC (DFE 1995) directive for pupils to respond to the 'work of others' necessitated a liaison between schools and gallery/museum education departments which, depending on the venue, included multicultural examples. Some of this may have been pragmatic or only tokenistic, visibility and celebratory tactics taking the place of a critical engagement with difference, but, at the time, teachers had to reconcile their personal ideals with external constraints. Under the current National Curriculum (DfEE 1999) the principles of inclusion underpin all subject areas and are provided as a generic introduction; it is important that you do not overlook this advice.

Elsewhere we have categorised current approaches to the art and design curriculum (11–18) (in Hickman 2004). The categories are not exclusive, they indicate:

> approaches that can be combined. For example, genre-based outcomes can be managed through formalist, perceptualist or expressive means: although in some schools a single approach dominates. However, there are two which are no longer evident: basic design, a cogent programme in the 1960s and 1970s, lost because its workshop-based strategies, if not its holistic philosophy, were subsumed by technology; anti-racist art education, promoted in the 1970s and 1980s, lost because it was no longer perceived as urgent in a post-apartheid era. However, with the publication of the *Macpherson Report* (1999) on the death of Stephen Lawrence, this was, with hindsight, premature . . .

1 perceptualist	mimetic procedures, a search for the 'absolute copy' reduction to appearances (Clement 1993);
2 formalist	a reduction to the visual elements, exercise driven, representational and/or abstract (Palmer 1989);
3 expressive	intuitive making through affective and/or material exploration: privileging the essential and individual (Witkin 1974);
4 genre-based	preconceived types perpetuated by teacher expertise and the imitation of exemplars, the successful work of past students, e.g. still-life, life-drawing, landscape, CD covers, ceramic figures;
5 pastiche	the imitation of canonic exemplars, occasionally assimilating the postmodern practice of parody (SCAA 1996 now QCA);
6 technical	the development of a succession of discrete technical skills: drawing followed by print-making, followed by batik, etc.;
7 object-based	a response to common – sometimes themed, often spectacular – artefacts in the form of a big still life/installation, e.g. natural and made forms; a multicultural potpourri (Taylor and Taylor 1990);
8 critical and contextual	an investigation of art as a means of social and cultural production privileging cognitive and analytical procedures (Field 1970; Dyson 1989; Taylor 1989);
9 Issue-based	an integration of the personal with the social, political and moral through responses to current and contentious issues (Kennedy 1995a);
10 Postmodern	promoting plural perspectives and approaches and embracing the new technologies (Efland *et al.* 1996; Swift and Steers 1999).

(Burgess and Addison in Hickman 2004: 18)

PGCE COURSES IN ART AND DESIGN

The art and design PGCE enables you to gain insight into the principles, processes and practice of art and design in education. It is founded on a model of partnership in which local schools and universities collaborate to provide a context in which the relationship between theory and practice can be investigated to inform learning and teaching in schools.

Practical teaching experience (PTE)

PTE takes place in partnership secondary schools where experienced art and design teachers are responsible for supporting and monitoring your development. You are introduced gradually to teaching through a sequence of observations and occasions for team teaching. You may also benefit from taking the opportunity to draw upon the expertise of staff in education departments in galleries, museums and other cultural sites.

Tutor groups

The tutor group, to which you are assigned at the beginning of the course, is your main support group. It is important that you view it as a vital resource for collaboration and actively participate in activities and discussions, sharing experiences, knowledge and skills.

OBJECTIVES OF THE PGCE COURSE

The PGCE course enables you to extend your knowledge and understanding of:

- your own creative processes to provide a basis for teaching;
- theories in the field of visual culture and how they can inform critical practice in art education;
- research in art education and how you can contribute to it;
- the status and position of art and design in relation to other curriculum subjects;
- cross-curricular collaborations and how they can enable an interdisciplinary pedagogy;
- the contribution of art, craft and design to the education of all young people in a plural society;
- planning for learning in art and design and how to monitor and assess pupils' progress;
- the organisation and use of space, tools and materials to ensure safe practice;
- the importance of communication skills in learning and teaching;
- rationales for learning in, about and through art and design and the ability to articulate an educational philosophy;
- what it means to be a committed professional, particularly in the PGCE year, and as you move through the Career Entry Development Profiles (CEDP), the Induction period and Continuing Professional Development (CPD);
- the Standards for the Award of Qualified Teacher Status (TDA 2006a) and their specific application within art and design.

HOW TO USE THE BOOK

The structure of the book is likely to correspond to the sequence of your PGCE course. It begins with the transition from artist to teacher, moves on to an investigation of learning, introducing you to planning, resourcing, management and assessment, before a consideration of issues such as attitudes to making and values in art education. Throughout the book, especially through tasks, we ask you to reflect critically on your experience, to question orthodoxies and evaluate your own and others' practice.

We ask you to engage with the associated fields of art education and visual culture in order to acknowledge changes in contemporary culture and to help you develop a reflexive practice. The book helps you to identify areas for development and suggests strategies to deepen your knowledge and experience of pedagogy.

A series of tasks is integral to each chapter. They function in various ways: collecting data through observation; asking questions and discussing issues with other students, your tutors, and colleagues in schools; planning and resourcing schemes of work and implementing and evaluating learning and teaching. It is important that you question the insularity fostered by those practitioners of art, craft and design who privilege notions of self-expression and originality to the expense of co-operation and partnership. Often we ask you to work collaboratively as a community of learners, for we believe that effective teaching is practised as a collective and reciprocal activity.

2 Making Connections between Subject Knowledge and Pedagogy: The Role of Workshops

Roy Prentice

In teaching we do not merely pass on a free-standing package of knowledge of the different periods, cultures and traditions in art, say, or the skills involved in working with different materials. What we do is rather to offer, however indirectly, a sense of the personal meaning which our curriculum has for us – its value, its relevance, its implications for us as particular human beings. In teaching . . . we represent not only our subject, more importantly, the stance we take towards it!

(Salmon 1995: 24)

INTRODUCTION

The purpose of this chapter is to support your growing understanding of the factors that help to shape the complex relationship between your expanding knowledge of art, craft and design and your developing skills and understandings as a teacher. A PGCE course provides a structure within which connections between personal knowledge of subject content and pedagogy can be explored in increasing depth and be strengthened and made explicit.

To assist you in arriving at a considered response, located in a theoretical framework, some key issues are identified and discussed within the following units:

Unit 2.1 the relationship between previous and present experience;
Unit 2.2 the nature of teaching and learning;
Unit 2.3 the role of reflective practice;
Unit 2.4 workshops: environments for enquiry.

OBJECTIVES

By the end of this chapter you should be able to:

- consider how to articulate the ways in which experience of professional practice informs the teaching of art and design;
- understand the significance of the wider, ongoing political and professional debates about educational standards and teacher effectiveness;
- bring more sharply into focus the nature and importance of the relationship between subject knowledge and subject application, particularly in the context of courses of initial teacher education;
- answer the fundamental question: how does what you know about art, craft and design, and the way you have come to know it, provide the basis for your professional development as a teacher and your evolving philosophy of art and design education?

UNIT 2.1 THE RELATIONSHIP BETWEEN PREVIOUS AND PRESENT EXPERIENCE

Whilst it is recognised that teachers, whatever their subject specialism, have in common a need to make genuinely creative and sustainable connections between their previous and present experience, you will find that for teachers of art and design there is sometimes insufficient acknowledgement of the impact subject-specific influences have on this process. It is useful for you to reflect upon the educational route which the majority of art and design teachers have followed in order to understand more clearly the nature of the issues involved. The pattern of national provision for undergraduate courses in art, craft and design is broad and complex. Degree courses differ enormously in their declared aims, content, structure and ethos. The majority of courses are highly specialised within a given area of activity, e.g. fashion design, fine art, three-dimensional design. However, modular courses offer students a wider range of experiences at the expense of working in a chosen area in depth. In addition, an increasing number of opportunities are available to students who wish to pursue more vocationally oriented routes or cross the boundaries of traditional specialisms and departments. Overall, they represent a continuum of professional practice in art, craft and design that constitutes the subject field. Such courses are predominantly practical, their nature being studio-workshop based. They aim, primarily, to develop students' creative capacities and technical skills as artists, craftspeople and designers – thus the emphasis is on making and increasingly on marketing that you are likely to have experienced.

For an artist or designer the decision to become a teacher raises fundamental and complex questions about professional integrity, creative energy, belief systems and self-image. Attitudes that influence responses to such questions include those which support a strong personal commitment to creative work. This is even more apparent when an increasing number of PGCE students have gained invaluable and often substantial experience of employment in a field of art, craft and design practice. It is significant that the majority of PGCE art and design students are motivated by a particularly strong subject allegiance and a well developed sense of personal identity. The recent availability of part time courses of initial teacher education in art and design has led to a deeper understanding of the reciprocal relationship between professional art practice and pedagogy when pursued concurrently.

As well as helping you to understand the complex connections between your previous and present experience it is essential at the outset that all those with whom you work, university tutors and school-based mentors, are equally aware of the unique combination of factors that have an impact on the transitional experience of intending teachers of art and design. Given the range of provision at undergraduate level, different courses provide students with different experiences of the 'same' subject. This, in turn, has a powerful influence on the stance adopted by secondary specialist teachers towards their subject:

> For young people in school, art and design is inseparable from the art and design teacher: the tasks, materials, goals of the lesson come infused with the teacher's personal identity . . . To teach is to reveal, both intentionally and unwittingly, what the curriculum really means to the teacher.
>
> (Salmon 1995: 24)

However, the formal curricula for art and design at school, undergraduate and post-graduate levels cannot be held solely responsible for the way the subject is perceived, valued, understood and ultimately taught. Teachers' constructions and reconstructions of 'what the curriculum really means' to them, are shaped by an amalgam of experiences encountered and revisited in a variety of formal and informal settings and relationships over an extended period of time.

Task 2.1.1 Identifying your personal position

In order to deepen your understanding of your personal position, identify and reflect upon the idiosyncratic baggage of beliefs, feelings, ideas, attitudes and skills you carry with you to the PGCE course. They underpin your present orientation towards art, craft and design and provide the growth points for your future development as an art educator.

UNIT 2.2 THE NATURE OF TEACHING AND LEARNING

It is apparent from the previous remarks that a model of art and design teaching based on a simplistic concept of transmission, e.g. of an acquired body of knowledge or set of practical skills, is flawed. Successful teachers demonstrate an ability to transform their knowledge of subject content into teaching material. They use appropriate representations of content to match the particular needs, interests and abilities of learners. Effective teaching of this kind relies on what Shulman (1986) refers to as professional understanding, a complex combination of knowledge of subject matter and knowledge of pedagogy.

From different perspectives an increasing number of contributors to the debates about teaching and teachers focus their attention on subject knowledge. No longer is it assumed that a teacher's competence is determined by pedagogical skills alone. The complex nature of teachers' knowledge of subject content has been investigated by Shulman and his colleagues at Stanford University (Wilson *et al.* 1987; Grossman *et al.* 1989). The ways in which teachers' personal belief systems about learning influence their approaches to teaching have been explored by Brophy (1991) and a further contribution to the discussion about what teachers should know is made by Aubrey (1994). Questions about the content of lessons, the issues they raise and the nature of teachers' responses are addressed by Shulman. In an attempt to demonstrate how teachers reveal different levels of subject knowledge through their classroom behaviour, Aubrey (1994) says, 'Where subject knowledge is richer, deeper and better integrated it is more likely that the teacher will be confident and more open to children's ideas, contributions, questions and comments' (p. 5).

While all teachers use representations, metaphors and analogies to transform subject content into teaching material, it is possible to identify qualitative differences based on depth of knowledge in the field. Predictably, if you have an understanding rooted in concepts, principles and underlying themes you are able to adopt a more flexible approach to your teaching and make more imaginative connections between different topics. This enables you to avoid the perpetuation of stereotyped and preconceived ideas about a given subject and to transform subject matter through carefully selected representations, using metaphor, analogy, illustration and demonstration.

In order to provide you with a theoretical framework within which the inter-relationships between three components of subject knowledge in a given discipline can be identified, the work of Grossman *et al.* (1989) is helpful.

1   Content knowledge includes:

- factual information;
- central concepts;
- organising principles and ideas.

2   Substantive knowledge includes:

- explanatory models or paradigms;
- conceptual tools used to guide enquiry and make sense of data.

3   Syntactic knowledge includes:

- relevant forms of methodology;
- ways of introducing new knowledge – justification and evaluation.

This model usefully informs a theoretical framework for a professional knowledge base for teaching. It embraces both subject content and pedagogy and includes:

1   General pedagogical knowledge

- knowledge of theories and principles of teaching and learning;
- knowledge of learners;
- knowledge of principles and techniques of classroom behaviour and management.

2   Subject content knowledge

- ideas, facts, concepts of the field;
- relationships between ideas, facts and concepts;
- knowledge of ways new knowledge is created and evaluated.

3   Pedagogical content knowledge

- understanding of what it means to teach a given topic;
- understanding of principles and techniques to teach a given topic.

As a teacher of art and design your understanding of pedagogical content knowledge determines how you approach, resource, teach and evaluate the art and design curriculum. This also helps you to understand how learners engage in learning with reference to subject-specific difficulties, attitudes, skills, requirements and misconceptions. Such pedagogical content knowledge is informed by knowledge of art and design and general pedagogical knowledge. Thus throughout your PGCE year it is important that you make creative connections between your developing knowledge of art, craft and design, your experience of art and design teaching and your wider awareness of the nature of teaching and learning.

Task 2.2.1 Identifying your knowledge base

With reference to the theoretical framework suggested by Grossman *et al.* (1989) identify your knowledge base in relation to the categories: content, substantive, syntactic.
In pairs examine which areas require extensive development. How can you ensure that this need is addressed during your PGCE year?

UNIT 2.3 THE ROLE OF REFLECTIVE PRACTICE

. . . reflection relies on ability to uncover one's own personal theories and make them explicit.

(Griffiths and Tann 1991: 86)

Through reflective practice it is possible to make explicit those aspects of professional practice in art and design and teaching that would otherwise remain implicit. The concept of the reflective practitioner proposed by Schon (1987) relies on a dynamic interrelationship between three phases of reflection:

- reflection-in-action;
- reflection-on-action;
- reflection *on* reflection-on-action.

The point is made by Schon that:

> . . . reflection-in-action is a process we can deliver without being able to say what we are doing. Skilful improvisers often become tongue-tied or give obviously inadequate accounts when asked to say what they do. Clearly, it is one thing to be able to reflect-in-action and quite another to be able to reflect *on* our reflection-in-action so as to produce a good verbal description of it; and it is still another thing to be able to reflect on the resulting description.
>
> (p. 31)

An example of reflection-in-action is a smoothly integrated performance by a group of accomplished jazz musicians. This is described by Schon as:

> Listening to one another, listening to themselves, they *feel* where the music is going and adjust their playing accordingly . . . Improvisation consists in varying, combining, and recombining a set of figures within a schema that gives coherence to the whole piece. As the musicians feel the direction in which the music is developing, they make new sense of it. They reflect-in-action on the music they are collectively making – though not of course in the medium of words.
>
> (*ibid.*: 30)

The development of a reciprocal relationship between an artist and the work-in-progress may be likened to the evolution of a fruitful conversation. Ideas and feelings are presented and articulated in ways that were unknown at the outset. Thus, there emerges through the interdependence of content and form 'a discovery in union' (Reid 1969: 279). Ben Shahn, a painter who, like other artists, has experienced this condition from the inside, says:

> From the moment at which a painter begins to strike figures of colour upon a surface he [sic] must become acutely sensitive to the feel, the texture, the light, the relationships which arise before him. At one point he will mould the material according to an intention – perhaps his whole concept to emerging forms, to new implications within the painted surface. Idea itself, many ideas move back and forth across his mind as a constant traffic . . . This idea rises to the surface, grows, changes as a painting grows and develops.
>
> (Shahn 1967: 49)

Beyond reflection-in-action, which is essential for professional practice, neither the jazz musicians nor the painter are required to articulate ideas about the nature of their creative process. As a teacher of art and design it is precisely this level of reflection on professional practice in art, craft and design and in education that you are required to develop. You are encouraged to think about each session you teach as a complex and subtle performance that is determined by your knowledge and understanding, skills and attitudes. As a reflective teacher you come to recognise the problematic nature of teaching art and design and systematically reflect upon your practice in order to improve it. In so doing you simultaneously become engaged in teaching and learning: a relationship that echoes the quality of creative activity in art, craft and design.

Such a view of teaching acknowledges the range of personal experience that you as a teacher – as well as pupils – bring to the educational enterprise in which you are involved and as Elliot (1991) points out:

> Learning to be a reflective practitioner is learning to reflect about one's experience of complex human situations holistically. It is always a form of experiential learning. The outcome of such learning is not knowledge stored in memory in propositional form, but *holistic understandings* of particular situations which are stored in memory as case repertoires.
>
> (p. 313)

Central to a creative model of professionalism, to which Elliot subscribes, is the view that the personal growth and professional development of teachers is inextricably entwined. As Day (1993: 84) argues, '. . . any attempt to improve children's learning depends upon some form of teacher growth'. A reflective teacher is valued for being a critical, resourceful and developing individual rather than someone who functions routinely in a predetermined role that merely reinforces the image of the teacher as an infallible expert.

However, the notion of the reflective practitioner in education has sometimes been discredited by critics who claim too much is attached to subjective judgements made in a vacuum and that a preoccupation with process and means is at the expense of concern for curriculum content and outcomes. Effective reflective behaviour requires you to subject your own practice to scrutiny within a conceptual framework that allows alternative theoretical interpretations and practical approaches to coexist. Traditionally the main focus for reflection has been on process, but, increasingly, the importance of reflecting on aspects of content is recognised in order to challenge assumptions about subject knowledge. Above all it is necessary for you to recognise and value insights into experience that can be gained through modes of human behaviour other than those that rely on patterns of processing that are linear, logical and rational.

Task 2.3.1 Reflection-in/on-action

With reference to the theoretical framework suggested by Schon (1987), discuss your experience of reflection within the creative process.

In pairs, consider how you might engage pupils in such reflection.

UNIT 2.4 WORKSHOPS: ENVIRONMENTS FOR ENQUIRY

Through your direct involvement in PGCE workshop studies you can discover opportunities to deepen your understanding of the issues raised in this chapter. The section which follows draws your attention to different dimensions of workshop activity. Throughout your course you are expected to maximise the potential of the different kinds of workshops in which you engage by making creative connections between your previous and present experience, theory and practice, art, craft and design and education. In order to explore ideas relating to subject knowledge and subject application simultaneously and 'in action', a strong case is made for the centrality of workshop studies in PGCE courses. It is to a detailed discussion of workshops that this unit is devoted.

Within the particular framework of each PGCE art and design course workshop studies are conceptualised, organised and approached in a variety of ways. Workshops may take place in the universities, schools, museums and galleries, local urban or rural spaces or in combinations of these locations and virtual sites. They may involve groups of students, students and teachers from partnership schools, groups of pupils and other artists or designers working in education through creative partnership schemes. It is likely that you will also be encouraged to engage in workshop activity as an 'individual', a necessary component of your private study or directed study time. You are invited to consider the implications for teaching and learning that such varied organisational frameworks may have, along with the value of workshops in which you are able to share your experience as a member of a group as opposed to those in which you work alone.

The stated intentions of workshop studies are also likely to differ from course to course, and the emphasis placed on the acquisition and development of skills and an exploration of ideas through issues-based work, for example, may shift within the same course. However, fundamental concerns and values are shared and these serve to underpin the importance attached to workshops and their role in the initial and continuing professional development of you as a specialist teacher of art and design.

Whatever the particular requirements of the workshops in which you participate, it is important that you exploit their rich potential as actual and virtual environments for material and digital enquiry. Through your engagement in different kinds of workshop activities you *become* able to enquire into the making of and response to art, craft and design and ways of teaching and learning in art and design. Workshops reaffirm the importance of active learning (learning through doing, practical knowledge). They provide 'a transaction with a situation in which knowing and doing are inseparable' (Schon 1987: 78). They foster learning modes that are experiential to ensure 'knowledge is not divorced from knowers' (Salmon 1995: 24). As Salmon says:

> Learning from experience calls for educational modes that are as far removed as they could be from traditional classroom teaching. Instead of didactic transmission of information such modes need to engage learners actively and purposefully in their own learning. In place of top-down knowledge pupils must construct things for themselves. And what is learned must go beyond

merely doing things; the learner must come to reflect on that practical experience, to articulate something of what it means.

(ibid.: 22)

Indeed, it is this growing capacity to articulate the personal significance of experience that is so empowering for all learners. It is through your developing capacity to reflect critically on practical experience that you gain greater insight into your own creative functioning as an artist, craftsperson or designer. This in turn informs your teaching and your ability to make finely tuned decisions about how best to structure learning experiences for others, in order to meet the individual needs of a wide range of pupils. This reflective dimension of workshop studies enables you to make and sustain vital and creative connections between your previous and present experience as a learner, on the basis of which your present and future approach as a teacher is shaped. Those aspects of your creative processes that have remained implicit in the past, become focal points of attention; they are made explicit. Their functions are articulated, shared, critically analysed, interpreted and made available for modification. Throughout your deepening awareness of your creative processes and preferred working methodologies it is important to recognise alternative ways of thinking and working in art, craft and design along with their implications for pedagogy. As a result it is possible to demonstrate how workshops in the field of art and design education can play a powerful role in helping you simultaneously to address subject knowledge and subject application.

Whatever the nature of the workshops in which you participate as a PGCE student, they provide you with opportunities to engage in practical art, craft and design activity and at the same time develop your critical awareness of the factors that influence the quality of learning. As a teacher of art and design you are required to create conditions that maximise the potential for pupils' experiential learning in the lessons, projects and courses for which you are responsible. To achieve this it is necessary to make numerous decisions about content and organisational structure. In order to provide an informed basis for such decision-making related to your teaching, you are asked to reflect upon and consider critically the content and structure of those PGCE workshops in which you participate as a learner. For this purpose it is recommended that you keep a detailed, annotated sketchbook, journal, log or diary to provide an ongoing record of your workshop experience, your reflections-on-action. By reflecting on your creative behaviour in this way you are able to identify the most significant factors that determine your preferred patterns of working. You are able, as a result, to articulate more clearly your understanding of the ways in which ideas are generated, shaped, supported, developed (or discarded) and refined. Strategies through which connections are made between concepts and skills, process and product, visual, written and verbal modes of communication can be explored. The insights gained through such explorations inform your growing awareness of the interdependence of skills, knowledge and understanding in the National Curriculum for Art and design. Additionally, when planning projects for teaching practice, you can investigate the far-reaching influences on creative activity of such ingredients as the nature of the brief, time, space and materials along with ways of dealing with variables of this kind.

In order to provide a framework for your reflections-on-action your attention is drawn to the following key factors on which you are invited to focus:

Time

Creative activity requires time so that it can be paced and structured in such a way that ideas have opportunities to evolve, to take shape through impact on material and, above all, time to explore alternatives through investigative research. Time is needed in order to become familiar with materials, tools, equipment and technical processes to acquire and develop practical skills through practice and to apply them in new situations. Through an application of such basic principles when making decisions related to ICT, opportunities for art and design teachers to transcend what Buckingham (2005: 12) calls, '. . . the *banality* of much new media use' can be demonstrated. Above all, workshops require sufficient time for art, craft and design activities to be engaged in, sustained and for personal satisfaction to be experienced through struggle. As Schon (1987) reaffirms:

> . . . nothing is so indicative of progress in the acquisition of artistry as the student's discovery of the *time* it takes – time to live through the initial shocks of confusion and mystery, unlearn initial expectations, and begin to master the practice of the practicum; time to live through the learning cycles involved in any design like task; and time to shift repeatedly back and forth between reflection on and in action.
>
> (p. 31)

Task 2.4.1 Addressing time

In tutor groups:
   Consider the basis on which time is allocated for art and design workshops and how it is organised and managed. You are constantly reminded that time is a precious commodity and not to be wasted. There is never enough of it for you to achieve what you know you are capable of achieving: if only you had more time. Compare what you, as a specialist, are able to achieve in a PGCE workshop and what a pupil might achieve within the constraints of their timetable. What are the implications for planning?

Widely available and advanced information and communication technology allows you instant, simultaneous and continuous access to a vast and rapidly increasing bank of information. The everyday expectation of an increasing number of individuals is that gratification is instant and that results are rapidly and easily won in their interactions with the material world and in their interpersonal relationships. However, information overload and clutter, speed and an inability to focus and reflect are increasingly cited as being detrimental to creative solutions being found to contemporary problems (Bergstrom 2005: 50).

   How can you enable pupils to question this expectation?

Structure

Task 2.4.2 Relating content to structure

In tutor groups:
   Consider the relationship between the organisational structure and the content of the workshops in which you are involved. Analyse their overall structure. Consider the underlying rationales on which they are based along with the appropriateness of their design to effectively achieve their declared intentions. Critically examine the main similarities and differences between your responses in these workshops and your experience of studio-based work during your degree course.

The organisational possibilities for workshop studies are infinite and it is important that you draw upon the range of workshops you have experienced to help you make informed decisions about how best to manage multiple variables in the lessons and projects that you plan for teaching practice.

   Some of the most challenging and far-reaching decisions that have to be made by both teachers and learners in art and design relate to *ways of getting started*. As a teacher it is necessary for you to address the advantages and disadvantages of different 'ways in' to each project you plan to teach. You may decide to introduce a whole class to a project through a common brief; alternatively it may be considered more appropriate for individuals to invent their own briefs, thus generating a range of starting points within an agreed framework. Ideas may be generated through the direct manipulation of media and processes, discussion, writing, brainstorming exercises, visits, the critical interrogation of artefacts or research material collected prior to the introductory session. Each approach requires of learners different kinds of skills. It is also important to be aware of the relationships between pupils' past and present experience. Different ways of introducing and developing projects help to challenge orthodoxies and disrupt routine ways of working and avoid uncritical and stereotyped responses. As work in art and design evolves it is necessary to make ongoing decisions about the management of variables in relation to time, content, media and working methods. In order to sustain a successful workshop project you need to structure it in such a way that an appropriate balance is achieved between freedom and constraint, skills and issues, individual and group endeavour, making and critical analysis.

Space

Task 2.4.3 The environment as determinant

In tutor groups:
   Consider the nature and organisation of the physical and virtual environments in which you work. How do they influence the kind of work you do, the way you approach it and the way you feel about it?

The setting in and through which your knowledge of and attitudes towards art and design, pupils, teaching and learning, are made explicit, has an unavoidable effect on the quality and mode of interaction between you and pupils, and between pupils and their art work. The degree of ease or difficulty with which connections can be made and explored between ideas, feelings, perceptions, media and processes is significantly influenced by the settings you construct and in which these elements coexist. A prescribed working area should operate as a coherent resource. The layout and visual, tactile and spatial richness of a room, for example, encourages and supports certain behaviours whilst it discourages and is unsupportive of others. The ways in which space is allocated and arranged by you in relation to specific activities and the nature and amount of mobility it allows and interaction with others it facilitates, all operate as controls over creative responses. So too does the availability of materials and equipment. The ways in which people perceive, think about and are motivated to use the resources in their immediate surroundings are influenced by the manner in which they are presented. Familiar things placed in unfamiliar contexts, and unfamiliar things placed in familiar contexts, provide powerful triggers for the release of creative energy.

Relationships

Task 2.4.4 Reflection-in-dialogue

In tutor groups:
   Consider how other members of your workshop group contribute to your subject knowledge and growing understanding about teaching and learning. Through your log or diary record the reflective dimensions of the workshop activities in which you have engaged.

Reflection-on-action can be likened to a conversation with yourself: an activity that helps you gain deeper insight into the conditions that favour creative thought and action. As a member of a group you are exposed to alternative views about art, craft and design rooted in different belief systems. Different ways of working coexist in the same workshop. To the extent that your subject knowledge base can accommodate different, sometimes conflicting, philosophies of art, through reflection-in-dialogue you are able to draw upon richer and deeper levels of understanding of subject content which allows you increasing openness to accept pupils' ideas.

   Through the opportunities that workshops offer you to share ideas with others, attitudes are modified. By shaping ideas to clearly communicate them; by making the implicit explicit they enter the public domain, become available for scrutiny. Above all, such interaction provides you with opportunities to reflect on reflection-on-action and reaffirms the value of exposure to alternative ways of seeing, thinking and doing.

CONCLUSION

The majority of teachers declare, verbally, that they encourage pupils to be creative in art and design lessons. Unfortunately, in practice, there is often an absence of those conditions which maximise opportunities for genuinely creative behaviour; for various reasons they are unacceptable in operational terms. Your verbal encouragement of risk taking, the flexible use of ideas, materials, time and space must be accompanied by personal behaviour which embodies these qualities and a climate which allows such behaviour to flourish.

Anyone who has not 'lived through' experiences in art and design of the same order as those in which pupils engage and struggle to make sense of the process through reflection, 'is not likely to ignite others with an intrinsic excitement of the subject' (Bruner 1960: 90). To achieve such ignition workshops have a central role to play in your experience of initial teacher education.

Further Reading

Buckingham, D. (2005) *Schooling the Digital Generation: Popular Culture, New Media and the Future of Education*, London: Institute of Education, University of London.

Salmon, P. (1995) 'Experiential learning', in R. Prentice, (ed.) *Teaching Art & Design: Addressing Issues and Identifying Directions*, London: Cassell.

Schon, D. (1987) *Educating the Reflective Practitioner*, San Francisco: Jossey Bass.

3 Learning in Art and Design Education

Nicholas Addison and Lesley Burgess

INTRODUCTION

This chapter explores learning and teaching in art, craft and design and introduces you to a range of learning theories, some of which are generic while others are specific to the visual and material arts. Issues relating to inclusion are addressed within each unit.

The units in this chapter examine:

Unit 3.1   definitions of learning and teaching (the didactic/heuristic continuum);
Unit 3.2   learning in relation to developmental psychology and theories of intelligence;
Unit 3.3   active and experiential learning;
Unit 3.4   the aesthetic;
Unit 3.5   complementary ways of learning;
Unit 3.6   language, motivation and learning in art and design;
Unit 3.7   enabling learning: transforming subject knowledge into pedagogical practice;
Unit 3.8   learning as a critical activity.

UNIT 3.1 DEFINITIONS OF LEARNING AND TEACHING

By the end of this unit you should be able to:

- understand the function and potential of different pedagogic methods and their effect on learning;

- use conceptual frameworks to focus your classroom observation and inform your lesson planning;
- consider how pupils' prior experience affects the way they learn.

As you begin to plan and teach schemes of work for art and design it is important to consider not only the content of lessons and the methods by which you aim to teach it but, more significantly, how it is that young people learn. After all, teaching in schools is a social process or set of practices through which pupils learn valued knowledge and skills with the guidance of experienced adults; teaching is the means not the end. On one level this focus on learning sounds as though it ought to be easy because what it is to learn seems self-evident; learning, it is often said, is the acquisition and assimilation of knowledge. But this neat definition focuses on the product of learning rather than ways particular people in specific contexts manage to learn. In this chapter therefore we encourage you to focus on learning as a process as well as the ways in which young people respond to different teaching strategies in the context of formal education. This does not mean that we ask you to neglect yourself as a learner; on the contrary it is necessary to reflect on your own learning, particularly in relation to art, craft and design, in order to understand the conditions under which it takes place. It is useful to carry out such reflection in discussion with your tutors and other beginning teachers so that you can learn from the commonalities and differences of your collective experiences. These experiences will help you to deploy a variety of teaching methods and strategies to construct motivational learning environments. But first it is worth spending time defining the learning process more clearly in order to provide a theoretical framework within which you can compare your experiences.

Learning is the natural process through which people, from infancy, interact with their environment and come to understand and negotiate their place and potential agency within it. Put simply, learning is a social process through which people make meaning from experience. As such, when children interact with their environment learning takes place in response to stimuli of all kinds, stimuli that are experienced as sensory encounters with the material world (a plant, furniture, toy, tool or electronic screen) and interactions with others (parents, carers, siblings, friends or pets). In this way children build up a store of knowledge that helps them to satisfy their unconscious drives and socially inflected, conscious intentions. In many instances a child's will and the will of others can be at odds and so a certain dynamic is set up that conditions the efficacy of such interactions; what is learned is, in this sense, always socially and culturally mediated and productive of power relations. The ways children learn and what they learn are therefore not just dependent on the faculty known as 'intelligence' but on their immediate environment, social relationships and historical circumstances. This upbringing naturalises local knowledge so that it appears to be common sense, and these common sense beliefs (what Bourdieu (1993) terms a 'habitus') remain embedded in each person and may limit the extent to which they can take on new ways of learning and new forms of knowledge. For example, if learning in secondary school sits uncomfortably with a young person's home or primary school experience then it is likely to be inhibited. It is therefore important for you to discover the ways in which your pupils have learnt to make and talk about art, craft and design outside schooling and in what contexts. These contexts do not only refer to different pedagogic relationships: parent/child; older/younger sibling; teacher/pupil etc., but to the different locations in which learning takes place.

These locations can be usefully categorised into three types: the home and local community (sometimes called informal learning) the everyday workplace and untutored forums such as global communication systems (non-formal learning) and the wider academic, religious and vocational education systems (formal learning). A learning activity, in distinction from an encounter, is a consciously constructed event designed for a specific purpose and, outside formal settings, it usually takes place in a social site pertinent to that purpose: a parent introduces a child into the process of cooking in the kitchen, giving them responsibility for one task within a sequence that leads to a meal. However, in schools, learning and the knowledge it provides can sometimes appear distanced from any immediate purpose, it becomes abstracted from use and instead becomes pure knowledge, a thing in itself. Although these formal events provide opportunities for learning, they also regulate it and as a result some types of knowledge and ways of learning become privileged over others. Which ways are privileged depend on the interests of those who have power to control the learning: e.g. parents, politicians, religious leaders, teachers. This is especially so in the secondary school where young people are inducted into the dominant cultural, social and political values of society; a rite of passage between childhood and adulthood.

Task 3.1.1 Learning in art, craft and design

Discuss in your tutor groups:
What activities in the home can be considered art, craft and design?
What skills are used in these activities?
Are they transferable to other activities?
Are the activities recognised in school?
How have you learnt in and through art, craft and design and what was the role of others in this process?

Learning in the secondary school has been characterised as increasingly technocratic and standardised, a process based on clearly defined aims delivered through centralised teaching methods that produce measurable outcomes fostering competition and achievement. An older tradition of progressive pedagogy would rather conceive education as a mode of inquiry that can develop creativity and collaboration (*ibid.*). In this respect, learning in art and design is potentially different to most school subjects in that it foregrounds ways of learning that acknowledge the role of the body as well as the mind and in which a relationship to the material world is as significant as a relationship with symbolic systems, of which language is the most privileged kind. Art and design therefore incorporates modes of learning that may appear marginalised in other parts of the curriculum and we aim to enable you to argue for their significance. But we would also like you to recognise ways of learning that pupils bring with them, ways of making and talking about art, craft and design in informal settings that are themselves often overlooked and marginalised. As identified here, the school subject is not just about making art products; art, craft and design are symbolic practices through which different social and cultural values are communicated and thereby attitudes formed and reproduced. Sometimes perceptions of what counts may need to be questioned; learning in art and design is therefore also about discussing visual culture in its many manifestations, especially the way image makers deploy representations to produce particular understandings of the world and the way designers shape the world in which people live, the domestic world and the built environment, to encourage specific forms of behaviour. We therefore need to consider a working definition for learning in art and design.

The didactic/heuristic continuum

Many theorists have supposed that learning is a social process of the acquisition, assimilation and application of knowledge (Capel *et al.* 2005: Unit 5.1). Developing cognition is a phrase that neatly describes this process, a process that includes acts of perception, intuition and reason. Acts of perception are the basis of experience and art, craft and design provide ways to represent and embody such experiences. Acts of intuition and reason are the basis of imagination: they are the processes through which perceptions are reworked to form concepts and precepts, in other words, are transformed into ideas and values. The processes of art, craft and design are transformative acts which produce and develop ideas and values in material, multimodal forms. These acts are, in themselves, evidence of learning. The result is a physical or virtual outcome that not only represents but also embodies knowledge, skills and understanding:

> The odd notion that an artist does not think and a scientific inquirer does nothing else is the result of converting a difference in tempo and emphasis into a difference in kind. The thinker has his [sic] esthetic [sic] moment when his ideas cease to be mere ideas and become the corporate meaning of objects. The artist has his problems and thinks as he works. But his thought is more immediately embodied in the object . . . The artist does his thinking in the very qualitative media he works in, and the terms are so close to the object that he is producing that they directly merge with it.
>
> (Dewey 1934: 15–16)

You might call this process understanding made concrete. It is therefore important to realise that process and product are interrelated concepts and that it is counter-productive to develop and assess one without reference to the other.

However, this definition of the learning process neglects the fact that learning is always a social process. The educational philosopher Vygotsky (1986) outlined a theory of learning which demonstrates how learning always takes place through interactions between two or more people within specific social/cultural situations. Such social processes ensure that what is learned and how it is learned are culturally mediated forms of knowledge and in this sense knowledge and the ways it is used mean different things to different people in different environments and times. This theory makes it clear that knowledge is something that is constructed rather than assimilated or acquired. Educationalists have applied Vygotsky's theories to a model of learning which is called co-constructivism (Watkins *et al.* 2003; Addison and Burgess forthcoming). Here, exploratory talk, productive dialogue, collaborative negotiation and mutuality characterise a process of reciprocal meaning-making leading to the construction of shared understandings. Often, constructivist theories focus on language and discourse as the primary mediator within interactions. However, we would wish to add aesthetic practices as vehicles through which shared understandings can be constructed; practices that are, in part, accommodated within activity theory (Wertsch 1998). From this theoretical position, the physical interactions with materials, environments and other people that characterise aesthetic practices can be understood to involve the body as well as the mind as co-productive of meaning. This has implications not only for the cognitive strategies you aim to deploy but also for the ways in which you organise the learning environment and its resources and the ways you enable pupils to work within such a codified space.

But before you begin to design learning activities for art and design it is important to investigate how knowledge in the field has been categorised and subdivided into different domains. Art education has a history and different ways of learning seem appropriate to different educational philosophies (Thistlewood 1992; Dalton 2001). For example, in Victorian England it was vital to train young people to copy exemplars accurately to ensure a compliant workforce capable of producing convincing drawings for industrial advertising, etc. After the Second World War Herbert Read (1950) argued that it was vital to develop a creative and compassionate population to ensure world peace. Today, critical skills have joined pragmatic and creative approaches not only to underscore the principles of democracy but also to provide a more flexible workforce. Three domains first developed in the 1980s remain influential:

a) A Conceptual Domain that is concerned with the formation and development of ideas and concepts.

b) A Productive Domain that is concerned with the abilities to select, control and use the formal and technical aspects of Art and design in the realisation of ideas, feelings and intentions.

c) A Contextual and Critical Domain that is concerned with those aspects of Art and design which enable candidates to express ideas and insights which reflect a developing awareness of their own work and that of others.

(SEC 1986a: 6–7)

The National Curriculum and public art and design examinations continue to relate these domains to classroom practice. Many earlier, specifically modernist models defined art education in terms of autonomous and discrete activities, whereas these later developments implicitly acknowledge art, craft and design as interrelated forms of social production. The NC states: 'Pupils should be taught about . . . continuity and change in the purposes and audiences of artists, craftspeople and designers from Western Europe and the wider world' (DfEE 1999: 20).

The SEC model should not be read as promoting three discrete areas that need to be taught separately. It represents three constituent parts of a complex and inter-dependent field. In order to integrate these domains you need to acknowledge the diversity of learning needs by employing different and changing teaching methods, from the didactic to the heuristic. Such a varied and systematic approach helps you to break down the oppositions between child and adult-centred, or progressive and traditional means of learning and allows you to satisfy the specific requirements of different tasks and the needs of individuals: (for an art and design specific advocacy of the Progressive Movement see Lowenfeld and Brittain 1987, for a critique see Abbs 1987: 32–46). Table 3.1.1 identifies the characteristics of the didactic/heuristic continuum. Extreme positions can be held regarding the efficacy of each method. We contend that no one method is entirely sufficient for all pupils or for every task. Therefore, it is important to maintain a balance between the methods in this and the other continuums presented and referenced below. Classroom observations during your course provide you with the opportunity to define the type of learning and teaching taking place.

The implications of this continuum for learning and teaching are explored in Unit 3.7 where examples of practice demonstrate how these theoretical positions are applied to the classroom. It is useful to refer to variations on this continuum such as 'The continuum of teaching styles' (Mosston and Ashworth in Capel *et al.* 2005: Table 5.3.2, pp. 286–287).

These types of learning and teaching have implications for inclusion and SEN. Warnock (1978) suggests that it is pupils' needs, not their disabilities, that should be identified when differentiating strategies for learning. She proposes 'common goals' and 'a common purpose' in education (p. 5). Common elements in the context of differing needs mean that the routes to achieve understanding have to be different. For example, pupils with literacy needs are likely to find activity-based learning (see Unit 3.3) more sympathetic in developing understanding of art, craft and design than in annotating drawings. In activities that demand written analysis, including self-assessment, these pupils require exactly the same literacy support that they receive elsewhere in the curriculum. In this instance, if specialist support is not available, it is necessary to build in opportunities for oral assessment so that they can try to articulate the understanding evidenced in their practice during group evaluations. However, if a pupil is unable to contribute to evaluation in either written or oral form it does not necessarily signify a lack of understanding. In the introduction to this unit we referred to the notion of embodied knowledge as understanding made concrete: it is important that you identify and record this, and acknowledge it in class discussions. Exceptionally able pupils require a different type of support, needing continuous challenges and tasks that stretch their abilities. In many curriculum subjects this group is identified with the most literate, logocentric pupils: it is often the same group who excel in art and design. However, there are cases where pupils with literacy and numeracy needs display highly developed mimetic skills, or advanced motor skills with which they can transform materials. Because they have been supported by a sympathetic curriculum in a particular department, they have managed to thrive.

Table 3.1.1 The didactic/heuristic continuum

centred	LEARNING	TEACHING	CHARACTERISTIC FORMS	pupil	teacher	JUSTIFICATION	DRAWBACKS
adult teacher	PASSIVE	DIDACTIC	• instruction • information • lecture • demonstration • closed procedures and structures	dependent memoriser imitator	expert provider (differentiation by testing)	• outcomes are certain (appropriate to factual syllabi) • introduces techniques, establishes a sense of shared beliefs and values • conditions pupils to become receptive, to observe, listen and record • encourages memory skills • confirms the expert status of the teacher	• can be authoritarian • single perspective • can alienate because it fails to acknowledge difference (abilities/backgrounds) • pupils may become dependent • results in conformity and normative outcomes • knowledge may be lost unless reinforced using other methods
	RESPONSIVE & ACTIVITY BASED	DIRECTED	• rehearsing and imitating activities • responses to given stimuli, e.g. still life, work of others, design brief • investigation (probable findings already known by teacher) • conditioned/ determined structures	responder	trainer director resourcer (differentiation by outcome and taste)	• provides common experience (eg core skills) **enables:** • continuity and progression • identification of pupils not on task • ease of assessment • efficient transfer of skills • activity: individual/ pair/group	• knowledge is given/fixed, determined by teacher's experience: often privileging making **neglects:** • pupils' prior knowledge • individual needs

(continued)

Table 3.1.1—continued

	LEARNING	TEACHING	CHARACTERISTIC FORMS	pupil	teacher	JUSTIFICATION	DRAWBACKS
centred	ACTIVE & EXPERIENTIAL	NEGOTIATED	• discussion/debate • collaborative work • purposeful investigation • critical evaluation • multi-faceted and flexible structures • interaction • reflexivity	contributor interactor	facilitator motivator guide negotiator supporter (differentiation by individual learning routes)	**provides:** • intelligent making • critical thinking • learning as social activity; art as social practice • mutual respect/trust **acknowledges pupils':** • prior knowledge • individual needs **enables pupils to:** • communicate ideas • evaluate their own and others' work • negotiate their own learning • cross boundaries	• time consuming • difficult to coordinate and resource • difficult to monitor and assess • teacher requires breadth of knowledge • teacher needs to be ready to relinquish a degree of control
	HEURISTIC ENGAGING	CO-CONSTRUCTED	• meeting needs • answering hypotheses • experimentation • unknown findings • discovery • problem solving • investigation	researcher self-motivator inventor discoverer	coordinator reciprocator (differentiation by role contribution)	**encourages:** • the application of knowledge to practical contexts • pupils as planners • divergent thinking • risk-taking • learner as teacher/teacher as learner	• pupils need to be ready to take on initiatives • difficult to resource • only works with pupil self-motivation • teacher may feel insecure • teacher needs to acknowledge self as learner
child pupil	OPEN	PUPIL-LED REDUNDANT	• self-determined structures, motivated by pupil interest • exploration	agent director	attendant technician	**appropriate for:** • highly motivated highly resourceful learners	• can be chaotic, unfocussed • lacks boundaries • can invite stereotypical responses and/or a rejection of learning

Task 3.1.2 The didactic/heuristic continuum

In your tutor groups discuss:
How these definitions relate to your observations of art and design teaching in your placement schools.

The visual/haptic continuum
(Lowenfeld and Brittain 1987: 356–368)

There is evidence to suggest that the way people learn about the world can be categorised into two extreme modes; the visual 'the observer, who usually learns about things from their appearance' and the haptic who 'utilises muscular sensations, kinesthetic [sic] experiences, impressions of touch, taste, smells, weights, temperatures, and all the experiences of the self' (Lowenfeld and Brittain 1987: 357). Lowenfeld believes that mental growth is only possible if pupils are allowed to interact with their environment on a sensuous level. He states that most people fall somewhere between these two extremes, but warns that art and design teaching can privilege the analytical and visual at the expense of the emotional and haptic.

Table 3.1.2 The visual/haptic continuum, characteristics of the two extremes

Visual	Haptic
spectator	participant
analytical	intuitive
detached	emotional
objective	subjective
abstract	concrete
mimetic	affective
optical	tactile and kinaesthetic
perceptual	synaesthetic

You must remember that although nobody is exclusively visual or haptic, particular Schemes of Work (SoW), and the way they are assessed, may isolate and promote activities which prioritise one or more of these characteristics. If the curriculum is geared exclusively to one extreme or the other, for example the visual, those pupils who learn effectively through the haptic mode will be inhibited in predominantly visual SoW. To accommodate all pupils you should:

- build in changes of activity that address the two modes;
- provide open routes so that pupils can choose the most appropriate mode;
- identify and target pupils who require additional support.

Haptic modes of learning are often associated with the way SEN pupils come to understand the world. Handling collections at museums can provide one forum in which the combination of direct contact with artefacts and discussion about their function and contexts can facilitate learning.

Task 3.1.3 The visual/ haptic continuum	
When planning SoW for your Teaching Placements: Look at the PoS for Year 7 in your placement schools and identify the extent to which SoW and their learning objectives relate to the categories in Table 3.1.2. Can you add to the table? Repeat the exercise for Year 10. Can you identify any differences between the two year groups? How will you ensure that your lessons are not weighted exclusively to any one side of this continuum? Discuss these issues with your school mentor.

The visual/haptic continuum bears some relation to more recent learning style theories, particularly those of Riding and Rayner (1998) for whom learners are divided into two cognitive styles: holistic–analytic and verbal–imagery, and Dryden and Voss (2001) where they break down into visual, auditory or kinaesthetic learners (see Capel *et al.* 2005: 253). But we would suggest that both theories are problematic on some levels: the first setting up a counter-productive opposition between word and image when in practice the relationship between the two is not oppositional but complementary and the second separating out three sensory means of reception, but ignoring others such as the olfactory and gustatory, as well as the possibility of synaesthetic learning, a fully embodied learning (Matthews 1999). Later we discuss the concept of multimodal learning (Kress *et al.* 2001) where such oppositions and separations are further critiqued. However, in the Lowenfeld and Brittain continuum the emotions are recognised as significant for the way people learn in art and design, and this can easily be overlooked in other models.

UNIT 3.2 LEARNING IN RELATION TO DEVELOPMENTAL PSYCHOLOGY AND THEORIES OF INTELLIGENCE

By the end of this unit you should be able to:

- evaluate how teaching in art and design is informed by theories of learning based on developmental psychology;
- consider the place of syncretistic vision in a curriculum which promotes analytical procedures;
- consider how theories of multiple intelligence define different types of learning and learners.

Many educational theorists subscribe to the notion of developmental stages which determine the ability of an individual to assimilate forms of knowledge at different moments in their psychological development: for example, it would seem inappropriate to ask a year seven pupil to produce a detailed design for a suspension bridge and provide all the technical data required for its construction. Clearly the degree of abstract and mathematical know-how combined with knowledge of specific materials and their behaviour would be outside the understanding of most 11 year olds. Developmental psychologists refer to the idea of 'readiness' as a necessary condition for assimilation and integration (see Capel *et al.* 2005: 247–249). For a learner to be 'ready' their educational experiences must match their level of understanding so that, with help, they can come to recognise and process information in such a way that it can be retrieved and applied to different contexts. Piaget (1962) formulated a theory that divides the maturing child's ability to reason into different, clearly defined stages. The first, the 'sensory-motor period', lasts into the second year and denotes a time when children, unable to conceptualise their experience, can only react to the world in response to sensations. From age two they begin to understand the world through interaction, by exploring their immediate environment actively, using all their senses, what Piaget calls 'concrete operations'. Only with time, and only once the child is ready, do they develop more abstract ways of thinking, 'formal operations', constructing and testing hypotheses, imagining the possible effects of actions not yet experienced. It follows from Piaget's theories that learning is not the same as knowing. Learning involves a process of understanding whereby knowledge is developed through interaction with an environment. His sequential stages signal different ways of understanding and each of these conditions the way people see. He coined the term 'syncretistic' to differentiate a child's vision from that of the adult's. For Piaget the syncretistic is a primitive way of seeing that is superseded by adult analytical processes. The syncretistic allows a person to see holistically and recognise objects through cues without the need to match part to part cumulatively as in analytical vision. Ehrenzweig (1967) is less dismissive than Piaget, suggesting that this way of seeing is dependent on a process of 'unconscious scanning':

> . . . unconscious scanning makes use of undifferentiated modes of vision that to normal awareness would seem chaotic. Hence comes the impression that the primary process merely produces chaotic phantasy material that has to be ordered and shaped by the ego's secondary process. On the contrary, the primary process is a precision instrument for creative scanning that is far superior to discursive reason and logic.
>
> (Ehrenzweig 1967: 5)

The child's syncretistic vision utilises these undifferentiated processes producing a view of the world which is unhindered by the analytical, conscious mechanisms which, Ehrenzweig believes, can inhibit creative thinking. In the artist or other creative practitioner, unconscious scanning is supplemented by more conscious processes so that those 'happy accidents' which appear to arrive from nowhere can be tested by empirical and analytical procedures at some later date. He considers in detail the implications of these ways of seeing for art education:

> The child's more primitive syncretistic vision does not, as the adult's does, differentiate abstract details . . . This gives the younger artist the freedom to

> distort colour and shapes in the most imaginative, and, to us, unrealistic
> manner.
>
> (*ibid.*: 6)

From about the age of eight children become more aware of adult expectations and
begin to analyse their own work against:

> . . . the art of the adult which he [sic] finds in magazines, books and pictures.
> He usually finds his own work deficient. His work becomes duller in colour,
> more anxious in draughtsmanship. Much of the earlier vigour is lost. Art
> education seems helpless to stop this rot.
>
> (*ibid.*: 6)

Unlike Piaget, Ehrenzweig does not believe that in passing from one stage to the
next the former stage is lost, merely that it is suppressed. Gardner supports the belief
that early forms of knowing are not eradicated or transformed: 'they simply travel
underground; like repressed memories of early childhood, they reassert themselves in
settings where they seem appropriate' (Gardner 1993: 29). Ehrenzweig does not
suggest that child art should be seen as a paradigm for art education, this would only
lead to aesthetics of regression. He suggests that anyone can retrieve their syncretistic
facility but usually in situations where analytical processes are bypassed, as in the case
of humour. He cites the example of caricature where perceived features can be
acutely distorted, yet the sense of likeness enhanced.

Ehrenzweig argues that the undifferentiated processes of syncretistic vision are
essential to creative thinking of all types (Ehrenzweig 1967: 32–46) and provides
evidence in the work of scientists and modernist artists. From the late 1960s he
influenced a tradition of process-led, conceptual practice, especially in the USA.
Robert Morris, acknowledging Ehrenzweig's concern with the whole visual field as
opposed to figure/ground differentiation shifted from using minimal geometric
forms, with their strong gestalt, to formless more heterogeneous materials like thread
waste, a byproduct of the textile industry used in packaging. Morris saw boundary-
crossing as essential to dedifferentiation and with Robert Smithson began to explore
the possibilities of ephemeral materials such as steam and time-based projects
recorded photographically. One effect of this shift was to move practice outside the
studio and gallery to any site or environment (Taylor 1995). This expanded field has
transformed contemporary practice, but has had little impact in secondary schools.

It is important for you to consider the implications of the change from syncretistic
to analytical perception for your planning and teaching, in general, but also in teaching
SEN pupils. When pupils first enter secondary school the 'rot' that Ehrenzweig
identifies may have already set in: how often have you heard the plea, 'but Miss, I can't
draw'. What the child is articulating here is their inability to draw in a particular way,
usually the analytical manner of observational, tonal drawing which is so ubiquitous
in schools. Pupils are undoubtedly aware of older pupils' work and may in comparison feel intimidated by their lack of technical and imitative skill. A limited sense of
what is good or right can be reinforced by your approach to practice, both as artist and
as teacher. Such rigid boundaries are quickly communicated to pupils:

Those teachers who were unable to tolerate their own spontaneity and the loosening up of their rigid planning could not tolerate the spontaneous and wilful reaction of their young pupils during their teaching practice either . . . But what has perhaps not been sufficiently realised is the close correlation between the two kinds of ego rigidity, the trainee's intolerance of the independent life of his [sic] own work of art and his intolerance of his pupils' independent contributions to his teaching programme. The unconscious fear of losing control underlies both.

(Ehrenzweig 1967: 101–102)

The art and design curriculum frequently privileges analytical modes of production suggesting they are of a 'higher order' than affective and emotional modes. Pupils with learning needs often display a syncretistic approach to representation and find it difficult to conform to the prevailing norms of a perceptualist tradition where likeness to appearance is the main criterion. This difficulty can manifest itself in a negative, 'primitive' self-image. It is therefore important that you discuss the potential of both syncretistic and analytical modes: provide reproductions of artists who work in affective and syncretistic ways (e.g. Cubists, Abstract artists, L'Art Brut) and include tasks in SoW that have affective as well as analytical criteria. Ensure that you examine analytical modes of representation by using them yourself. This enables you to break down such activities into realisable steps and helps you to explain the process to pupils who have not developed the mimetic strategies that other pupils may already demonstrate without your assistance.

Task 3.2.1 Recognising and valuing the syncretistic

With your art and design mentor:

- look at the reproductions of artists' work on display throughout the department and identify their syncretistic and analytical elements;
- go through the same process with examples of work by Year 7 pupils;
- discuss ways to explain to pupils the value of the syncretistic in their own and others' work;
- devise a lesson which enables pupils to approach a task in either a syncretistic or an analytical mode;
- consider the implications of syncretistic expression for assessment.

Theories of intelligence

The emphasis in the secondary curriculum on factual knowledge and measurable outcomes privileges subjects and approaches that promote the ability to reiterate given knowledge. This is believed to provide a knowledge base from which pupils can construct arguments, solve problems and organise experience. The critical curriculum advocated by this book recognises the importance of cognitive processes for developing visual and aesthetic literacy but invites you to question whether logical and sequential processes and abstract reasoning hold all the answers to the creative curriculum. The dominant culture of accountability has driven teachers back to the certainties of traditional, tried and tested, pedagogic methods. Systems of assessment that measure abstractions through tests, like the IQ, are still applied in educational contexts but are recognised as biased and woefully inadequate for determining the potential of pupils to contribute to society (see Capel *et al.* 2005: 197). More inclusive theories of intelligence do exist and in particular Gardener's theory of multiple intelligences (1983) has influenced practice in schools. He proposes that other faculties besides reasoning can be defined as intelligences and categorises them into seven types:

> linguistic
> musical
> logical–mathematical
> spatial
> bodily-kinaesthetic
> interpersonal
> intrapersonal

This more inclusive taxonomy embraces 'ways of knowing the world' that recognise artistic processes, although, at first sight, only spatial intelligence specifically belongs to art and design. However, when you read Gardner's definitions, 'bodily-kinaesthetic' intelligence is described as 'the use of the body to solve problems or to make things' (Gardner 1993: 12) and clearly also belongs to the subject.

Gardner (1993: 6–7) points out that there are different types of learners: intuitive, traditional and disciplinary. He describes the intuitive learner as natural, naive and universal; the traditional student as scholastic, one who works comfortably within school systems; the disciplinary expert as one who can successfully apply their knowledge to new contexts. He feels it is essential that teachers and pupils should be willing to take risks, including the possibility that they might fail, rather than reiterate the safe formulae known to produce standard outcomes. He suggests that 'such a compromise is not a happy one, for genuine understandings cannot come about so long as one accepts ritualised, rote, or conventionalised performances' (*ibid.*: 150).

More recently Gardner has extended his taxonomy to include naturalist intelligence (by which he means to refer to the environment), although he has also considered the possibility of spiritual and existential intelligences (Gardner 1999). There is no doubt that Gardner's work has been deployed to support the teaching of arts subjects in schools, but within the research community his theories are criticised for being based on his personal intuitions and reasoning rather than on full empirical data and testing. One difficulty we have with his taxonomy is that it looks as though it could proliferate into many more categories; why not visual, cultural, material, etc.? (White 1998), but also with the way in which social interaction is relegated to merely one category, the interpersonal.

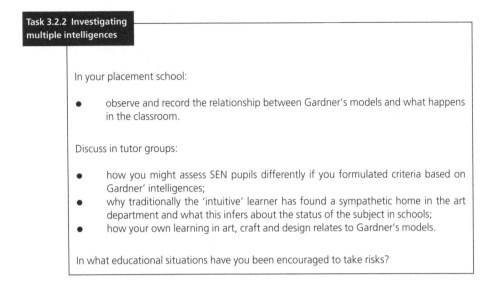

Task 3.2.2 Investigating multiple intelligences

In your placement school:

● observe and record the relationship between Gardner's models and what happens in the classroom.

Discuss in tutor groups:

● how you might assess SEN pupils differently if you formulated criteria based on Gardner' intelligences;
● why traditionally the 'intuitive' learner has found a sympathetic home in the art department and what this infers about the status of the subject in schools;
● how your own learning in art, craft and design relates to Gardner's models.

In what educational situations have you been encouraged to take risks?

UNIT 3.3 ACTIVE AND EXPERIENTIAL LEARNING

By the end of this unit you should be able to:

● define active and experiential learning;
● develop strategies to facilitate active and experiential learning.

It is important to recognise that making in art and design lessons is nearly always active in the sense that pupils are engaged in doing something, whether self-generated or in response to direction. However, doing is not everything. It is wrong to assume that because pupils are busy they are learning something, they may only be reinforcing existing knowledge: activity is not active learning. Passive learning, that achieved through listening, has traditionally been suspect: I hear, I forget; I see, I remember; I do, I understand (anon, Chinese proverb). Lowenfeld and Brittan suggest that activities such as drawing enable pupils to transform passive into active knowledge:

> The child draws only what is actively in his [sic] mind . . . A child knows
> a great deal more in a passive way than is included in the drawing. Part of a
> teacher's responsibility is to make this passive knowledge more active.
> (Lowenfeld and Brittain 1987: 36)

The implication here is that you should not spoon-feed pupils with ready-made formulae as this can repress their personal interests and inhibit their motivations. Instead, you must find ways to encourage them to take on increasing responsibility for their learning.

Similarly experiential learning is commonly defined as learning by doing. This definition is partly true, but from the point of view of construct theory (Kelly 1955;

Rogers 1969) experiential learning is recognised as a reflexive activity, where action and reflection are coexistent, both interdependent and interactive. Only where pupils are 'engaged actively and purposively in their own learning' (Salmon in Prentice 1995: 22) is the term 'experiential' appropriate. Usher and Edwards (1994) suggest:

> Experience is most often accorded importance as the 'authentic' representation and voice of the individual. Experiential learning has been constructed as a progressive and emancipatory movement in education, a shift away from the learning of canons of knowledge which, it is argued, marginalises the majority of learners by not giving value to their voices and thereby demotivating them.
>
> (p. 187)

Developing strategies to enable pupils to learn experientially requires effort. Pupils do not assimilate knowledge as presented, as though it were something separate and 'out there', information to be recorded and retrieved at will. They interpret it in relation to their existing conceptual models. Experiential strategies include acknowledging:

a)   prior knowledge and learning, both in and outside school;
b)   social and cultural backgrounds.

It follows that knowledge is actively and differently constructed by each individual. Therefore it is important for you to recognise these differences by valuing pupils' ideas and contributions and by providing opportunities for them to:

- take responsibility for their own learning;
- devise and give presentations;
- reflect, discuss and evaluate;
- negotiate meanings.

To achieve this the teacher needs to rethink their traditional role as knowledge provider by making the relationship between teacher and learner more reciprocal (Freire 1985; Giroux 1992), although you must remember this relationship is always 'asymmetrical' (Reid 1986). However, it is essential that you help pupils to understand dominant signifying systems so that they can negotiate their position within them. The National Curriculum (DfEE 1999) requires pupils to be 'taught about: codes and conventions and how these are used to represent ideas, beliefs and values in works of art, craft and design' (p. 20). The representation of space is an accessible example. To a Year 7 group you might be tempted to explain this didactically, as a body of given knowledge, by comparing oblique and axonometric projection in Japanese Ukiyo-e prints to one-point perspective in painting from the early Italian Renaissance (Willats 1997: 37–69). The mathematical concepts implicit in this explanation are unlikely to be remembered if delivered in this way. To develop a more experiential approach you need to devise questions inviting pupils to compare and contrast the spatial organisation in these different traditions. For example:

- How can you tell that a figure or object is in the background or the foreground?
- Where do the lines leading you back from the foreground finish?

Pupils can then formulate rules governing these differences in spatial organisation and apply them to their own representations. Not until this point should you provide a mathematical explanation against which pupils can test their findings. This could be extended by asking pupils to investigate the use of these systems in contemporary forms, e.g. computer games and Manga comics (Schodt 1993; Adams 2000). In addition pupils can explore alternative spatial systems, for example aerial and oblique perspective (Willats 1997; Cole 1992).

As a teacher you need to consider your own position as far from neutral (hooks 1994: Ch. 6). You bring with you your own preconceptions which are embedded in everything you say and do. Your beliefs about the appropriate role of teachers and pupils affect the interactions that take place. A statement such as: 'this class is never responsive' says as much about the way you have devised your lesson as it does about the pupils themselves. It may be that pupils are not usually expected to be responsive, in which case it will take you time to gain their trust and convince them that responsiveness is beneficial to their learning.

Setting up situations and environments which encourage pupils to take ownership of ideas, and thus their learning, is equally demanding. Pupils are more likely to be motivated if the learning environment is stimulating and well resourced. This is not to be confused with an environment of pure spectacle manufactured in the hope that an overwhelmed audience will be stunned into reverential compliance, e.g. one consisting solely of the virtuoso displays of artefacts selected by teachers or work by star pupils from the past decade. Pupils are more likely to be motivated if they contribute to the environment themselves, one they have helped to construct. Such contributory practice is particularly useful in building self-esteem. In this context learning is not the transmission of knowledge from expert to novice but an active and productive partnership where meanings are questioned and negotiated to help construct a learning community (Lave and Wenger 1991; Watkins 2005). This is not to suggest that teachers should surrender their responsibility to provide new and different forms of knowledge. A complete reliance on pupils' prior knowledge and experience would clearly limit the art and design curriculum in ways which would leave it open to accusations of insularity and introspection.

Task 3.3.1 Differentiating between activity-based and active/experiential learning

Look carefully at the lesson plans you have observed or taught in your first teaching practice.

Identify those activities that encourage responsive, active and experiential learning.

Ensure that you differentiate between activity-based and active/experiential learning when revising existing SoW and planning new ones.

Further reading

hooks, b. (1994) *Teaching to Transgress*, New York; London: Routledge.

Giroux, H. (1992) *Border Crossings*, New York; London: Routledge.

Salmon, P. (1988) *Psychology for Teachers: an Alternative View*, London: Hutchinson.

UNIT 3.4 THE AESTHETIC

By the end of this unit you should be able to:

- define the aesthetic;
- identify its significance in art and design.

The aesthetic refers to the way in which all experiences are mediated through the body, a multisensory and multimodal process. Art and design activities are particularly appropriate for developing sensory awareness and an understanding of how materials can be transformed into objects of use and beauty. Aesthetic practices are often central to the way young children's making activities are researched and understood (Gentle 1988; Matthews 1999). However, in research into secondary education it is often the more obviously cognitive processes of problem-solving and critical thinking that become priorities and the role of the body and the senses in learning is marginalised or overlooked. We think it is important that you remind pupils of the ways curiosity and inventiveness, so natural in childhood, go hand-in-hand with decision making and an induction into symbolic systems such as social codes and conventions. Aesthetic learning is embodied learning where an individual not only expresses ideas about the world and contributes to the built environment, but also engages in a social exchange, for, as Meskimmon (1996) argues 'the body is the site at which the social and the biological/psychological meet. It is not just a natural "given", but a constructed web of meanings and subject positions' (p. 201). Aesthetic learning in art and design is therefore a way to ensure that pupils retain a way of understanding the world that refutes the mind/body split of the logocentric curriculum.

Given the fundamental role of aesthetics in learning it is interesting to see how it is often most associated with the reception of art, an aesthetic experience, particularly as it is explored within philosophy (see Chapter 12). The aesthetic experience is often regarded as beyond words. It is a very complex phenomenon; any generalisations are as likely to obscure as to illuminate. However, it has been associated with the following terms: auratic, sublime, intense, illuminating, intuitive, emotionally gratifying, uplifting, enervating, sensual, significant, spiritual, liminal, disturbing, and can therefore be seen as something beyond the ordinary.

It is defined by the *Pocket Oxford Dictionary* (1992) as:

> aesthetic -adj. **1** of or sensitive to beauty. **2** artistic; tasteful.
> -n. (in pl.) philosophy of beauty, esp. in art.

What role did aesthetics play in your own art education?
Was the term referred to during your degree course, if so in what context?
Record your answers and compare them with other members of your tutor group.
Try and identify any common experiences.
Did the following have any influence: degree specialism, year of degree?

Parsons (1987) and Taylor (1992) are noted for promoting the aesthetic dimension of art and design education. Parsons proposed a cognitive developmental account of aesthetic experience which brings together Piaget's stages of scientific thought with Kolhberg's stages of moral judgement (pp. xii–xiii). Parson's stages are summarised under the following headings:

- favouritism
- beauty and realism
- expressiveness
- style and form
- autonomy.

Like Piaget and Kolhberg, Parsons sees development as progressive or incremental: a sequence of steps, each one a new insight toward a mature understanding of art, in his terms, a 'more adequate' understanding. Taylor (1992) promotes Abbs' concept of a dynamic aesthetic field in which responding, evaluating, making and presenting form a 'highly complex web of energy linking the artist to the audience' (p. 3). He devises a framework for pupils to engage with artworks: content, form, process, mood. Taylor asserts that this 'provides an invaluable means of empowering young people so as to enable them to enter effectively into the aesthetic field' (*ibid.*: 69). Taylor was influential in the 1980s for promoting a resource-based approach to art and design education. However, in line with the NC he continued to promote an approach which, beginning with Fry's 'significant form' culminated in Greenberg's formalism (Frascina and Harrison 1982). This is based on the assumption that responses to high art and some natural forms are universal, that art's intrinsic properties can profoundly affect the viewer on an emotional and even transforming level. Taylor's work with the National Gallery (1999) on universal themes can be seen as a continuation of this somewhat limited method. He also builds on Hargreaves' 'traumatic theory of aesthetic learning' to promote the related notion of 'illuminating experience' and 'aversive experience' (Taylor 1986: 18–34).

Hargreaves (1983) makes the distinction between the incremental and the traumatic theories of learning in art. The traumatic accounts for those aspects of aesthetic learning which cannot be explained by the incremental. The viewer finds their response to the artwork intense, even disturbing, and this has a powerful impact on their learning and long-term memory. Although it may be possible to learn from such an experience it is not one that can be taught or in any way accurately predicted.

Such 'traumatic experiences' occur only occasionally and many people may never respond in this way. It is a reaction reserved for the work of others; reception, not one experienced through engagement with materials and processes; production.

Abbs (1987) reiterates the arguments outlined in the influential Calouste Gulbenkian Report (Robinson 1982) reinforcing the notion of the aesthetic as a distinct way of knowing. He agrees with Fuller (1993) that aesthetic experience begins with the sense responses of individuals and then radiates out to become a 'shared symbolic order'. Critics of this view assert that aesthetic values can be construed as elitist and should be replaced with a more critical, sociological analysis. Abbs (1987) defines the term:

> Aesthetic derives from the Greek word *aesthetic* meaning *things perceptible through the senses*, with the verb stem *aisthe* meaning: to feel, to apprehend through the senses. Here in this small cluster of words: perception, sensing, apprehending, feeling, we begin to discern the nature of the aesthetic mode.
> (p. 53)

This cluster makes it clear that for Abbs the aesthetic is a fully cognitive mode, parallel but distinct from logical and discursive modes.

In differentiating between aesthetics and logic you have to consider how these two ways of knowing differ and relate. A percept is the mental product of perceiving and is peculiar to a specific experience. Aesthetic response is the means to differentiate between percepts that are, or are not, pleasing: it is a process of synthesis that appears to be immediate, total. A concept is the mental product of conceiving, an abstraction, a system of classification which assists in forming patterns of predictability. Logic is analytical and cumulative, it organises concepts to solve problems. However, it is not the only means, for example you have already encountered the process of unconscious scanning (Unit 3.2). Aesthetics defines the taste for things; logic defines the application of concepts. Both ways of knowing are liable to inform decision making in the production and reception of the arts. We call this 'aesthetic literacy' which is grounded within the field of visual culture rather than within some notion of high art.

Grounded aesthetic

Willis (1990b) coined the term 'grounded aesthetics' in an attempt to retrieve the term 'aesthetic' from its traditional western fine art preoccupations with 'abstract or sublimated qualities of beauty'. He recalls how a particular pop song can suddenly evoke and come to represent an intense personal experience and how a dramatic episode on a television 'soap' can parallel and illuminate personal problems and dilemmas. Such moments can make personal experiences more understandable, more controllable. Willis attempts to lift aesthetics off its pedestal. He insists that young people need to engage with aesthetics not only because of its intrinsic values but because of its capacity to produce social meanings. Grounded aesthetics ensures that learning takes place within an expanded field of visual and material culture. In the past, the hierarchical and canonic view of 'Art' has marginalised if not dismissed the home culture of most pupils. It is important that you question this separation by investigating issues of taste.

Taste

What is taste? The dictionary definition (*Shorter Oxford* 1993) links it closely with judgement and aesthetics: 'mental perceptions of quality, judgement, discriminative faculty . . . aesthetic discernment'. Pateman (1991) offers three other possibilities:

1 Taste is a faculty of mind, innate and equal in virtually everyone, which is capable of education, and by means of which we distinguish (or are struck by) the difference between the beautiful and the ugly, the tasteful and the tasteless.
2 Taste is the faculty of the mind, innate and very much unequal, which is capable of education, etc.
3 Taste is a set of socially inculcated dispositions or culturally transmitted abilities to respond to and distinguish objects, events, etc. as beautiful or ugly, tasteful or tasteless. What counts as beautiful is itself socially and culturally constructed as part of the construction of taste, and may vary without any evident restriction. Taste is always differentially distributed: some people get it, some don't. It thus can act as a social discriminator or distinguisher, and is used by individuals to define and distinguish themselves socially.

(p. 175)

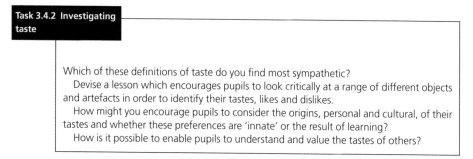

Task 3.4.2 Investigating taste

Which of these definitions of taste do you find most sympathetic?
Devise a lesson which encourages pupils to look critically at a range of different objects and artefacts in order to identify their tastes, likes and dislikes.
How might you encourage pupils to consider the origins, personal and cultural, of their tastes and whether these preferences are 'innate' or the result of learning?
How is it possible to enable pupils to understand and value the tastes of others?

Critical aesthetics

The well-known phrase 'beauty is in the eyes of the beholder' suggests that aesthetic appreciation is something personal and possibly inexpressible. However, many within the field of aesthetics refuse to accept that aesthetic experience is beyond articulation, arguing that such experiences require reflection, deliberation and justification if they are to become socially meaningful. For example, Crowther (1996) suggests that the search for objectivity in 'critical aesthetics' can facilitate the deepening of aesthetic experiences. Lack of familiarity with the artistic conventions of a culture may make it difficult to fully appreciate and understand its material and visual artefacts; judgement may be rooted in personal taste based on different and conflicting criteria. This is not to suggest that the codes and conventions of producers are the only criteria for aesthetic judgement, rather that judgements of taste are contingent upon personal experience, exposure to and knowledge of familiar discourses (acculturation). A richer understanding is afforded by knowledge of motivations, intentions and art as social production.

The claim that taste is universal is no longer tenable, not only in relation to determinants such as social class, race and ethnicity but also in relation to gender and sexuality. Abbs (1987: 31) defines feminism as a method of literary and artistic criticism, a movement away from the primary aesthetic engagement and encounter. But it is important to recognise that a cultural object addresses a gender specific audience and at the same time contributes to the 'gendering' of that audience (McRobbie 1994).

It is all too easy to claim that the subjective nature of our responses to art makes it difficult, if not impossible, to teach aesthetic appreciation. Davey (1994) insists that it is important for aesthetics to have a place within art education, but that first teachers have to address some of the 'strangulating preconceptions about philosophical aesthetics' and understand:

> . . . when art works are either historically or culturally juxtaposed . . .
> questions of aesthetics no longer involve the study of pure appearances but
> an understanding of how art facilitates the coming into appearance of
> meaning, an epiphany, which simultaneously reveals and binds us to the
> cultural horizons we are within.
>
> (p. 74)

Davey's article provides a compelling argument for the inclusion of aesthetics in the art and design curriculum, one that replaces the Kantian notion of 'pure perception' with a phenomenological account of 'significant perception'. He suggests that aesthetic experiences are highly relational or associative and only evoke a response when they are personally significant.

As you can see, aesthetics is a difficult concept to define. A fixed, traditional definition confines it to the history books. Rather than resort to this, you need to consider its place in relation to contemporary practice by reflecting on its relation to both reception and production taking into account contradictions and possibilities.

Further reading

Abbs, P. (1987) *Living Powers*, London: Falmer Press.

Pollock, G. (1988) *Vision and Difference*, London: Routledge.

Taylor, R. (1992) *Visual Arts in Education*, London: Falmer Press.

UNIT 3.5 COMPLEMENTARY WAYS OF LEARNING

By the end of this unit you should be able to:

- define tacit learning and intuition;
- identify their significance in art and design;
- consider the extent to which tacit learning and intuition have informed your practice in art, craft and design.

Tacit learning

Polanyi's theory of cognition (1964) successfully overcomes the traditional dichotomy between making and thinking. His explanation of tacit knowledge, the kind of knowledge that we cannot fully articulate, is commonly used by artists to account for their difficulty in explaining how they have worked their materials and the skills they have deployed in making. Cognition, according to Polanyi, constitutes a continuum between tacit and, its opposite, explicit knowledge.

In formal education, explicit knowledge is highly valued. Explicit knowledge can be articulated conceptually and in most areas of the curriculum it is thought to be the only kind of knowledge there is. In contrast, tacit knowledge arises in and through what Polanyi calls 'indwelling', a kind of empathetic participation: learning through example, learning by trial and error. Learning a tacit skill involves going through a series of 'integrative acts' by which a pupil grasps the full meaning of a process. This requires imitation, practice, repetition and complete immersion: it takes time. Heidegger (1954) talks about the relationship between making and thinking: 'Every motion of the hand in every one of its works carries itself through the element of thinking, every bearing of the hand is rooted in thinking' (p. 16).

How then can this be taught? The constrained timetable at KS3 militates against prolonged engagement with making. Tacit learning is allied closely to the apprenticeship model of learning, a model that bears some relationship to aspects of vocational education in FE and may increasingly return in the form of applied diplomas in schools (see the DFES website www.dfes.gov.uk/14–19/index.cfm?sid=3 for further information). Unless you can provide some continuity in practice, learning with materials is liable to be superficial and quickly forgotten. There are voices in art education arguing that it may be more productive to limit the range of materials and processes so that pupils come to know a few, and know them well (Mason 1995).

Task 3.5.1 Tacit learning

Look at Schemes of Work used in your placement school, identify:

- those practices that can be associated with tacit learning;
- the extent to which continuity is considered both in terms of prior knowledge and available time;
- in what ways tacit learning can enable progression.

It is essential that you recognise and build upon the tacit knowledge pupils have developed in other curriculum areas and outside the school. For example, many pupils have particular skills in ICT which can prove a foundation from which to develop art and design specific exploration. We predict that it will take some years before the majority of teachers' ICT skills match those of their pupils: here is a particular case for reciprocity. There is a middle way between tacit and explicit learning. The terms 'intelligent making' and 'practical thinking' (Burgess and Schofield 1998) provide this bridge.

Intuition

Art and design is one of the few subjects in the school curriculum that has recognised the significance of preconscious, cognitive processes for the development of the imagination. The American philosopher John Dewey (1934) emphasised the central position of intuition for creativity; although he would concur with the notion that creative acts are 90 per cent perspiration and only 10 per cent inspiration:

> 'Intuition' is that meeting of the old and the new in which the readjustment involved in every form of consciousness is effected suddenly by means of a quick and unexpected harmony which in its bright abruptness is like a flash of revelation; although in fact it is prepared for by long and slow incubation.
>
> (p. 266)

Bastick (1982) explores intuition by identifying a series of associated properties which he also relates to the term 'insight':

1. Quick, immediate, sudden appearance
2. Emotional involvement
3. Preconscious process
4. Contrast with abstract reasoning, logic or analytic thought
5. Influenced by experience
6. Understanding of feeling
7. Associations with creativity
8. Associations with egocentricity
9. Intuition need not be correct
10. Subjective certainty of correctness
11. Recentring
12. Empathy, kinaesthetic or other
13. Innate, instinctive knowledge or ability
14. Preverbal concept
15. Global knowledge
16. Incomplete knowledge
17. Hypnogogic reverie
18. Sense of relations

19 Dependence on environment
20 Transfer and transposition.

(p. 25)

> Which of Bastick's properties do you relate to your practice as an artist?
> To what extent do you think intuition is antithetical to critical practice?
> Identify in your teaching, actions which you would ascribe to intuition.

Bruner (1960) contrasts analytical and intuitive thinking, suggesting that they are complementary in nature:

> . . . intuitive thinking characteristically does not advance in careful, well-defined steps. Indeed, it tends to involve manoeuvres [sic] based seemingly on an implicit perception of the total problem. The thinker arrives at an answer which may be right or wrong with little if any awareness of the process by which he [sic] reached it. He rarely can provide an adequate account of how he obtained his answer, and he may be unaware of just what aspects of the problem situation he is responding to. Usually intuitive thinking rests on familiarity with the domain of knowledge involved and with its structure, which makes it possible for the thinker to leap about, skipping steps and employing short cuts in a manner that requires a later rechecking of conclusions by more analytical means whether deductive or inductive.

(p. 58)

In an age of accountability in which clearly identified learning objectives are a prerequisite, the notion of intuition may appear unhelpful because definitions are both vague and all-embracing. The properties that Bastick identifies can easily be confused with such concepts as the imagination and creativity, and Bruner's explanation bears a close resemblance to Polanyi's definition of tacit learning.

The ability of the mind to process information in ways other than cognitive reasoning is frequently overlooked or marginalised in the school curriculum. 'Intuition' 'is abused by ordinary people who want to avoid thinking, by philosophers who without much examination dismiss it as mere subjective hunch opposed to reason' (Reid 1986: 28). Intuition is often used as a catch-all phrase to describe pupils whose learning does not easily correspond to the logocentric curriculum: pupils who may find themselves being diagnosed as having learning difficulties. We wish to propose a definition of intuition as a synthesis between the preconscious and unconscious processes identified by Bruner, Ehrenzweig and Polyani. Ehrenzweig advocated art educators recognise the importance of unconscious processes for creativity; Bruner's reference to 'the implicit perception of the total problem' can be related to Ehrenzweig's 'syncretism' which he describes as pre-analytical; Bruner's reference to 'familiarity with the domain of knowledge involved' can be related to

Polanyi's 'indwelling'; Bruner's reference to 'skipping steps and employing short cuts' equates with Ehrenzweig's notion of unconscious scanning.

UNIT 3.6 LANGUAGE, MOTIVATION AND LEARNING IN ART AND DESIGN

By the end of this unit you should be able to:

- identify different methods and strategies to motivate learning in art and design;
- consider the role of language for learning in and through art and design.

It is evident from the diverse approaches and methods outlined in the previous units in this chapter that learning is a multi-faceted process. Within the limitations of the school timetable how is it possible to ensure a range of methods to differentiate for pupils' learning needs?

Motivation

Motivation, or the will to do, is central to education: without it any learning is liable to be short term or superficial. Until the 1960s theories of motivation were dominated by two schools, the psychoanalytical and the behaviourist. For Freud motivation was conditioned by the tension between the primary, innate drives of survival, e.g. hunger, sex, communication, and the secondary, rational processes of the ego which modify instinct in relation to social and cultural conventions. The emphasis here is on internal drives constrained by external factors, the social possibilities of satisfying the pleasure principle. Behaviourists suggest that motivation is essentially a reaction to external stimuli and that these provoke certain patterns of behaviour. Thus, habits of learning can be formed through such processes as repetition and imitation and are reinforced positively or negatively according to past experience. Since the 1960s these deterministic models have been layered and questioned by theories that promote more interactive processes. Social psychology proposes that an individual's personality interacts with variables such as class, race, geography, culture, so that motivation is conditioned by social environment. Cognitive psychology also recognises the social environment as a constructive element but emphasises the formation of knowledge through a dynamic interaction between it and an individual's innate cognitive processes. The educational implications of these general theories can be found in Capel *et al.* 2005 Unit 3.2: Table 3.2.1, pp. 123–124).

There are many ways in which you can motivate pupils to learn: the following list indicates some of the most effective:

- *Differentiate* individual needs to promote personalised learning.
- Create a *safe environment* in which *risk* can take place.

- Infect through your *enthusiasm*.
- Communicate your *high expectations* to pupils.
- Enable pupils to share and take *ownership of ideas*.
- Develop *critical reflection* so that pupils themselves can effect change.
- Develop pupils' competence in subject specific and transferable *skills*.
- Provide pupils with constructive *feedback*.
- Acknowledge successful learning through *praise* and *positive reinforcement*.
- Catch pupils *doing things right*.
- Recognise the significance of *self-* and *peer-group esteem*.
- Allow for the possibility of *pleasure*.

Language for learning in art and design

Learning in art and design mostly requires working with objects that communicate through non-verbal means. It is something of a paradox therefore, that you often have recourse to words in order to communicate clearly and efficiently. The mediation of language is central to a critical approach to learning and your command of verbal skills not only benefits your teaching but your advocacy of the subject within your placement schools and the wider educational community.

For Vygotsky (1986) the potential for learning is revealed and realised in inter-actions with 'more knowledgeable others'. One of his main contributions to the understanding of learning is the concept of the 'zone of proximal development' which refers to the gap between what an individual can do alone and what can be achieved with the help of 'more knowledgeable others'. For Vygotsky, the foundation of successful learning is co-operation and the basis of that success is communication and language. He believes that children solve practical tasks with the help of their speech as well as their eyes and hands. This includes inner speech, thinking things through for oneself, as well as explaining to others.

The recognition of the important role of language for learning in art is well established. Field (1970) insists that art is more than making, it is also about appreci-ating. He notes that art educators find it extremely difficult to grasp the 'inwardness of the aesthetic experience mode' without experiencing it for themselves, without 'making' art. Field goes on to insist experience is not enough. Art and design educa-tion should not merely enable pupils to 'think' in the subject, to approach it from within, but also help them to see it in a wider context. To do this a pupil needs to learn to become articulate 'to be able to discuss the nature of art experience, the criteria for art, the purpose of art' (p. 111). Field believes this must go beyond private conversations between pupils and their teachers to the formation of a common language, a language which pupils must learn to apply in formal discussion.

Right from the start, you need to consider the language you use in the classroom and find a means to communicate clearly without diluting the specificities of the subject. The NC (DfEE 1999) provides a framework. However, what at first sight may seem clearly defined is open to interpretation.

Task 3.6.1 Interpreting the language of the National Curriculum

In small groups, define NC terms or phrases.
  In your tutor group, compare the explanations. Try to come to some consensus.
  Write down your agreed terms and consider whether they are accessible to a KS3 pupil. If not, 'translate' them. Test their efficacy in the classroom: record pupils' responses and usage.

Try to be consistent in the way you use and explain NC terms. You can determine whether the 'official' language has been assimilated by pupils when you analyse the language they use to evaluate their own and others' work. You must begin from this critical point of view and expect your pupils to do likewise. Encourage them to question at all times. If they have not understood your explanation do not dismiss their concerns but work with them to construct definitions.

Developing language

The KS3 Strategy for Literacy (DfES 2001a) recommends that art and design teachers consider some of the ways that work in art, craft and design develops speaking, listening, writing and reading skills, alongside ways in which language can enhance an understanding of the subject. The emphasis here is on transferability; a one-way process. Eisner (1998) has pointed out the dangers of justifying art and design by the way it serves other subjects. He warns teachers against the temptation to defend the subject in ways that indicate its subservience to core skills. You need to consider the reciprocal relationship between word and image, number and artefact. Literacy and numeracy can serve art and design just as effectively as it can serve them.

Osbourn (1991) has suggested that learning to respond to visual stimuli is 'a two way process in which language is used to communicate perceptions and understanding while perceptions and understanding serve to stimulate language' (p. 33). She challenges a traditional view of art appreciation, one that involves a 'silent viewer, cut off from reality and wrapped in his or her own thoughts' (*ibid.*). Instead she insists that responses to works of art should be discussed and debated and that by encouraging pupils to do this you develop their ability to discriminate, analyse, scrutinise, interpret and communicate understanding.

Franks (2003) looks at the relationship between language as a mediating tool for teaching making in art and design. He reminds you that all interactions are multimodal and therefore any utterances cannot be divorced from the actions that accompany them or that are the result of instruction, description or suggestions. He claims: 'So for example, it becomes a problem if meaning-making is reduced to a circumscribed definition of "literacy". This kind of reduced plan cuts across and shapes and squeezes into conformity the multimodal capacities shown by young people in their imaginative play' (p. 83). One way in which you can ensure that language is not only used as a regulative tool of behaviour management and instruction is to develop a culture of questioning where you and pupils continuously clarify, challenge and negotiate meanings and practices. This culture of questioning can have its formal moments by adopting the habit of bringing pupils together in plenaries, whether these are at the beginning, midway through or at the close of lessons (DfES 2004a).

a) Forming questions

You have already introduced pupils to a subject-specific language and explained and or modified it depending on their responses. Use it consistently in your introductions, demonstrations, class discussions and reviews. You need to reinforce these terms by highlighting key words and phrases in your planning and class reference materials and by using them in oral and written communication, including assessment. If you use unfamiliar jargon or a laissez-faire approach to questioning, pupils can feel threatened. It is therefore necessary to prepare questions carefully, aligning them to the learning objectives you have identified for your pupils (Capel *et al.* 2005: Unit 3.1).

Questions can be defined as:

- open: open to experience and interpretation, e.g. How do you think women are represented in advertising?
- closed: there is only one answer, often factual, e.g. What material is this sculpture made from?
- pseudo: there is only one answer and it is known to the questioner, e.g. What is my name?
- framed: open to interpretation within a given framework, e.g. Among the contemporary crafts-work we have investigated, which artefact bridges the fine/applied art divide most effectively?

When instigating discussion and debate or when questioning pupils about their own and others' work, teachers often use questions such as: 'What do you think?' 'What can you tell me about this?' or 'How does it make you feel?' (Meecham 1996: 72). The pupils' response is usually an unqualified value judgement: 'I like it', or, 'It's rubbish'. The further question, 'why?' can yield embarrassed inarticulacy or evasion.

> Given particular subject matter or a particular concept, it is easy to ask trivial questions or to lead the child to ask trivial questions. It is also easy to ask impossibly difficult questions. The trick is to find the medium questions that can be answered and that take you somewhere.
>
> (Bruner 1960: 40)

Initially, most pupils respond more readily if questions are concise, focused and related to their own experience and prior learning. These do not have to be questions of known 'fact' (pseudo) but should encourage pupils' descriptive, analytical, deductive or speculative skills. Open questions allow for individual, subjective answers. These have their function, but only once pupils feel confident that their opinions will be taken seriously, and used to extend the debate. However, on some occasions it may be necessary for you to challenge pupils' preconceptions. You should avoid being confrontational. Refer to Chapter 10 for examples of the types of question you can use to investigate works of art; refer to Chapter 7 for examples relating to the evaluation of pupils' work.

b) Enabling discussion

To promote a critical curriculum in art and design it is essential that you develop the classroom as a site for dialogue, a place where talk is productive and meanings negotiable. Small group discussion is likely to facilitate exploratory talk, particularly when you are introducing new concepts. In contrast, whole class discussion may sometimes be intimidating and inhibit pupils from contributing. You need to acknowledge that making meaning through talking takes time, just as when writing, several drafts are required. Initial discussion is likely to be searching, and can include half or incomplete sentences, awkward syntax, even muddled or contradictory ideas: this is particularly noticeable when pupils are asked to provide personal responses or information. It is important as a teacher that you recognise and value such hesitant, tentative, half-formed ideas. The teacher doesn't always have to know the solution; in art and design there doesn't always have to be a right answer. You need to provide pupils with the opportunity to discuss and test out new concepts in the safety of small peer or friendship groups before they feel confident enough to present ideas in a wider forum.

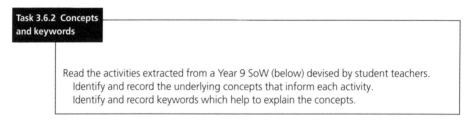

Task 3.6.2 Concepts and keywords

Read the activities extracted from a Year 9 SoW (below) devised by student teachers.
Identify and record the underlying concepts that inform each activity.
Identify and record keywords which help to explain the concepts.

Scheme of Work: Beyond weaving

Aims

- pupils consider and define craft practice;
- pupils experiment with and develop weaving techniques in relation to three-dimensional portraiture;
- pupils respond to given materials;
- pupils use assessment and evaluation to develop work.

Activities

Research and investigation

- mind-mapping craft practice;
- discussion around selected representational images;
- introduce weaving and construction materials and techniques;

- in groups investigate initial responses and experiment with materials by categorising, comparing, contrasting and considering their representational possibilities;
- introduce sketchbooks as a research resource;
- homework: select found materials to incorporate in weaving.

Developing ideas through making and evaluation

- experiment further to identify materials suited to a representational function;
- construct a framework for three-dimensional weaving;
- begin weaving;
- pupils review their work by assessing it with a partner and evaluating it in their sketchbook;
- the teacher introduces further resources and contextual information each session, inviting pupils to question boundaries between fine art, craft and design;
- as the SoW progresses pupils present their work to larger groups: final crit. in the form of a whole class evaluation.

(PGCE students 1998)

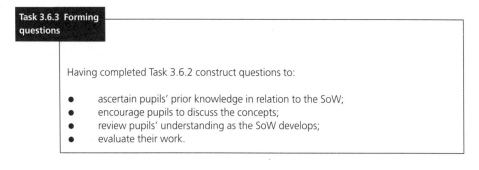

Task 3.6.3 Forming questions

Having completed Task 3.6.2 construct questions to:

- ascertain pupils' prior knowledge in relation to the SoW;
- encourage pupils to discuss the concepts;
- review pupils' understanding as the SoW develops;
- evaluate their work.

Mind-mapping and concept mapping

Mind-mapping is usefully deployed as a whole class or group activity in order to explore ideas for practical and critical investigation. It involves choosing a key word to represent a theme or issue. A key word is presented to pupils as a trigger to invite related ideas. It should be a quick activity, almost a word association game, so that ideas come thick and fast. The key concept is written in the centre of a flipchart and associated words recorded around it: this is often referred to as a spider diagram (Figure 3.6.1). Where one association derives from a word other than the key, it is connected to its originating term. You may find that initial responses take the form of cliches and stereotypes: these should be acknowledged but you can encourage pupils to extend their frames of reference by repeating the activity without duplicating words from the first version.

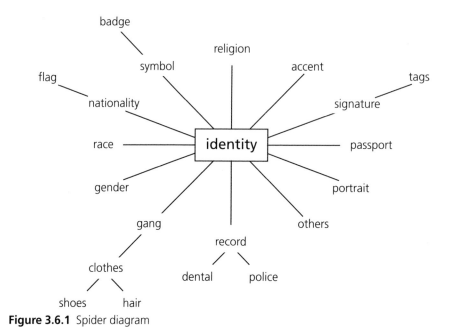

Figure 3.6.1 Spider diagram

Buzan (1982) provides a useful practical guide to these and similar processes although it is not art and design specific. Creber (1990) discusses Buzan's methods: '[Mind-mapping] offers what Witkin might call "a holding form" without the interference caused by a premature inner demand for explicitness and order in what should be the crucial, thinking, stage of the operation' (p. 49).

Concept mapping takes this process a stage further by grouping words into categories and defining them in terms of priority and possible developments. Figure 3.6.2 groups the associations made in Figure 3.6.1 into four categories:

Group	Personal	Records	Signs
others	gender	passport	flags
nationality	shoes	dental	badge
religion	tag	police	tags
gang	signature		signature
race	clothes		symbol
gender	hair		shoes
shoes	passport		clothes
badge			hair
passport			
clothes			
hair			

Figure 3.6.2 Identities

This grouping helps pupils identify an area for investigation. They might consider why some terms fall into more than one category and why others are closely identified with just one.

Task 3.6.4 Mind-mapping and concept mapping

In pairs, Mind-map the theme 'sense of place'.
Exchange your diagram with a neighbouring pair and translate it into a concept map.

Task 3.6.5 The role of language in art and design

Consider the following quotations by artists, critics and educators concerning the complementary and conflicting role of language in art and design education.
Which one most succinctly represents your view?
Which one do you find most problematic?
Compare your thoughts with other students.

The initial appraisal of images is therefore a complex cognitive process which integrates the visual and the verbal questioning in such a way as to contribute extensively to the development of cognitive skills. The cognitive value of instruction in the arts is seriously underestimated. By promoting talk and thereby making understandings and responses more public, this view can be effectively changed.

(Osbourn 1991: 54)

The published statements of artists are often found to be elliptical, contradictory, evasive and rhetorical . . . yet it is rare to find an inarticulate artist . . . there is a way in which the artist too is an onlooker, a beholder of his or her work, and suffers, no less than any other member of the audience, from the problem of defining just what has been thought, made and achieved. A mixture of fantasy and conjecture, anecdote and metaphor are the likely companions of a working process which is as much concerned with concrete things as it is with words. It is often the case, too, that the interpretation and the theoretical position of the critic, the interlocutor, is at some variance with what the artist believes; they stand at different sides of the artwork. The work itself – seeming all too often in danger of disappearing under a superfluity of words – is where the artist and audience meet. This is where the discussion begins.

(Searle 1993: 1)

Meanings belong to culture, rather than to specific semiotic modes. And the way meanings are mapped across different semiotic modes, the way some things can for instance, be 'said' either visually or verbally, others only visually, again others only verbally, is also culturally and historically specific. . . . But even when we can express what seem to be the same meanings in either image-form or writing or speech, they will be *realised* differently. For instance, what is expressed in language through the choice between different word classes and clause structures, may, in visual communication, be expressed through the choice between different uses of colour or different compositional structures. And this will affect meaning. Expressing something verbally or visually makes a difference.

(Kress and Leeuwen 2006: 2)

I think of words as an invisible material; manipulating punctuation is like focussing the attention on certain points in the painted surface of a pot.

(Britton 1991: 4)

Turning art experience into language also carries its dangers. Words are prisons as well as searchlights and pigeonholes, for what we see and comprehend. So a vocabulary which is provided by the teacher rather than invented by the pupil may constrain and regiment our seeing and interpreting. The rate at which new and specialist vocabulary is added to pupils' existing personal vocabulary needs sensitive and teacherly monitoring, so that the elastic between the new to the familiar is never stretched to breaking.

(Stibbs 1998: 203)

Further reading

Franks, A. (2003) 'The role of language within a multimodal curriculum', in N. Addison and L. Burgess (eds) *Issues in Art and Design Teaching*, London: RoutledgeFalmer.

Kress, G., Jewitt, C., Ogborn, J. and Tsatsarelis, C. (2001) *Multimodal Teaching and Learning: The Rhetorics of the Science Classroom*, London: Continuum.

UNIT 3.7 ENABLING LEARNING: TRANSFORMING SUBJECT KNOWLEDGE INTO PEDAGOGICAL PRACTICE

This unit explores ways in which you can reinforce, extend and develop your subject knowledge. It is important that you reflect upon your own learning in art, craft and design before attempting to transfer, wholesale, practice that may be inappropriate for learning in a classroom context. However, it is possible that, although your practice is seen as falling outside the orthodoxies of 'school art', it corresponds to the NC and examination learning objectives. You should consider the range of practices endorsed by the NC and examination syllabuses and how your own practice can inform and develop it.

By the end of this unit you should be able to:

- recognise the origins of existing orthodoxies, their uses and limitations;
- consider ways to transfer knowledge and understanding into good pedagogic practice;
- investigate methods and strategies and consider ways to use them in your teaching.

The following list, although not exhaustive, highlights productive and receptive practices which have become marginalised or neglected in art and design:

Productive:

- audience- and site-specific problem-solving
- visualisation and memory training – mind and word picturing
- using metaphor
- meaning making and material exploration
- synaesthesia
- associative, automatic and chance processes.

Receptive:

- comparison
- semiotic analysis.

You are asked to question the current validity of these practices and consider their application to the curriculum.

Audience- and site-specific problem-solving

When you direct pupils' attention to the needs of others you help them to understand the social functions of art, craft and design. This encourages pupils to consider the relationship between their own needs, tastes and preferences and those of potential audiences. For example, setting the task of designing packaging for household goods intended for the visually impaired ensures that pupils investigate the haptic, tactile and spatial properties of materials.

For a site-specific task you can ask pupils to design a playground for pre-school children. To do this pupils have to take into account: materials, safety, ergonomics, economics, aesthetics, etc., survey local playground provision and identify the needs of users, both children and adults. Following this research, they feed back findings before mind-mapping possible solutions. You can differentiate tasks and allocate them to individuals and small teams. More ambitious projects of this kind might involve pupils working with designers, including work placements, or with community artists on funded public art initiatives. A project of this type, to develop a crazy-golf course, attracted design submissions by such well-known yBa's as Gavin Turk and Damien Hirst (Millar 1998: 8). Pupils could be asked to develop their own design solutions and then compare and contrast their efforts with professional schemes. See also the 'Experimental Playground Project' (http://www.snugandoutdoor.co.uk) which is a partnership between architects, artists, contractors, engineers and other professionals with the users of playgrounds both primary pupils and carers/teachers.

Task 3.7.1 Site-specific SoW

Devise a SoW in which pupils design and construct a public monument (or model) for their school grounds. The monument should commemorate a significant event in the history of the school or local community.

Visualisation and memory training

Working from observation is one of the major orthodoxies in British secondary schools. This has its roots in a number of traditions; the academic, with its emphasis on emulating canonic exemplars; the perceptual, in particular Ruskin's advocacy of working directly from nature, and vocational education with its reliance on copying to facilitate technical competence. However, there is an alternative tradition that promotes the practice of working from memory. Horace Lecoq de Boisbaudran (1862) wrote a treatise propounding a method that was influential for Whistler's painting. His friend, T. R. Way, recalled the significance of this approach:

> I shall never forget a lesson which he gave me one evening. We had left the studio when it was quite dusk, and we were walking along the road by the gardens of Chelsea Hospital, when he suddenly stopped, and pointing to a group of buildings in the distance, an old public house at the corner of the road, with windows and shops showing golden lights through the gathering mist of twilight, said, 'Look!' As he did not seem to have anything to sketch or make notes on I offered him my notebook, 'No, no, be quiet,' was his answer; and after a long pause he turned and walked a few yards; then, with his back to the scene at which I was looking, he said, 'Now see if I have learned it', and repeated a full description of the scene, even as one might repeat a poem one had learned by heart. Then we went on, and soon there came another picture that appealed to me more than the former. I tried to call his attention to it, but he would not look at it, saying, 'No, no, one thing at a time'. In a few days I was at the studio again, and there on the easel was the realisation of the picture.
>
> (Spencer 1989: 106)

Dependence on working directly from the object, graphically or by using a camera, can blunt the power of the memory to recall sights and filter-out incidental details. As you devise programmes for image making you are advised to consider the difference between recording from observation, with its scientific and documentary credentials, and recording from memory, which is more selective. Your teaching should include opportunities for both to be developed.

In the sphere of children's education Marion Richardson explored methods of visualisation that depend, to some extent, on memory training. She was particularly keen that pupils develop imaginative responses to different forms of stimuli:

> Mind-Picturing involved the learner closing the eyes and allowing images of any type – figurative, non-figurative, ornamental, etc. – to appear in the 'mind's eye' whereas Word-Picturing consisted of carefully worded descriptions of actual events or paintings recalled by Richardson, or poems read by her, acting as stimulants for pictorial work. As in mind-picturing, the product was not predetermined, indeed the word-picture might or might not relate to the described image.
>
> (Swift 1992: 118)

The way in which mind-picturing is described here suggests a process similar to a sort of willed day-dreaming in which memories, in the form of images, are allowed to flood the mind. Such images can either be recorded immediately, using spontaneous methods, or developed over time using premeditated processes. In word-picturing the learner has to recall or imagine forms associated with or equivalent to given text.

Task 3.7.2 Using memory and imagination

Recall and record: the front elevation of your home, the face of a 'significant other'.

Attempt to conjure images using mind-picturing techniques: what materials have you made available to record your sensations/memories as they occur?

Choose a brief text, song, piece of music or smell as a stimulus and apply word-picturing techniques.

Metaphor

Metaphor and its close relative metonym are linguistic terms which denote a process of explaining one thing by relating it, respectively, to something similar (similarity in difference; her fortitude is explained with reference to an oak), or, something with which it is associated (a connection or attribute; throne stands in for royalty). Kress and Leeuwen (2006) suggest that all representation is a process of metaphoric transformation; in the following explanation they reference a child's drawing of a car (in the form of several circles/ellipses):

> . . . the sign-maker's interest at this moment of sign-making settled on 'wheelness' as the criterial feature of 'car'. He constructs, by a process of analogy, two metaphors/signs: first, the signified 'wheel' is aptly represented by the signifier 'many wheels' to make the motivated sign 'car'. . . . The sign is thus the result of a double metaphoric process in which analogy is the constructive principle. Analogy, in turn, is a process of classification: x is like y (in criterial ways). Which metaphors (and, 'behind' the metaphors,

which classifications) carry the day and pass into the semiotic system as conventional, and then as naturalised, and then as 'natural', neutral classifications, is governed by social relations of power. Like adults, children are engaged in the construction of metaphors. Unlike adults, they are, on the one hand, less constricted by culture and its already-existing and usually invisible metaphors, but, on the other hand, usually in a position of less power, so that their metaphors are less likely to carry the day.

(p. 8)

Visual and material metaphors and metonyms are analogous to linguistic ones in that they are representations: an image of a basket of fruit stands in for fertility; an abundance of hair animality. Traces of the making process can also be recognised as metaphoric/metonymic: erratic mark-making can stand in for agitation; associations can be invited between applied and rendered surfaces and attractive or repellent experiences. Metaphoric processes, while layering and enriching an idea, 'can simultaneously grasp the familiar and make it strange' (Davison and Dowson 1998: 253). Hercules' lion-skin cloak reinforces his physical strength but the neutral garb of a Victorian angel obscures his Biblical gender.

The following task suggests ways for you to explore the metaphoric potential of the properties of materials without any preconceived outcome in mind (see Plates: 1, 2, 3, 4).

Task 3.7.3 Exploring metaphor through material resources

A scrap bank has provided you with: rolls of telephone wire, damaged sponges, different sizes of cardboard tubes, skeins of wool, buttons, hooks, eyes and zips. These examples suggest types of material that might come your way.

In tutor groups collect a similar array of found materials and work through the following exercise:

Consider each material separately: What do they signify by their look? What do they signify by touch? What do you associate with them? Record your sensations and thoughts.

Place the materials in pairs and other multiples: follow the same process, see whether these combinations suggest different meanings.

Investigate each of the materials individually, testing their resilience, durability, malleability, etc.

Explore the materials in different combinations so that they interact, e.g. penetrate, envelop, entwine, intervene by cutting, distressing, piercing. etc. Record your associations.

Use these experiments as a basis for designing and constructing a garment or body adornment which is intended to either attract or repel an identified audience.

Devise a similar task or adapt this one for use with GCSE pupils.

In the above task the focus is on recyclable materials collected from scrap schemes and resource banks. These are sites where industry offloads surplus materials. For an annual fee schools can have regular access to these materials as long as they can provide their own transport. The result of this can be that you have a selection of arbitrary materials that do not obviously relate to your existing SoW. The task should enable you to take advantage of this situation to help pupils learn.

Synaesthesia

Synaesthetic approaches are concerned with the correspondence between different forms of sensory experience, the way that sights, sounds, surfaces, tastes and smells relate, or seem to relate, to one another. Synaesthetic responses are types of cross or multi-sensory metaphors in which one sense experience can suggest or even realise another: for example, a particular colour, deep purple, finds its metaphoric equivalent in a low chord on a cello, and the feel of thick velvet. The way young children learn about the world is mostly through sensory exploration and they begin to form associations between different sense experiences:

> Size, colour, texture, temperature, weight and plasticity are all aspects of the objects which a child can sensorily enjoy . . . This sensory awareness and discrimination in a young child is the basis of an intelligence which helps the child to bond to the earth and all the perplexing variety of sensations which life provides.
>
> (Gentle 1988: 37)

This notion of correspondence became very important to artists in the nineteenth century. In their respective fields Gautier, Baudelaire, Wagner and the Symbolists experimented with synaesthetic evocation and affect, Seurat with the relationship between angles and vectors and their related colours; Whistler gave his paintings musical titles; Debussy provided his Preludes for piano with pictorial ones. For many artists in the twentieth century, synaesthetic experiment was a key method for developing 'non-referential' modes. Kandinsky describes his synaesthetic theories amid the dubious metaphysics of his spiritual ruminations (Chipp 1968: 152–155) and more recent abstract artists such as Gillian Ayres have continued such experiments (see *The Sixties Art Scene in London* exhibition at the Barbican in 1993). Synaesthetic correspondence can be one way to make sense of the ambiguous combinations of image and material in 'surrealist' art, with its deployment of automatic processes and strategies of defamiliarisation: from Ithell Colquhoun's *Sea-Star 1* (1944) to Cathy de Monchaux's *Evidently Not* (1998) (Plate 21). Film, and more recently video, have been ideal vehicles for synaesthetic expression, although the emphasis on the spoken word in commercial cinema has limited its development. However, pop video and multi-media DVDs provide an ideal outlet for synaesthetic correspondence, for example, that between visual and musical rhythm or colour/graphic fields and ambient sounds (e.g. Tsukamoto's film *Tetsuo: the Iron Man*). To what extent these correspondences can be seen as universal, culturally specific or personal and contingent is open to debate: for instance how might Helen Chadwick's *Chocolate Fountain* (1994) encourage synaesthetic responses?

Task 3.7.4 Synaesthesia

1   Devise exercises for your pupils which explore the relationships between sound, shape and colour; texture, smells and colour, or any other combination.

Pupils can extend these exercises by exploring the correspondences between sense experiences and other affective states such as moods and emotions. If you or your pupils desire, representational elements can be introduced.

Compare and discuss the results with fellow students.

2   Devise a lesson introducing the concept of synaethesia to a KS3 class to help them respond to a text, a memory or a sense experience.

3   Consider how you might introduce synaesthetic elements into an animation project using multi-media.

Associative, automatic and chance processes

An immediate and accessible way to motivate pupils, particularly those who feel insecure in their technical abilities, is to introduce automatic and semi-automatic techniques as a starting point. We do not propose this strategy to subvert analytical procedures and aesthetic practice in a nihilistic way like some of the Dada artists, but more in the manner of the Surrealists for whom chance processes acted as a method for freeing up the imagination and loosening the hold of codes and conventions. As early as the fifteenth century Leonardo in his *Treatise on Painting* suggested that artists look at chance configurations on stained walls to elicit images. Even as controlled an artist as Degas is said to have been frightened by fresh sheets of paper and he would stain them with coffee grounds before beginning a drawing.

Ernst often used rubbings, the process of 'frottage', to form chance or arbitrary configurations into which he would see or project images: his *Histoire Naturelle* (1925–26) is entirely dependent on this process. Wolheim (1987) discusses the concept of 'seeing in' theorising its central place in human perception:

> Seeing-in as I have described it, precedes representation: it is prior to it, logically and historically. Seeing-in is prior to representation logically in that I can see something in surfaces that neither are nor are believed by me to be representations . . . I can, for instance, see headless torsos in clouds ranged against the vault of the sky. And seeing-in is prior to representation historically in that surely our remotest ancestors engaged in these exercises long before they thought to decorate their caves with images of the animals they hunted.
>
> (pp. 47–48)

Whether seeing-in precedes representation or is a type of transformative mental activity that is itself a form of representation, is open to debate. What is significant is that seeing-in appears to be an innate faculty and is therefore an accessible method for all to use. Inviting pupils to apply this faculty to their immediate surroundings can convince them that even the most prosaic and banal of environments may hold the potential for imaginative transformation. The school environment can be used as a resource to stimulate imaginative responses beginning with the rubbing or casting of surfaces such as wood-grain, stone and brick. The process of seeing-in can be applied to any circumstance. Ithell Colquhoun's view of herself lying in a bath (Chadwick 1991: 104), transformed from a nude into a rocky but nonetheless gendered seascape, is an amusing and troubling work, the process of which many pupils will both readily understand and be able to apply to their own work.

Another favoured method practised by Dominguez, Ernst, Leonor Fini and Ithell Colquhoun was decalcomania (Chadwick 1991). You can follow this process by spreading paint or other pigments over a non-porous surface such as glass or perspex. You then press either paper, card, canvas, etc. against this surface, pulling it off to reveal a tracery of interconnecting lines. The resulting forms are not unlike the veins on a leaf or certain fan corals. These forms can be used either as stimulus for 'seeing-in', as with Ernst's *Europe after the Rain* (1940–42), or as discrete elements for use in a collage or decorative schemes.

A more fully automatic process is the common practice of doodling; the type of drawing you produce when your conscious mind is engaged in another activity such as phoning a friend. Masson, Dali, Fini, Matta, Bacon, are particularly noted for using this process, although, once again they use it as a stimulus for seeing-in. The Abstract Expressionists developed a doodling-like process as a means to bring to the surface unconscious imagery, a process realised through painterly action. This strategy, at its limits, was used by artists like Pollock to subvert the figural potential of painting, to bypass representation (Clark 1999: 299–369). In schools the wholesale imitation of the 'look' of such work, without the contexts and motivations of its making, is an absurd spectacle. Such strategies may however provide pupils with ways of working that they can use to make their own metaphors.

An engaging Surrealist practice, the 'Exquisite Corpse' is more familiarly known in the UK as the party game 'Consequences'. Recall this game and relate it to the Surrealist activity. The sort of chance encounters that are the result of this process can be similarly activated by choosing images at random and juxtaposing them to form unlikely, amusing, contradictory or troubling combinations. This can be developed as a more conscious activity and used to construct meanings which require the viewer to make or seek connections which may not, at first sight, appear logical or coherent. Man Ray and Meret Oppenheim produced notable objects which subvert the utility of domestic objects, respectively *Object to be Destroyed* (1923) and the *Fur-lined Object* (1936). Ernst conjured unexpected, dreamlike images using photomontage, Hannah Hoch brutally satirised bourgeois conventions using popular sources to subvert its proprietous tastes, Heartfield juxtaposed contrasting and opposing images to undermine the credibility of Nazi ideology (Ades 1976). More recently, artists have taken up the decentring and dislocation of Dada photomontage and the Surrealist object to construct phobic, abject installations and objects; e.g. Matthew Barney, Louise Bourgeois, Robert Gober, Mona Hatoum, Damien Hirst, Sarah Lucas and Rona Pondick.

Task 3.7.5 Chance and automatic techniques

Try out some of the chance and automatic techniques cited.
Go through the process of elaborating the outcomes by seeing-in.
Devise a SoW which incorporates one or more of the techniques as a starting point.

Comparison

Comparative methods can prove useful in engaging pupils' attention and inviting them to investigate the work of others (Dyson 1989: 129–132). You may wish to set up a comparative task by presenting pairs (or more) of reproductions designed to focus on points of similarity and difference. For example, Dyson proposes six types of comparison:

- art objects/every-day objects;
- different art objects with the same subject matter;
- pupils' own work/appropriate art objects;
- artefacts of different periods;
- objects, texts, etc. of the same period;
- art objects of a particular school or period.

Task 3.7.6 Making comparisons

In your tutor group:
think of additional combinations and their implicit questions;
note these and decide in what context and for what reason (learning objective) you would use them in a lesson.

Chapter 10 develops these ideas and provides additional methods.

Semiotics

Semiotics, or the science of signs, is increasingly recognised as an important field for art education. In recent years it has been associated more with theories of communication (Fiske 1982) and in schools, Media Studies (Buckingham and Sefton-Green 1994). However, most critical practice in art, craft and design can be recognised as a form of semiotic inquiry because it involves processes of analysing and decoding visual and other sensory signs (Addison 2005). Although increasingly there is good critical practice in art and design it is often dependent on individual teachers' interest and training and can lack a coherent theoretical basis. Kress and Leeuwen (2006) have formulated a method for analysing the objects of visual culture and claim that any

2-D image from the west can be analysed by applying their system. Although the authors emphasise the differences between visual and verbal modes of communication, their method is analogous to linguistic models. However, it manages to avoid the hierarchical dangers of some art historical investigation because it treats all manifestations of visual culture in the same way; it does not suggest a different method for investigating a special 'expressive' practice called art. Some critics have suggested that while Kress' theory of social semiotics is convincing when applied to media images, advertising, illustrations and to a lesser extent painting, it neglects three-dimensional forms, craft and most contemporary art practice. You should always approach universalising methods with caution but this book is well worth investigating.

See Chapter 10 for a development of this issue.

UNIT 3.8 LEARNING AS A CRITICAL ACTIVITY

We close this chapter by introducing you to a set of strategies for critical learning based on the work of Grossberg (1994) and others, which we offer up as a way to question those orthodoxies in art and design that have sedimented over recent years into the predictability of school art. Such questioning does not encourage you necessarily to reject traditional forms of learning but rather to evaluate it and renegotiate a place for it in a dialogue with critical, contemporary practices.

Critical learning

Grossberg (1994: 16–21) identifies four models of a critical and progressive pedagogy which he describes under the headings: Hierarchical, Dialogic, Praxical, Articulation and Risk. The 'hierarchical' addresses problematic bodies of knowledge. The teacher retains a traditional role in a position of authority, determining the frames of reference by which an issue can be discussed. Grossberg suggests that such an approach can assist 'emancipationary struggles' but that it is the teacher who defines what is right. The 'dialogic' shifts the balance of power, giving pupils and students a participatory voice. Because this approach acknowledges differing positions it questions the authority of teachers and thus asks them to reflect on their own position. However, unless pupils and students are empowered with critical and discursive tools it is ineffectual. The 'praxical', while demanding dialogue, is not content to leave it there. It requires that the educational community effects institutional change by challenging the structures of power. It can only do so by forming alliances with the wider emancipatory community and joining in their political struggle. The difficulty with this model is that once the tools which secure emancipation have been established they can become the hierarchical tools of a new orthodoxy. The 'pedagogy of articulation and risk' in art and design marks a subtle shift, bringing together the political idealism of the dialogic and the praxical while acknowledging that the expanding field of cultural practice requires people to make connections between what was once considered separate and ineffable and what is now theorised as social and personal. Grossberg defines this synthesis as an:

> . . . affective pedagogy, a pedagogy of possibility (but every possibility has to
> risk failure) and of agency. It refuses to assume that even theory or politics,
> theoretical or political correctness, can be known in advance. It is a pedagogy
> which aims not to predefine its outcome (even in terms of some imagined
> value of emancipation or democracy) but to empower its students to begin
> to reconstruct their world in new ways and to rearticulate their future in
> unimagined and perhaps even unimaginable ways.
>
> (Grossberg in Giroux and McLaren 1994: 18)

The implication here of a wholesale abandonment of authorial control by the teacher
has led critics to question the validity and practicability of this form of critical pedagogy. Ellsworth (1989) suggests it is all well and good in theory but it 'has developed
along a highly abstract and utopian line which does not necessarily sustain the
daily workings of the education its supporters advocate' (p. 297). Todd refutes this:
'One gets the sense [from Ellsworth] that critical pedagogical discourse has remained
virtually unchanged and the assumptions never questioned thereby encouraging
unself-reflexive teaching practices' (Todd 1997: 73).

As Hall suggests, 'a theory is only a detour on the way to something more important' (in Giroux and McLaren 1994: 17). This notion of an 'affective pedagogy' corresponds to hooks' 'engaged pedagogy' (1994) which is one of continuous flexibility
and flux. She insists that 'engaged pedagogy' does not offer a blueprint, rather it:
'recognises each classroom as different, that strategies must constantly be changed,
invented, reconceptualised to address each new teaching experience' (hooks 1994:
10).

How, as art and design teachers, can you make sense of this rhetoric so that it has
some impact in the classroom? Throughout this book we recommend methods and
strategies to enable experiential learning and interactivity (Unit 3.3), discursive and
critical inquiry (Chapter 10), the use of inclusive resources (Chapter 5) and a culture
of reciprocity (Chapters 3, 5, 12). You may have observed and used one or more of
these practices on your course, but it is unlikely that you have consistently used them
all. If you find ways to connect them so that they converge, albeit it at unexpected
nodal points, this collective and collaborative pedagogy can radically alter the culture
of the classroom.

Strategies for a critical pedagogy

Interventions

Pollock (1988) asserts:

> The structural sexism of most academic disciplines contributes actively to
> the production and perpetuation of a gender hierarchy. What we learn about
> the world and its peoples is ideologically patterned in conformity with the
> social order within which it is produced.
>
> (p. 1)

In Chapter 5 Burgess identifies ways in which the 'repressed question of sexual difference' can be raised so that 'the differentiation which is so manifest and so symbolic [is seen to be] produced' (Pollock in Addison and Allen 1997). This awareness impacts on other areas of difference, whether those of class, race, disability, age, sexuality, by using strategies of visibility and intervention. In addition to external interventions; institutional: galleries and museums; human: artists, craftspeople and designers, local communities; publications: teaching-packs, slides, DVDs, you must remember that the contribution of pupils, particularly in bringing in their home cultures, can act as an intervention into the official curriculum, what Shohat and Stam (1998) refer to as 'affective investments' (p. 14).

The work of Adrian Piper can be cited as an example of multiple intervention. Her video art, a medium that is in itself a technical intervention within the traditional art and design curriculum, confronts viewers with issues that raise repressed questions of identity: class-based, racial, sexual. We do not suggest that these issues are unproblematic, on the contrary, they may make the classroom a place of challenging interaction. To manage this it is essential that teachers construct a safe environment within which to orchestrate the classroom dynamic using systems such as debating structures, small group and whole class discussion and presentations (Unit 3.6). We do not advise you to attempt this type of interventionary strategy until you know pupils and their contexts well and have a confident knowledge of the work and issues under discussion. To begin, you could try these strategies with classroom teachers and experienced artists/educators. They are the sort of strategies that could usefully be deployed in a curriculum development project towards the end of your course.

Mediations

The preceding section makes claims for the position of contemporary practice in art education. This emphasis on looking across practice at any one given time encourages an interdisciplinary and dialogical approach. However, it is also important to consider histories, the means by which the past is preserved, presented and perpetuated. Material and visual culture is increasingly used as a primary form of evidence in the construction of genealogies, histories, trajectories and notions of essential identities. Paley (1995) recommends the use of 'mediations' recognising that the past can be a powerful base for developing thinking about the contemporary, but at the same time, insisting that you need to:

> Simultaneously honour and reformulate experiences associated with the historical without being trapped in either it or its objectification . . . don't position history as an absolute to be memorised and consumed in the manner of formal educational exercises, but as a point of 'mediation' that requires continuous sustained rereading, reworking and reconstruction.
>
> (p. 178)

Chapter 12 asks you to question essentialist constructions used to preserve the power of vested interests which have become naturalised as 'truths'. Gretton (2003) for example, warns that, 'Uncritical recourse to canonical works and figures will reproduce the canon as something natural and inevitable, not as an institution in which social structures, relations of cultural power, are reproduced and legitimised' (p. 181). He proposes what he terms a critical pragmatism and argues that 'we need to find acceptable ways of dealing with it [the canon] in our teaching, ways which neither worship the sorts of cultural power the canon represents nor stare down cultural relativism with cultural authority' (p. 179). Giroux (1992) reminds you:

> The radical educator deals with tradition like anything else. It must be engaged and not simply received. Traditions are important. They contain great insights, both for understanding what we want to be and what we don't want to be. The question is: in what context do we want to judge tradition? Around what sense of purpose? We need a referent to do that. If we don't have a referent then we have no context to make sense of tradition.
>
> (pp. 17–18)

The difficulty in art and design is to manage the wealth of traditions afforded by museum culture. Material and visual culture not only 'represents' the past, it is a part of the visual field of the present, as much available for configuring the future as it is for reconfiguring the past. We ask you to be aware that appropriation can all too easily decontextualise difference, past and present, because it is one of the primary means for reproducing stereotypes (see Chapters 5 and 11). As a strategy, we suggest that in extending your historical knowledge you focus on points of cultural interaction. This way, you discover that plurality is not entirely new; cultures are rarely discrete and autonomous, rather they are fluid and dynamic, mediated through continuous interaction, whether confrontational or collaborative.

Transformations

The beginning of a new century marks a symbolic point of transition, a time for you to choose between the centring and essentialist strategies of traditional pedagogies or the uncertainties of critical transformations, what Giroux (1992) describes as 'border pedagogy':

> Border pedagogy necessitates combining the modernist emphasis on the capacity of individuals to use critical reason to address the issue of public life, with the postmodern concern with how we might experience agency in a world constituted in differences unsupported by transcendental phenomenon or metaphysical guarantees.
>
> (p. 29)

Border pedagogy decentres as it remaps. The terrain of learning is inextricably linked to shifting parameters of place, power, identity and history. This terrain can only be traversed by crossing borders, can only be understood by questioning the forces by which its borders have been generated, can only be transformed by creating new borderlands. To be effective, pupil and teacher must become 'border crossers', they must learn through interaction, link arms together, as it were, and skip along the 'yellow brick road'. However, this is not Hollywood. Do not always anticipate being in step, nor singing in unison or conventional harmony: the travellers may each be singing their own tune; common goals, different needs. Like Dorothy in the *Wizard of Oz* (1939), who only comes to self-understanding by helping others, engaging the pupil in border pedagogy requires that you reflect on your own position and practices and reconsider them in relation to the narratives of others (including pupils). Border pedagogy can thus be considered a process, not of transmission, but of imaginative transformation: ultimately the borders are not out there they are in you. As Giroux (1992) asserts: 'What border pedagogy makes undeniable is the relational, constructed situated nature of one's own politics and personal investments' (p. 35).

Further reading

Giroux, H. and McLaren, P. (eds) (1994) *Between Borders*, London: Routledge.

hooks, b. (1994) *Teaching to Transgress*, London; New York: Routledge.

Paley, N. (1995) *Finding Art's Place: Experiments between Contemporary Art and Education*, New York; London: Routledge.

4 Curriculum Planning

Lesley Burgess and
David Gee

INTRODUCTION

What is a curriculum?
What constitutes an effective art and design curriculum?
What do I need to do before discussing the curriculum with colleagues in my placement schools?

This chapter introduces you to ways of planning an art and design curriculum for pupils at Key Stage (KS3). A brief examination of the development of the National Curriculum (NC) Art Orders helps you to identify the implications for teaching and learning as prescribed in the new NC Order for Art and design (see http://www.qca.org.uk/curriculum).

This chapter is designed to help you look critically at curriculum planning and consider a number of methods used by secondary school art departments to decide the content and delivery of the curriculum. You are introduced to some key concepts to build your understanding. This enables you to contribute to existing schemes of work (SoW) and devise new ones.

You are encouraged to identify, understand and, where necessary, challenge prevailing school art orthodoxies (Downing and Watson 2004) by developing SoW and lesson plans to promote ways of working which give full consideration to pupils' prior learning, interests, social and cultural capital.

The aims and content of the curriculum need to be discussed with others before planning SoW otherwise this can be a bewildering and isolating experience. Sharing curriculum planning with others, and knowing what and how you can contribute through your teaching, empowers you in the classroom and gives a clear message to your pupils about your purpose and professionalism.

OBJECTIVES

This chapter should help you to:

- develop a working definition of the curriculum;
- become conscious of the variety of aims and values informing the curriculum;
- understand how individual art departments plan their curriculum;
- contribute to the curriculum when planning SoW, lessons and home-work;
- contribute to the development of a socially and culturally inclusive curriculum.

UNIT 4.1 WHAT IS THE CURRICULUM?

By the end of this unit you should be able to:

- define the term 'curriculum';
- examine the development of the curriculum since the Education Reform Act (ERA) 1988 to understand the development of the NC Art and design, Programmes of Study (PoS), SoW, and lesson plans;
- develop SoW and plan lessons in which the learning objectives are carefully formulated and differentiated to take into consideration the specific needs of the pupils involved;
- be aware of innovations in the way that schools are organised and the curriculum is delivered, e.g. the 'integrated curriculum' with its inter-disciplinary studies and thematic teaching being trialled by a number of the new Academies.

Put simply the curriculum is a plan or framework for that which is taught and/or learned. Pupils gain access to it through the ways they are taught and the conditions in which they learn.

The ERA (1988) introduced the NC to provide a 'common learning experience' for the majority of pupils who attend state schools. PoS describe the 'matters, skills and processes' which are required to be taught for each subject. Teachers use these to plan the content of SoW and individual lessons.

The curriculum in its broadest sense includes all the conscious and unconscious influences on pupils' learning. Learning does not take place in a vacuum. The ethos of the school, its rules, regulations, shared values, the individual beliefs and interests of teachers all form part of the curriculum. Schools plan some of these aspects through agreed aims and whole school policies; however, increasingly these are dictated by government agencies: the DfES, TDA and QCA. As Steers (1998) points out, 'demands for greater accountability from the teaching profession lead inexorably to ever tighter control, if not specification, of the curriculum and its assessment and, through these mechanisms, to control of teachers' (p. 2).

Since the introduction of the NC teachers have found it increasingly necessary to discuss and plan their delivery of the curriculum and share responsibility for devising and resourcing SoW. Departments are required to record and review these to provide evidence that they are managing resources effectively and raising pupils' achievement.

When taking on responsibility for classes it is essential to refer to previous SoW and curriculum frameworks in order to develop what has gone before and avoid needless repetition. It is dispiriting to hear pupils moan, 'not portraits again', as a result of you not knowing that this topic has been taught before. Consulting written records also helps you to identify the previous learning objectives and ensure some continuity and progression of learning. It is also important that you compare SoW and lesson plans with pupils' art production (work on display, in folders, etc.) in order to relate learning intentions with visible outcomes and discuss these with your SCT.

Where can I find the curriculum?

The following list identifies some of the key inputs influencing the art and design curriculum:

- NC (http://www.qca.org.uk);
- The art and design department's aims (spoken and written);
- SoW;
- Lesson plan;
- Archives and displays of pupils' work;
- Selections of resources/equipment/materials;
- Art and design department and whole school policies, e.g. homework;
- Teaching methods and styles;
- Department/school ethos;
- Collective memory of department members;
- Range of teachers' subject knowledge/specialisms;
- Department discussion/meetings.

Task 4.1.1 Where can I find the curriculum?

Ask a number of art and design teachers what the 'curriculum' means and where it can be found. Check their replies against the list provided above. Add to the list where appropriate.

The National Curriculum Art Order

Introduction

The publishing of PoS in 1995, their subsequent revision in 2000 from two discrete yet interdependent Attainment Targets (ATs) to one, obliged all state schools in

England to reconsider how they planned, monitored and assessed both what was taught and learned. Most departments now endeavour to provide a balanced programme of art, craft and design work with knowledge and understanding of artists, craftspeople and designers developed alongside practical work.

During the 1990s many educators including Meecham (1996) Cunliffe (1996) Hughes (1998a) and Swift and Steers (1999) suggested practice based upon the NC Order rarely challenged the prevailing orthodoxy or 'school art' which prioritises a perceptual or formalist, modernist approach.

> As we move towards the Millennium, we are still delivering art curricula in our schools predicated largely upon procedures and practices which reach back to the nineteenth century – practices and procedures which cling to the comfortable and uncontentious view of art and its purposes. As a result, secondary Art & Design education in England and Wales is, in general, static, safe and predictable.
>
> (Hughes 1998a: 41)

According to more recent research (Downing and Watson 2004) this situation has been slow to change and many departments are reluctant to develop a curriculum which reflects the ideas and technologies of contemporary visual culture. In fact Rogers and Bacon (2002) claim art rooms for the most part, are still designed for the traditional mediums of drawing and painting on paper and not for 3-D and installation work and, as Callow (2001) points out, multimedia approaches require sophisticated and expensive equipment and require teachers to be trained to use it.

It is your responsibility as a student teacher to ensure you understand the rationales and constraints underpinning existing practice so that you can both value and build upon its strengths, recognise its limitations and identify areas of potential development.

Task 4.1.2 Planning SoW

- Find out how SoW are planned and recorded in art and design departments in your placement schools.
- If the department uses set schemes, find out when, how and who compiled them.
- Find out how often schemes are reviewed and new ones introduced.

The PoS outlined in the NC Order covers the range of experiences and opportunities which schools should make available to pupils. It does not specifically set out the manner in which this is to be done, nor place them in any particular hierarchy. When planning SoW or lessons it is possible both to use the PoS to stimulate ideas for SoW and as a means of checking whether you are working within its framework. As you become more familiar with its terms, and can readily identify practical applications, it becomes less of an 'order' and more of a conceptual and organisational framework on which you will build your practice.

A background to the NC Art Order

The proposals for the NC Art Order (DES 1991a) claim to provide a broad and flexible framework to enable teachers to develop their own SoW. It was developed by a Working Party representing teachers, academics, artists and industrialists who identified the main concerns and aims for art and design education. These were identified in the NC Art Working Group Interim Report (DES 1991b) as: 'visual communication, aesthetic sensibility, sensory perception, emotional and intellectual development, physical competence and critical judgement' (p. 7, 3.6).

The following aims were drawn up by the Art Working Group and formed the basis of the PoS:

> We take the view that from age 5 to 16 art education should:
>
> - enable pupils to become visually literate: to use and understand art as a form of visual and tactile communication; to have confidence and competence in reading and evaluating visual images and artefacts;
> - develop particular intellectual and technical skills so that ideas can be realised and artefacts produced;
> - develop pupils' aesthetic sensibilities and enable them to make informed aesthetic judgements in art and design;
> - develop pupils' design capability;
> - develop pupils' capacity for original thought and experimentation;
> - increase pupils' capacity to enjoy and value the visual, tactile and other sensory dimensions of the natural and made environment;
> - develop pupils' ability to articulate and communicate ideas, opinions and feelings about their own work and that of others;
> - develop pupils' ability to respond thoughtfully and critically to ideas, images and objects of many kinds and from many cultures.
>
> (*ibid.*: 13, 4.1)

Task 4.1.3 The NC aims for art and design education

In your tutor groups consider the following:

- Are all the aims equally important?
- Are some aims are more relevant to pupils at KS1 and 2 than KS3?
- How do you prioritise the aims in relation to pupils at KS3?
- What are the implications for planning KS3 SoW?

Task 4.1.4 Reflecting on the aims of your own education

Speculate on and make a list of the aims which governed your art and design education. Compare your list with the Art Working Group's aims.

The legacy of recording from observation

High expectations were placed within schools by the Art Working Group (DES 1991b) on the role of 'observation and recording of visual images'. It was thought that this would enable pupils to express feelings and emotions, transform materials into images and objects, plan, visualise and design. Recording, and especially drawing from primary and secondary sources, was, and still is, a major feature of pupils' art and design education. It is promoted as the primary means for developing visual communication. Recording also has a part to play in the acquisition of formal, technical and conceptual languages, as well as the assessment of technical and critical skills. This view is reiterated in the NC Art Order (DFE 1995) and the NC Order for Art and design (DfEE 1999).

Task 4.1.5 Observing and recording visual images

Identify the extent to which observing and recording visual images plays a part in the SoW in which you have been involved or observed.

Discuss *other* ways of recording and communicating ideas and feelings and how these could be incorporated within a SoW. Refer to contemporary practice and a range of cultural/artistic traditions.

Embedded values within the Art Orders

The earlier NC Art Order (DFE 1995) was written in a particular place by a particular group at a particular time and cannot help reflecting attitudes and values which may to some extent have shifted. The PoS and attainment targets depend, for their coherence, on shared assumptions about child development and the role of art, craft and design in education and beyond. These assumptions are culturally determined and may be different for other societies and indeed, within them, perceived differently by various groups.

The UK is often construed as a liberal democracy. Within this ideological framework early childhood is seen as a time for self-discovery and development. The 'self' is seen as autonomous and generative and if provided with an appropriate environment 'naturally' is motivated to create, construct images, experiment with materials and seek out stimulus and challenges. Evidence for this is easy to find through observing very young children at play who become stimulated by drawing, painting and assembling things. Most activity of this kind is assumed to be performed for *intrinsic* reasons: adults only become involved when they show interest or intervene.

The formation of the art and design curriculum in the early years

From the first occasion when adults assist or direct activities or attempt to provide meanings or interpret children's intentions, two sets of conflicting motives come into play.

Child's motives: intrinsic/inherent

- to seek sensory gratification;
- to seek mental stimulation;
- to explore the environment;
- to begin to construct personal meaning.

Adult's motives: extrinsic/instrumental

- to aid social awareness/development;
- to develop skills;
- to set and solve problems;
- to provide means of production/exchange;
- to transfer or exchange meaning.

Mixed motives!

The role of the curriculum is seen by many as a means of directing pupils' interest towards experiences which they can communicate and share with others. The arbitrary nature of subjective and personal experience is made coherent through developing a common language, e.g. formal, technical and conceptual languages must be systematically learned if skills, knowledge and understanding are to develop in a coherent way. In this way intrinsic and extrinsic needs are addressed simultaneously.

Task 4.1.6 What motivates you?

Reflect on your own art, craft or design activity and compare your attitude and motivation when working on personal projects with those prescribed by others, e.g. commissions, examination criteria, OFSTED.

How can you include these motivational factors in your planning?

Discuss with other students.

Breadth or depth? The dilemma for art and design departments

Arriving at a consensus about what a 'broad and balanced' art and design curriculum might contain is not easy. The sometimes conflicting needs of teachers, pupils and society have to be resolved. Whatever is stressed gives rise to criticism of neglect in other areas: if the focus is on measurable outcomes, the individuality of pupils can be overlooked or highlighted in a negative way; if too finely differentiated on the specific needs of individuals, it can be impossible to make valid comparisons between pupils or establish common expectations or standards.

With the limited number of periods for teaching the subject in most schools, departments are obliged to use their time wisely and think carefully about what they can or cannot include in their SoW.

They often favour one of the following strategies:

- to limit the range of activities, materials and resources and rerun and refine existing SoW. In this instance pupils are motivated by developing specific skills in response to tried and tested, high-quality resources supported by exemplars; *or*
- to provide a broad range of activities, materials and resources to promote experimentation and innovation. In this instance pupils are motivated by '*novelty*' and the opportunity to take responsibility for the development of their ideas.

Task 4.1.7 Breadth or depth? Real or false choice?

- Discuss the above strategies with art and design educators.
- Decide which best describes their practice.
- Consider whether the strategy adopted is age/Key Stage related.
- Identify and discuss alternative models.

The Manifesto for Art (Swift and Steers 1999) proposed 'a postmodern solution for postmodern situation' recommending a shift from the existing school art orthodoxy towards an art curriculum that promoted 'difference, plurality and independence of mind' with an 'emphasis on negotiation of ideas which arise from asking pertinent questions, and testing provisional answers, rather than seeking predetermined ones' (p. 1). Consider the recommendations from Swift and Steers' Manifesto (available from http://www.nsead.org.uk). How do these compare with your findings to date?

Planning by departments

Departments vary in their ability to discuss and plan the curriculum together. This is often made easier where there is informal day-to-day contact, a difficulty when schools are on split sites or employ part-time teachers. Approaches to curriculum planning vary. The following examples characterise some alternatives, but are by no means exclusive.

Example department A

Teachers devise separate SoW, all using the NC as a guide. Each attempts to provide a broad and balanced programme independently.

Advantages:

- teachers are free to develop SoW which build upon their interests and areas of expertise;
- teachers take responsibility for broadening their own subject knowledge and are personally accountable for pupils' achievement;
- a variety of teaching styles is almost guaranteed.

Limitations:

- possible duplication of resources;
- static teaching styles: self-review may not be sufficient to encourage variety or development;
- isolation: lack of common themes or purpose, the department becomes 'invisible', lacks identity/presence; strengths of individuals are not maximised for departmental good.

Example department B

The department meets before each term or year and decides on a theme, either for each or all year groups, from which SoW are planned.

Advantages:

- themes provide both content and context for whatever material and process is used;
- resources can be shared and built up collectively;
- common themes provide the department with a strong identity within the school.

Limitations:

- an orthodoxy of method, style, delivery can establish itself: shared practice can become self-justifying and closed to internal or external criticism;
- only those involved in the original discussion can fully claim ownership of the theme: much of the content of the discussion is not recorded and is thus inaccessible to other teachers.

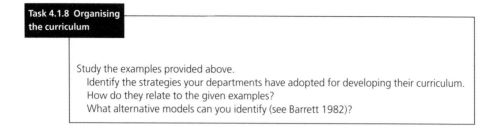

Task 4.1.8 Organising the curriculum

Study the examples provided above.
   Identify the strategies your departments have adopted for developing their curriculum.
   How do they relate to the given examples?
   What alternative models can you identify (see Barrett 1982)?

A guide to planning SoW

Introduction

The ability to plan, record and review schemes of work and individual lessons is increasingly regarded as a key requirement for effective teaching and learning.

Teachers vary in the way they plan: some record in detail the content, resources and methods to be used, others prepare simple checklists of inputs and activities. Obviously not everything that occurs during lessons can be planned for or predicted: indeed, pupils sometimes respond in ways very different to those you anticipate. Therefore, your planning should not be too rigid. Good planning 'builds in' flexibility and allows for serendipity. Planning provides both a structure and a set of objectives with which to guide events and also a means to measure your pupils and your own performance.

Task 4.1.9 How do teachers plan?

Ask art and design teachers how they plan SoW and lessons.
   How do they allow for flexibility?
   Compare their methods with those used by teachers from other subjects.

Your first SoW

When planning SoW for the first time you need to consult with the subject teachers with whom you are working. Identify and discuss the area of the PoS you are expected to cover. It is useful to 'brainstorm' ideas with the subject teacher, school tutor/mentor or another student teacher. Conduct your research at a level suited to your classes. Remember, although you may be aiming for depth and some degree of complexity, you are planning for secondary school pupils; you need to give careful consideration to both the 'pitch' and the 'pace' of your projects.

Decide exactly what your aims are; think about a 'way in'. Ask an experienced teacher whether your aims are suitable and if they align with the NC and/or examination syllabuses.

Divide your ideas into sections. Find: resources, websites, books, posters, display materials, etc. If using techniques and materials which are new to you, practise these yourself using the school materials and equipment so that you are aware of their qualities and limitations! Decide how you are going to set tasks and activities and plan a homework programme to provide a range of differentiated tasks.

Getting your aims right!

Establishing aims for SoW is not straightforward. Pupils' competence and maturity can vary enormously within the same group. With new technology, schools have instant access to detailed information about each pupil including their levels of achievement in each subject, expected grades at examination alongside more personal information. Teachers use these more and more to develop 'personalised learning' plans (Miliband 2004) and to plan activities, assess students' performance and give them feedback and targets. How realistic, ambitious and comprehensive can you expect your aims to be? Can you expect all children to meet them or should you make exceptions? In general, the more clearly your aims are differentiated to the specific learning needs within the group the more chance they have of being realised. The way you word and use aims in your SoW can either assist or hinder you. If appropriate, they provide a helpful focus; if inappropriate, they become a distraction.

Aims about aims

Think clearly about the support you offer and the resources you need to realise your aims. Bearing in mind that lowering expectation becomes self-fulfilling decide whether you expect:

- all pupils to meet your aims to a comparable standard;
- all pupils to meet your aims at some level;
- the majority to meet your aims and make provision or adapt the aims for those who do not;
- only a few pupils to meet your aims.

To teach confidently you need to be able to justify your aims educationally and ethically. It is easy to corner yourself unintentionally by not fully thinking through your aims. In the same way that the NC requires pupils to review and modify their work as it progresses, you may find it necessary to change the way you plan your SoW or adjust them as they progress.

Teaching for diversity

As a general rule the more homogeneous the ability and experience of the group, the easier it is to narrow the focus of your aims. Within a wider range of pupils wider terms of reference are often needed. Often, teachers regard aims as reminders of what they should provide. If some choice or variety of outcome is built into a SoW, many of the problems associated with too narrow an aim or focus can be avoided. A tension always exists between consistency and flexibility. You can tell whether aims are appropriate or not by monitoring the response of your pupils.

Using pupils' responses to help differentiate needs and monitor the effectiveness of your aims

If you focus too much on your aims without having thought through the support needed to achieve them:

- some pupils make a half-hearted attempt to meet them, are not fully engaged and look for the easiest solutions;
- others feel daunted from the start; they are be unable or unwilling to make any effort and 'play up' accordingly.

If insufficient emphasis is given to your overall aims and the SoW is allowed to completely alter course:

- enterprising pupils make good use of the opportunity to adapt resources to their own purposes;
- others pretend to do this, but with the sole intention of exhausting your supply of materials;
- the more insecure feel that you have let them down in your duty to direct, stimulate, entertain them and will switch off and stop working.

Checklist for planning SoW

Consider the following before planning:

- the number of lessons available;
- what you want pupils to learn;
- the areas of the PoS you are covering;
- pupils' prior learning;
- the materials, processes, skills and techniques you need to gather, organise and rehearse;
- the critical, contextual and historical resources needed to support, guide and reinforce learning;
- the way you balance and integrate the requirements of the Attainment Target: skills, knowledge and understanding;
- how to make each lesson memorable and significant;
- assessment and evaluation.

Look for opportunities to:

- build on pupils' existing knowledge, skills and interests;
- stimulate and maintain motivation through a variety of tasks, materials and teaching methods;
- develop technical skills through instruction;
- arrange pupils to work in pairs and groups as well as whole-class teaching;
- provide opportunities for group discussion and peer appraisal;
- involve pupils in discussion about the content and direction of their work; provide a degree of choice;

- broaden pupils' frames of reference;
- refer to a range of traditions and examples of contemporary practice;
- identify key concepts and terms and build pupils' art, craft and design vocabulary;
- set a variety of types of homework, e.g. gathering information, practice of techniques, writing up methods, etc.

Conclusion

There are many variables in teaching which make it difficult to find a perfect formula for planning SoW. A lesson idea which works well with one group can be a disaster with another. The weather, time of day, incidents from previous lessons, etc. can all affect the pupils' receptiveness to your aims.

Some student teachers find that they have been too elaborate in their planning or have complex aims which they cannot communicate to pupils. In an effort to provide breadth, you may try to cover too many strands of the PoS without allowing time to establish connections between them. You may feel that every piece of work needs to be accompanied by reference to an artist or that all work should start with drawings from observation in order to fully comply with the four strands of the Attainment Target: exploring and developing ideas; experimenting with and using media; reviewing and adapting work as it progresses; investigating others' work and applying knowledge and understanding. As a result a veil of predictability descends as the formula tightens and is perpetuated!

Variety of input and outcome is one way of guaranteeing the involvement of the largest numbers of pupils. It should be possible when planning your SoW to change the focus from time to time, sometimes concentrating specifically on technique, sometimes investigating traditions and changing cultural forms (including youth culture) through discussion and written tasks, sometimes providing opportunities for pupils to make discoveries of their own and at other times directly instructing them and transferring knowledge.

In Chapter 3 it is noted that pupils have many different ways of learning. Consequently you need to find different ways of presenting instructions or stimuli and adapt your teaching method and style accordingly. This does not, however, remove the need for the security provided by consistent routines, rewards and sanctions.

Lesson planning

Much of the advice in the previous section on SoW also applies to lesson planning. When planning lessons you should set specific learning objectives for each lesson, a skeletal outline of this should form part of the SoW. However, individual lessons are best planned in response to and after careful reflection on what has been achieved and can be developed from previous lessons. It can prove useful to devise a *proforma* for planning lessons. This encourages you to develop effective working habits and balanced, purposeful lessons. The layout and content of your *proforma* may need

Table 4.1.1 Scheme of work

Framework for Scheme of Work

Title:			
Class: Key Stage: Teacher:	Date: Period: Room:		No in group: SEN: EAL: IEP:

Aims: • • •		Session:	Structure and Content:
		1	
		2	
		3	
		4	
		5	
		6	
Links to NC KS3 Strategy and Examination Criteria: • • •		7	
		Key Concepts & Terms: • •	
Assessment Strategies: *How and when will evidence be gathered, recorded and communicated?*		Resources: *Materials and Equipment - H&S. Visual Resources: Stimulus materials/critical and contextual references. Outside Agencies: visits to galleries, museums.*	
Personal Learning across the Curriculum: *Identify ways in which this will be addressed.*			

Table 4.1.2 Constructing a lesson

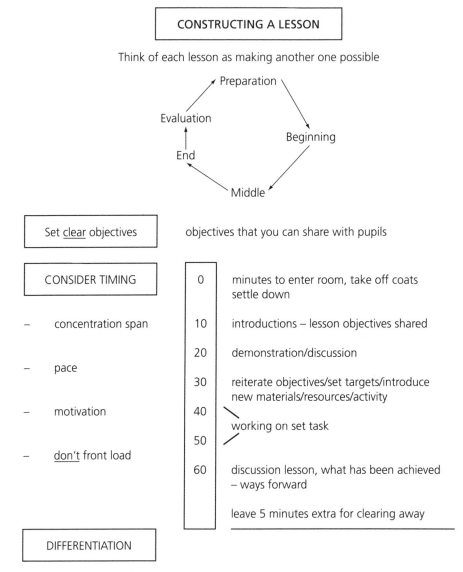

CONSTRUCTING A LESSON

Think of each lesson as making another one possible

Preparation

Evaluation

Beginning

End

Middle

| Set clear objectives | | objectives that you can share with pupils |

CONSIDER TIMING

	0	minutes to enter room, take off coats settle down
– concentration span	10	introductions – lesson objectives shared
	20	demonstration/discussion
– pace	30	reiterate objectives/set targets/introduce new materials/resources/activity
	40	
– motivation		working on set task
	50	
– don't front load	60	discussion lesson, what has been achieved – ways forward
		leave 5 minutes extra for clearing away

DIFFERENTIATION

English as an Additional Language EAL; Special Educational Needs SEN

Check with class teacher re special needs
Special needs also apply to gifted children
Have extension work prepared

visual
motor
hearing
emotional
behaviour
etc.

Table 4.1.3 Lesson plan (a)

LESSON PLAN

OBJECTIVES

By the end of the lesson, students/pupils will be able to:

State ....

Describe ....

List ....

Identify ....

Prioritise ....

Solve ....

Demonstrate an understanding of

Objectives within a lesson

1   to learning   ⟶

2   to behaviour   ⟶

Work collaboratively/in pairs/cooperatively

Work independently

Sustain effort – concentrate (an adults concentration span is 20 minutes maximum!)

Maintain an orderly environment

Table 4.1.4 Lesson plan (b)

Lesson Plan

Title:		Lesson No: *Eg. 3/6*
Class: Key Stage: Teacher:	Date: Period: Room:	No in group: SEN: EAL: IEP:

Learning Objectives:
What the lesson enables pupils to learn.

-
-
-

Strategies for personalised learning:

Resources:	Key Concepts & Terms:

Activities

Times:	Teacher:	Pupils:

Homework:

revision after a few lessons in order to reflect and assist your emerging teaching style and methods.

Task 4.1.10 Defining aims and objectives

- Write down your definitions for the following terms: aim; objective.
- Compare and discuss your definitions with other student teachers.
- Check them against dictionary definitions.

Consider the following checklist:

Subject knowledge

- Do lesson objectives help towards achieving the overall aim?
- Are the concepts, resources and tasks suitable for the age and range of achievement of the pupils?
- How does the lesson build on individual pupil's skills, knowledge and experience?
- How does the lesson build on the achievements and challenges of earlier lessons?
- At the end of the lesson, what *new* knowledge, understanding and skills do you hope pupils have gained? (i.e. the learning objectives)
- What new vocabulary do you want pupils to learn and how can it be communicated and reinforced?

Teaching and class management

- What teaching methods can best meet your objectives? e.g. whole class, group work, pair work, discussions, talk/demonstration, etc.
- What is the most appropriate way to arrange the tables/equipment, etc.?
- How do you establish or support existing routines for distributing materials, work, etc.?
- How do you sustain pupils' interest, e.g. by balancing discussion with practical activities, taking pupils off task to discuss progress, changing activities, addressing different ways of learning, reviewing and summarising achievements in order to set future targets?
- What strategies do you use to help pupils compare methods, approaches and outcomes and engage with the work of others?
- How do you ensure that all pupils benefit from the lesson? How do you differentiate individual needs? In what ways do you cater for the SEN pupils, including gifted and talented pupils? What extension tasks are provided for pupils who successfully complete work before the rest of the class?
- How do you communicate concepts and tasks to pupils for whom English is an additional language (EAL)?
- How and at what point in the lesson do you set homework?

Assessment

How do you ensure:

- pupils know if they have achieved the learning objectives?
- feedback is provided: group/individual?
- pupils understand the criteria used for assessing progress and achievement?

Homework

Homework can extend pupils' art and design education by many weeks over their schooling. Reasons for setting homework are to:

- consolidate the skills, knowledge and understanding taught during lessons;
- practice techniques and media;
- gather information and resources;
- investigate histories, issues, artists, craftspeople and designers, etc.

Task 4.1.11 Homework survey

Consult a copy of your department's homework policies.
   Discuss with other teachers the various ways and reasons for setting homework. Add to the list provided above.

UNIT 4.2 EQUAL OPPORTUNITIES: AN OPPORTUNITY FOR THE ART AND DESIGN CURRICULUM

By the end of this unit you should be able to:

- examine equal opportunities and identify ways to develop strategies which acknowledge difference and plurality;
- adopt approaches for developing and interpreting visual images and artefacts which recognise that interpretations may be changed, by reference to gender, race and class to give widely different meanings.

Introduction

The range of what we think and do
is limited by what we fail to notice.
And because we fail to notice
that we fail to notice
there is little we can do
to change

until we notice
how failing to notice
shapes our thoughts and deeds.

(Goleman after R.D. Laing 1985: 24)

Vital lies, simple truths; the psychology of self-deception

Equal opportunities policies attempt to foster the conditions whereby pupils can fulfil their potential irrespective of their gender, race, ethnicity, religion, nationality, social class, sexual orientation or level of ability. Any serious engagement with visual and material culture must tackle issues of representation and identify ways in which meanings are constructed. In order to engage with these you need to acknowledge that meaning is not fixed but constantly changing and needs to be negotiated within a range of forces: historical, social, cultural, political, and economic.

Equal opportunities does not imply the assimilation of 'other' cultures or groups by blurring distinctions, in other words encouraging minorities to radically alter their identity in order to conform to the majority. Neither does it involve ignoring the background of pupils by assuming or pretending that they share the same cultural traditions and history: rather, it means acknowledging all the groups who make up contemporary society in curriculum planning and in the *content* of PoS. To this end teachers of art and design need to develop a broad subject knowledge informed by an understanding of changing traditions and ensure that this is reflected in their SoW. This is happening increasingly as more teachers and pupils develop plural or multiple identities through travel and migration. The art and design curriculum presents many opportunities for teachers and pupils to understand, appreciate and inform the diversity and hybridity of traditions and shifting practices.

The writers of the NC Art are concerned to make the Order applicable to as wide a range of pupils and communities as possible. Initial concern regarding a bias towards European traditions was modified for the NC Art (DFE 1995) in order to ascribe equal value to cultures 'Western and non-Western'. Meecham (1996) aptly describes the 1995 Orders as 'an almost multi-cultural approach, with an eye to the National Heritage' (p. 75). However, the PoS for KS3 identifies a list of named periods in the Western tradition as examples: 'Classical, Medieval, Renaissance, post-Renaissance through to the nineteenth and twentieth centuries' (DFE 1995: 6). Although this was not meant to suggest a chronological approach, it has resulted in teachers feeling obliged to include the given examples to the exclusion of more inclusive work, whether contemporary or drawing on a range of cultural traditions. In the NC Art and design (DfEE 1999) this list was revised to include 'pupils investigate the diverse roles and intentions of artists, makers and designers in the community, from past and present and from familiar and unfamiliar cultures'. Neither document mentions race, gender or social circumstance. Children with SEN are only provided for by means of general 'access statements', none being specific to art and design. No strategies are offered for teaching gifted pupils although 'Exceptional Performance' is recognised in the Attainment Target Levels. In fact, the Order is so skeletal that while a broad, informed, differentiated, intercultural, and issue-based approach is possible, it is not specifically required. However, it is your responsibility to ensure that the curriculum you develop is inclusive and gives full consideration to equal opportunities.

Equal Opportunities: checklist for planning

Gender

Foster mutual respect and achievement between the sexes:

- Investigate the roles of men and women as makers, spectators, participants, consumers and critics of art, craft and design.
- Investigate the cultural/economic reasons for the predominance of one sex in some traditions/practices, e.g. the 'great masters' of Western Art, the batik craftswomen of Nigeria.
- Investigate the causes of bias and stereotyping within art, craft and design practice.
- Introduce pupils to activities which may once have been the preserve of one sex.
- Highlight women's achievements, through your choice of references, books, exhibitions, collections, etc.

Race/culture/religion/class

- Investigate the different status and definitions and relationships between art, craft and design in different communities.
- Investigate the different roles and functions that art, craft and design have for the different communities.
- Aim to plan SoW which reflect culturally and socially diversity, both within and outside the UK, and question orthodoxies and hierarchies.
- Identify the main cultural and religious traditions represented by the groups/individuals in your school; recognise the contribution pupils can make to the department's range of references and resources.
- Incorporate pupils' backgrounds and frames of reference in SoW.
- Be aware and respectful of pupils' religious beliefs and customs.
- Seek examples of achievements from a broad range of activities and social groups, e.g. include vernacular styles, folk traditions, popular culture.

5 Resource-Based Learning (RBL)

Lesley Burgess

The crucial success factors in achieving breadth and depth for all are the resources available, most of all the teacher.

(QCA 2006)

INTRODUCTION

This chapter introduces you to the central role of RBL in art and design education. It encourages you to reflect critically on the resources used by yourself and others in order to determine whether they are appropriate and, if not, how they might be developed. You are encouraged to think carefully about the ways in which outdated attitudes and values can be perpetuated through the uncritical use of existing resources. You are alerted to ways in which you can all too easily reproduce traditional power structures by working with existing exemplars of 'good' practice and/or promoting cultural heritage uncritically and without reference to context.

Nearly ten years ago the SCAA (1997) *Survey and Analysis of Published Resources for Art (5–19)* found that resources are selected on the basis of:

- teachers' interests, motivation and training;
- what happens to be accessible;
- finances available.

Specific areas identified as difficult to resource were: non-Western art, sculpture, women artists and craftspeople, designers, design education/design history, local artists, contemporary art/artists. Other areas mentioned were: photography, environmental art and architecture. More recently a research report by the National Foundation for Educational Research (NFER) authored by Downing and Watson (2004) suggests that, although there is 'evidence of slow-changing orthodoxy' in the main, this situation has not changed. This is reinforced by OFSTED's (2005) overview and trends in secondary art since 1997 which points out that even though 'the proportion of very good and excellent art teaching is higher than in any other subject . . . the innovation and experimentation that characterise these lessons are often absent elsewhere' and that 'in too many departments, including some where teachers are given every opportunity to take ownership of the curriculum, the choice of media is largely traditional and the artists referred to are remarkably predictable'. These gaps in provision provide the focus for this chapter with the exceptions of sculpture, craft and design which are covered in other parts of this book.

OBJECTIVES

By the end of this chapter you should be able to:

- resource schemes of work (SoW) giving careful consideration to the impact that your selection, use and presentation of resources can have on pupils' reception, enjoyment and understanding of the subject;
- identify a wide range of resources available for teaching and learning in art, craft and design;
- recognise the central role of new technologies in accessing information and locating resources;
- reflect critically on the range of resources currently used in schools and HE, including those accessed through www and the Internet, and consider their suitability for different pupils;
- identify opportunities to work in partnership with outside agencies, including galleries and museums, in order to develop the range of resources currently at your disposal;
- appreciate the need to constantly update and reconsider resources to ensure that they reflect contemporary practice as well as changing traditions.

UNIT 5.1 DEFINITIONS

By the end of this unit you should be able to understand:

- why the term RBL was introduced in the 1970s;
- its reception in art and design education at a time when it was being denigrated in some other areas of the curriculum;
- the changing definition of RBL in response to curriculum developments and new technologies.

So much pupil learning in schools is alienated in the sense that it consists of other people's knowledge purveyed in transmissional mode. Pupils have no share in the knowledge, nor any control over the learning processes. In

addition, it is difficult to see the relevance of such learning for their own
interests.

(Woods 1996: 127)

Woods (1996) suggests that RBL was promoted in the 1970s in an attempt to provide
the antidote to 'alienated' learning. He identifies its genesis in child-centred progres-
sivism which promoted: learning by discovery, making teaching relevant to pupils'
concerns, integrated knowledge, democratic decision making. Critics of RBL
claimed that, in practice, it became: a restraining rather than a liberating culture; a
coping rather than a teaching strategy; operating in the interests of social control
rather than pupils' interests (*ibid*.). However, despite this criticism and RBL's limited
reception in other areas of the curriculum, it has had a significant impact in art and
design. Rather than dismiss RBL as problematic, many art and design educators
successfully related it to the pedagogic principles underpinning critical practice and
intelligent making.

> The concept of 'Resource-Based Learning' (a significant feature of GCSE
> art and design), developed from the successful teaching and learning methods
> which were based on extensive provision of an appropriate range of learning
> resources. In the main, these consisted of collections of objects and reference
> materials used both as stimulus for practical work and 'research' as work
> progressed.
>
> (Binch and Robertson 1994: 112)

The term RBL was initially used in art and design to signify the promotion of work
from 'direct experience' using primary sources. This signalled a clear rejection of the
ubiquitous practice of inviting pupils to respond creatively to a given topic using only
their imagination or images (usually gleaned from the Sunday newspaper supple-
ments). Many educators acknowledged the need to move away from such restricted
practices. They recognised that sheer self-expression required no artistic form and
mere transcriptions of photographic reproductions required no understanding of
content. They responded by insisting that the main stimulus for all pupils' work must
be close observation of collections of 'real' objects. This gave rise to the 'big still life',
large classroom installations of objects and artefacts selected in response to a given
theme: Natural Forms; Reflections; Celebrations.

Kennedy (1995a) explains why he considered it to be an essential part of the
curriculum:

> Students are encouraged to work from real things for two main reasons. The
> first is a matter of quality control – there is less of a credibility gap between
> concept and realisation in representational terms if students work from direct
> observation rather than a secondhand image. The second concerns
> empowerment – the student is in control of the basic elements when
> working directly and can determine angle, view, scale, rendering, etc.
> Working from secondhand imagery frequently means that the student's own
> intentions have to be compromised by compositional decisions made by

others. Generally, in one scenario the student is active and learns more, in the other the role is more passive with the student potentially disempowered.

(p. 9)

However, RBL was not without its problems, the extent to which some teachers rigidly adhered to this 'perceptual' approach proved limiting; pupils fed on an exclusive diet of close observation tended to develop a narrow definition of art. Critical and contextual resources used to support this approach drew exclusively upon the nineteenth- and early twentieth-century western still-life tradition. The perceptual approach fostered the development of the 'school art' orthodoxy that Hughes claimed was 'static, safe and predictable . . . divorced from contemporary ideas in the spheres of art practice, critical theory, art history or museology' (Hughes 1998a: 41). By restricting the range of practices, emphasising 'basic skills' and tech-niques and reducing drawing to accurate recordings of things seen, it avoided the difficulties of assessing work which was issue-based, innovative, idiosyncratic or from cultural origins other than western.

Simpson (1987) insisted:

> The emphasis has been on direct observation and recording – the object being the 'environment' . . . drawings of twigs, shells sections of any fruit or vegetable you care to name, sculls, bits of bikes' lamps and those tedious pencil drawings of crushed coke cans and toothpaste tubes; all now definitely passé.
>
> (p. 255)

By the mid-1980s use of the term had expanded to include galleries, museums and other outside agencies; artists, craftspeople and designers in education. Art and design teachers began to collect resources to fulfil the GCSE requirement to develop pupils' critical and contextual skills. The proposed introduction of the National Curriculum Art Order (NC) (DES 1992) with its requirement 'to study the work of others' resulted in a rush to locate suitable resources. In deliberations about the NC Art, RBL extended beyond concrete examples (object, material, human) to include abstract concepts such as time and space. At an NSEAD conference in 1992 the then HMI for Art, Colin Robinson, claimed that: 'the most valuable resource any art and design teacher has is time'.

Time

On average one hour per week is allocated for Art and design at KS3. However, given that each pupil will probably be ill for at least one lesson, and out for school trips, choir, orchestra, steel drums, sports day and rubella injections you can expect any one child to attend 30 timetabled hours each year. At KS4 time allocation is typically two and a half hours per week, but with reductions for work experience, field trips and mock exams the likely total is 75 hours per year. Many art and design teachers find it necessary to provide extra-curricular opportunities for pupils to develop their studies.

Prentice (Chapter 2) insists that student teachers need time to become familiar with material, tools, equipment and processes; the same claim can be made for pupils. Time restrictions are often cited as the reason why some teachers offer a limited range of practices; they claim to provide depth in the basics rather than introduce a broad range superficially. Similarly, restrictions in equipment, materials, room size and available storage space are cited as the reasons why pupils do not develop work in three dimensions or ICT. It is important to take these issues into consideration when developing SoW. However, the opportunity provided by your ITE course to visit a number of art and design departments introduces you to different ways to overcome such hurdles.

Rooms/equipment/hardware/resources/finances

> Many secondary pupils are being denied opportunities to realise their creative potential because of lack of resources, time and expertise in art & design in school.
>
> (NSEAD 2001: 14)

Surveys and reports by RSA (1998), NSEAD 2001, OFSTED (2005), (QCA 2006), reveal that there continue to be strong contrasts from school to school in the range of activities undertaken and the resources available. Some schools have a wealth of equipment: airbrushes, cameras (video, digital, still) computers (monitors, scanners, colour printers, Internet access, whiteboards, data projectors, etc.), looms, silk-screens and kilns; others have to make do with hoghair brushes, powder paints, recycled boxes and a few blunt lino-cutting tools. It is useful to determine how often different resources are used. You should explore the ways teachers in your HEI and placement schools organise resources; are they stacked in neatly labelled boxes or do they litter the department in a cacophony of conflicting colours and textures? Do pupils have open access, or are they closely guarded for fear they might disappear? Are they catalogued, if so, what headings are used? Find out where you can obtain or borrow resources. Some local education authorities have extensive loan collections, as do a number of galleries and museums. Most regions have 'scrap banks': collections of recycled/industrial waste, offcuts of paper, cloth, wire, metal, plastic, all made available for educational use at giveaway prices. The pressing need for most art and design teachers to be thrifty was highlighted in a survey by Artworks/NSEAD (2001). This survey identified that the annual spending, through capitation, on art and design consumable materials had deteriorated over the preceding six years to an annual average rate of just £2.68 per pupil on roll (this ranges from a minimum of 60p to a maximum of £7.30). In contrast the average annual spend per pupils in the independent sector was found to be £12.40. Only one third of state schools were able to supplement this with additional funding. Interestingly, the most common source of additional income came from taking on initial teacher education (ITE) students!

Your ITE year is an opportunity to identify what is available and to develop a comprehensive bank of critical and contextual resources.

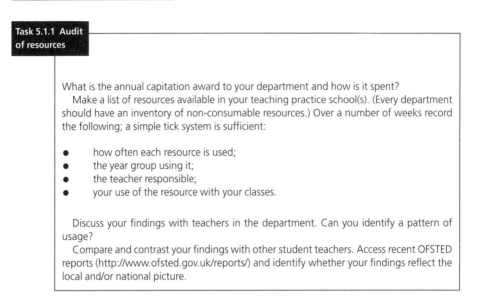

Task 5.1.1 Audit of resources

What is the annual capitation award to your department and how is it spent?

Make a list of resources available in your teaching practice school(s). (Every department should have an inventory of non-consumable resources.) Over a number of weeks record the following; a simple tick system is sufficient:

- how often each resource is used;
- the year group using it;
- the teacher responsible;
- your use of the resource with your classes.

Discuss your findings with teachers in the department. Can you identify a pattern of usage?

Compare and contrast your findings with other student teachers. Access recent OFSTED reports (http://www.ofsted.gov.uk/reports/) and identify whether your findings reflect the local and/or national picture.

More recently the term RBL has been reintroduced in mainstream education and has taken on a new, extended meaning; the pedagogic use of new technologies or ICT. RBL can now be defined as 'diverse course delivery methods using non-traditional modes'. It takes place in Learning Resource Centres – quiet study spaces with access to virtual learning environments (VLE) and other computer-based networks and resources (CD-Rom, DVD, Internet, Intranet). In FE and HE these are already well established and teachers from all curriculum areas are expected to be actively involved with 'learning assistants' ensuring that course design, support, assessment, review and evaluation are available on the Internet and the Intranet. The terms 'library' and 'librarian' suddenly seem quaint and old-fashioned. RBL has shifted from the periphery to the heart of the learning process. Tomlinson (1996), Kennedy (1997), Dearing (1997) and Miliband (2004) have argued for accessibility, inclusivity, flexibility, transferable skills; greater opportunities for students to manage their own learning. The implications for teaching in secondary school are rapidly becoming apparent.

Further reading

Downing, D. and Watson, R. (2004) *School Art; what's in It? Exploring the Visual Arts in Secondary Schools*, Slough: NFER.

OFSTED (1998) *The Arts Inspected*, Oxford: Heinemann.

OFSTED (2005) *The Annual Report of Her Majesty's Inspector of Schools 2004/5: Art and Design in Secondary Schools*, Online. Available HTTP: <http://www.ofsted.gov.uk/publications/annualreport0405/4.2.1.html> (accessed 8 October 2006).

Rogers, R., Edwards, S. and Steers, J. (2001) *Survey of Art and Design Resources in Schools*, Artworks/NSEAD. Online, Available HTTP: <http://www.artworks.org.uk> (accessed 28 October 2006).

SCAA (1997) *Survey and Analysis of Published Resources for Art (5–19)*, London: SCAA.

UNIT 5.2 OWNERSHIP

By the end of this unit you should be able to:

- consider who or what decides or dictates the types of resources you use;
- understand whether selection is based upon *your* subject specialism, interests and enthusiasm or prescribed by others;
- ensure your use of resources provides an inclusive education informed by cultural and social issues;
- recognise the value of including the work of contemporary artists in your teaching.

Task 5.2.1 The naming game

> Name six artists, give their dates and identify their art form.
> Take no more than two minutes to complete this task.
> Once a name has been written down you must not change it or cross it out.
> Do not confer with anyone else.

Education has been named as one of the major ideological state apparatuses – that is, not just a place of learning, but an institution where, as in the family, we are taught our places within a hierarchical system of class, gender and race relations.

(Pollock 1996a: 54)

The important role that you play in constructing pupils' understanding of 'what counts' in art and design cannot be overlooked. Your choice and use of resources provides pupils with a clear statement about what is to be valued. Before you address issues of marginalisation and invisibility you have to question your 'own ready implication in a discourse of mastery' (Usher and Edwards 1994: 81); to recognise that, often unwittingly, you can reinforce orthodoxies. Even when they think they are being 'inclusive', research (Stanworth 1987; Spender 1989) has proved that teachers continue subtly, albeit unintentionally, to reinforce 'bias' and 'otherness'.

Task 5.2.2 Reviewing personal choices

> Look back at the list of six artists produced in Task 5.2.1.
> Analyse them using the following questions as prompts:
>
> - Does your selection represent: exhibitions you have recently visited; artists with a high media profile; artists who have been influential in the development of your own practice; your contemporaries?
> - If you had been asked to list six artists, craftspeople or designers, would your list be different?
> - If you were completing this list in a contemporary art gallery would this alter your selection?

- Is there a gender balance?
- Is cultural diversity considered?
- Have you included yourself?

Long established metanarratives are powerful forces which are difficult to disrupt; this is as true in art and design as it is in other areas of the curriculum. However, it is worthy of note that in successive versions of the NC Art and design Order, unlike any other NC Order, teachers have been actively encouraged to 'explore different codes and conventions'; 'recognise ways in which works of art, craft and design reflect the time and place in which they were made' (DES 1995), investigate 'art, craft and design in the locality, in a variety of genre, styles and traditions, and from a range of historical, social and cultural contexts' (DfEE 1999: 21). Far from being an exhortation to sustain a 'school art orthodoxy', these statements insist pupils should engage with a range of social and cultural issues.

Throughout this unit you are encouraged to think carefully about the way the critical and contextual resources can conceal a hidden curriculum. You are asked to query the 'modernist' trap which reproduces traditional hegemonic power structures and reinforces dominant western patriarchy. Hughes (1998a) calls for a total reconceptualisation of the art and design curriculum and, like Efland *et al.* (1996), he insists that merely 'tinkering on the edges' is insufficient. Swift and Steers (1999) in their Manifesto for Art called for 'a postmodern solution for a postmodern situation' and make a convincing case for an art curriculum based on the principles of difference, plurality and independence of thought, outlining the need for teachers to replace orthodoxy with innovative and imaginative approaches, and espousing the need to encourage qualities such as empathy, playfulness, ingenuity, risk-taking, curiosity and individuality. Perhaps such demands are needed to effect the smallest changes in practice. However, as a student teacher you need to think carefully about your role in developing pedagogies. If art and design education is to evolve, to reflect new ideas and ways of working, then you must see yourself as both an advocate for change and an agent of change. It is important that you consider what a 'reconceptualised' curriculum might comprise and what resources you need to support it. However, until you understand the existing system tread carefully. To 'boldly go where no one has gone before' would be foolhardy. The approach you are asked to adopt is one of interventions (Pollock 1988) or what Spivak calls 'strategic essentialism' and Hall refers to as 'arbitrary closure' (in Shohat and Stam 1995). Through interventions to destabilise, to question, to identify why invisibilities have come about. Through strategic essentialism/arbitrary closure to understand that although you need to adopt a position or approach in order to practise, in time it is likely to be replaced. It follows that you need to recognise that to permanently prescribe (fix) resources and their usage in school is a redundant project. Some of the resources you use with conviction today are likely to be spurious/irrelevant in the future. Shohat and Stam (1998) insist that any study of visual culture should be a 'provocation', to cause new questions to be asked, 'to interrogate the conventional sequencing of realism/modernism/postmodernism' (p. 31). Spivak insists, 'it is the questions we ask that produce the field of inquiry and not some body of materials which determines what questions need to be posed to it' (Rogoff 1998: 16).

While it is impossible to start from scratch it is important that you do not only adopt resources produced by others. You need to intervene in 'packaged pedagogies' by challenging the authority of stereotypical resources, turning statements into questions and by introducing artists, objects and artefacts which refuse to comply.

Task 5.2.3 Visual resource sheets

You probably have acquired or produced visual reference sheets and reproductions early on in your course and are using them in your introductions, demonstrations and displays. How are you selecting these resources?

Are they there to produce exemplars, canonic or otherwise, for pupils to emulate?

Do they include diverse cultural perspectives?

Have you selected them to act as a stimulus for discussion?

What questions might they provoke?

Do you provide contextual information to complement choices made for formal reasons?

At what stage in a SoW should you bring in critical and contextual resources?

What is the function of your references in each SoW?

Are you producing these resources to promote learning as a sign of your effort and enthusiasm, a form of spectacle?

Would it be more beneficial if pupils investigated an issue of interest and produced their own referenced presentations?

Pupil interventions – pupils as a resource: gatherers and collectors

> Every student transferring to secondary education at the age of 11 brings with them a unique canon of visual literacy ranging from, in most cases a subliminal expertise of sophisticated televisual constructs (absorbed osmosis – fashioned from many thousands of hours of home viewing) to conscious and articulate appraisals of what one Year 10 student termed TCCC Twentieth Century Cultural Clutter.
>
> (Kennedy 1995a: 8)

Ownership of lesson content merits careful consideration. Edwards and Furlong (1978) suggest that RBL is conditional upon teachers' prevailing knowledge; that it rarely takes into account pupils' interests. The National Curriculum (DfEE 1999) infers that the responsibility is not just that of teachers: 'Pupils should be taught to research, select and organise a range of visual evidence and information' (Loeb 1984: 6), and notes that when pupils are encouraged to become researchers and are 'given permission' to become contributors to, as well as clients of, the education system, a rich and sometimes unexpected resource becomes available.

Willis (1990a) insists teachers need to broaden education's traditional, restricted notion of art and design in order to embrace 'moving culture'; to recognise that children are already engaged in imaginative, expressive and decorative activities that are grounded in the needs and functions of everyday life.

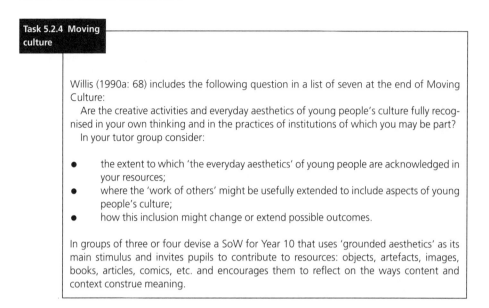

Task 5.2.4 Moving culture

Willis (1990a: 68) includes the following question in a list of seven at the end of Moving Culture:

Are the creative activities and everyday aesthetics of young people's culture fully recognised in your own thinking and in the practices of institutions of which you may be part? In your tutor group consider:

● the extent to which 'the everyday aesthetics' of young people are acknowledged in your resources;
● where the 'work of others' might be usefully extended to include aspects of young people's culture;
● how this inclusion might change or extend possible outcomes.

In groups of three or four devise a SoW for Year 10 that uses 'grounded aesthetics' as its main stimulus and invites pupils to contribute to resources: objects, artefacts, images, books, articles, comics, etc. and encourages them to reflect on the ways content and context construe meaning.

As an art and design graduate you are familiar with the impact context can have on the reading and reception of objects and artefacts. You need to consider how this applies to pupils' contributions. An object, image or statement which one pupil values for highly personal reasons (memory, association) may be perceived by another as promoting stereotypes (race, gender, social class, age, sexuality). Troyna and Hatcher (1993) have suggested that while both racism and anti-racism feature in pupils' culture 'the existence in every class of children who have a clear anti-racist commitment is potentially the most powerful resource' (p. 203). Pupils can provide perceptive and sometimes unexpected insights into a wide range of issues from ecology to exploitation and, given the right environment, will voice their concerns about censorship, marginalisation and prevailing hierarchies. This encourages both teachers and pupils to question their value systems and to develop classroom relationships and 'communities of learning' (Watkins 2005), on the basis of equality of treatment and an ability to acknowledge difference and value the viewpoints of others.

Pupil interventions, both material and theoretical, should be acknowledged as an important resource for teaching and learning in art and design. However, just as pupil participation is significant, non-participation can also be significant; it can conceal a desire not to be exposed (hooks 1994). Stanley warns us that 'pupils may well have good reasons to be suspicious of an invitation to share their cultural heritage with the teacher and with other pupils not otherwise exhibiting much tolerance, let alone respect, for minority cultures' (Stanley 1986: 177). He reminds you that you do not start with a cultural 'tabula rasa' and you may be perceived as patronising or overbearing (*ibid.*). Buckingham and Sefton-Green (1994), Gillborn and Mirza (2000) and Youdell (2004) identify the 'identity traps' and resulting 'subtle resistances' by pupils who are unwilling to examine issues which impinge on their own identity. Such pupils contribute only what they think fits the status quo rather than their own beliefs and values.

The decision to proceed or not to proceed, to intervene or not to intervene, to take up issues or to leave them unchallenged are decisions the teacher is compelled to make as an authority – both real and imagined – in the class. This is not to say that students cannot intervene or challenge, but that students' actions have very different consequences because of their very different relation to power.

(Todd 1997: 72)

Plugging the gaps

As already noted, research by NFER (Downing and Watson 2004) and findings by OFSTED (2005) claim that there are still significant gaps in provision in the content of resources for art. These findings confirm that while the use of male, European artists (predominantly painters) remains prevalent, those resources relating to cultures and traditions from the more remote past, in contemporary society, by women artists and from non-Western traditions continue to be overlooked by a number of art and design teachers in their planning and teaching of the programme of study. Where 'readymade' inclusive packs are not available then it is up to you to ensure that you adapt existing resources or develop new ones to redress this imbalance. To disregard the past in favour of the present is not the solution. Similarly, any suggestion that aspects of race, gender, class can be viewed as discrete entities is erroneous.

Task 5.2.5 Resource development plan

Constraints in terms of resources is the first excuse teachers make in defence of these omissions so you must try and plug these gaps from the outset.

You have already produced an audit of the resources available in your placement school; do the same for your HE institution and your own collection.

Consider these in relation to the canon of 'plentiful resources' outlined above.

What are you making visible?

Are there any gaps?

What do you need to form a more comprehensive and inclusive set of resources and reference packs?

On the basis of these omissions produce a resource development plan for the year. Don't feel that you have to cover each aspect separately but find ways of combining them.

What further resources can you recommend to your host department and their school library?

Cultural diversity
(see also Chapter 11: Towards a Plural Curriculum)

In a Keynote lecture presented at the Commonwealth Institute, London in 1991, Brandt expressed his disquiet about the way art and design education seemed to be promoting cultural diversity through the use of cultural artefacts. He claimed that they were defined by their place of origin; there was no reference to the way 'images and objects presented were not value-free but value laden; not politics-free but politics laden, not context-free but context-bound'. He insisted education is not simply about making pupils aware of diversity or providing exotic or celebratory examples, it must also confront negative stereotyping. Brandt reminds us that art is not divorced from society, 'it is not simply an aesthetic expression, but also a sociopolitical and cultural statement which has added to a legacy of heritage of representation and speaks loudly to contemporary society'. Brandt insists educators move beyond providing cultural artefacts or adopting 'ethnic techniques' by ensuring that they are grounded in the right cultural contexts.

Multiculturalism has been replaced and/or subsumed by a succession of terms, 'New Internationalism', 'Global Art', 'World Art', 'Cross-cultural Studies', 'Cultural Hybridity'. More recently the notions of cultural diversity and interculturalism have encouraged educators to recognise that most earlier definitions are inadequate and limiting. Perminder Kaur, a British artist who is Sikh, reiterates this point when she claims:

> It's very difficult to make statements about particular things. Issues about race and colour are very complex now. In my work there is no longer direct polarisation between two distinct cultures. The work still contains questions concerning identity but on a more subtle level.
>
> (Proctor 1996: 11)

Proctor insists that Kaur's work does not suggest that 'home is "no-place", but rather "an-other" place, somewhere just over the next hill, in the next land . . . and any belief that we have found it is just an illusion' (*ibid.*). Meera Chauda (Plate 14) addresses similar issues in her work when she introduces traditional Hindu images into collages which ironically represent her place in contemporary British society.

The gap between theory and practice, between rhetoric and reality remains wide. It exists on a number of levels. Perhaps the most significant here is the difference between what teachers suggest should happen in the classroom and what can be observed. Aware of the complexity of the hidden curriculum, and the unspoken messages given to others by our choice of resources, it is tempting to play safe rather than take risks. To get it right is time consuming and involves research. Resources to support an inclusive curriculum are rapidly increasing. The following organisations and publications are worthy of note:

Resources

The Institute of International Visual Arts (InIVA): an arts organisation which promotes the work of artists, academics and curators from a plurality of cultures and cultural perspectives. It has a reference library and archive which includes reference sources such as books, exhibitions, catalogues, slides, CD-ROMs, video and audio tapes. It is an excellent source of information conferences, residencies, useful websites and online lectures (InIVA www.iniva.org). Its education packs on Landscape and Portraiture introduce British artists from a number of cultural backgrounds confirming a broad and rich notion of British art practices. They position Hogarth and Bacon alongside Sonia Boyce, Chila Kumari Burman and Vong Phaophanit with Andy Goldsworthy.

The Diversity Art Forum; formerly the African and Asian Visual Art Archive (AAVAA): Learning Resources Centre, University of East London, is an important resource centre for slides and publications, dissertations, audio and video tapes. Diversity Art Forum has inherited archive material of Black, African and Asian artists living in Britain. To enhance that material the archive has now included new material from artists outside Britain and western Europe. Diversity Art Forum has new material from artists working in sound art, Internet as well other visual media (http://www.aavaa.org.uk/).

Cahan, S. and Kocur, Z. (eds) (1996) *Contemporary Art and Multicultural Education*, New York, London: Routledge. This provides theoretical foundations, good colour plates, lesson plans and a useful annotated bibliography. Its focus on the USA should not be seen as a deterrent in this context. It deals with interrelationship between contemporary art, multiculturalism and social class. It promotes issue-based work within an expanding field, and includes contemporary concerns such as ageism, Aids, gay rights and racism.

Shohat, E. and Stam, R. (1995) 'The Politics of Multiculturalism in the Postmodern Age', in *Art & Cultural Difference: Hybrids & Clusters, Art & Design Magazine*, 43. A collection of theoretical papers supported by visuals.

Mason, R. (1995) *Art Education and Multiculturalism*, Beckenham, Kent: Croom Helm. This includes case studies of multicultural programmes in education, discussion re the interrelationship between aesthetics and social learning, art criticism and anthropology.

Loeb, H., Slight, P. and Stanley, N. (1993) *Designs We Live By*, Corsham: NSEAD. This provides cross-cultural resources for design work including West African, Celtic, Japanese and Islamic examples.

Third Text: an international journal covering 'third world perspectives on contemporary Art and Culture' provides a critical forum for discussion and appraisal of artists marginalised by racial, sexual and cultural differences. For further information write to: Third Text, PO Box 3509, London NW6 3PQ. See also the selection of key articles in *The Third Text Reader on Art, Culture and Theory* (2002) edited by R. Araeen, S. Cubitt and Z. Sardar, London: Continuum.

Lloyd, F. (1999) *Contemporary Arab Women's Art: Dialogues of the Present*, London: WAL.

198 Gallery: is 'a centre for the production & exhibition and interpretation & participation in contemporary art representing a range of diasporan identities'. 198 Railton Road, Brixton. www.198gallery.co.uk.

The October Gallery: www.octobergallery.co.uk/homepage.shtml, is an art gallery dedicated to the appreciation of art from all cultures around the world. The Gallery exhibits and promotes art of the *transvangarde* – or trans-cultural avant-garde – that is to say, the work of artists who, whilst working at the forefront of their own respective cultures, assimilate into their work elements from other cultures as well.

International Journal of Education Through Art: which began publishing in 2005, seeks to publish articles that present 'an international perspective on fundamental issues concerning education through visual arts'. As part of its remit it forefronts recent trends such as globalisation and the impact of the educational reforms associated with multiculturalism and cultural identity. For example, in volume 1, Soganci discusses the unease about figurative representations in Islamic culture in an article entitled 'Mom, why isn't there a picture of our prophet?'

Gender

> Since the mid-1970s there has been a legal requirement for equality of
> opportunity of treatment of boys and girls . . . One approach to this is to
> present pupils with examples of work, of artists, craftworkers and designers
> of both sexes. In this way, both boys and girls can grow up knowing that
> the full spectrum of media techniques and skills is open to them.
>
> (DES 1991a: 60)

> Ignorance does not just mean not knowing women's names or being able
> to identify pictures, sculptures, photographs, films or videos by women. It is
> much more complex. It is about an invisibility of meaning that arises from
> the indifference and indeed hostility of the culture to where these works
> come from, what they address and why they have something to offer that
> realigns our understanding of the world in general. If I call the work of
> women 'different' I immediately fall prey to the deadly paradox: to name
> what makes it interesting to study art by artists who are women is to
> condemn the artists to be less than artists: women.
>
> (Pollock 1996b: xiii)

While Borzello claims that feminist art criticism has 'just touched the national
curriculum with its fingertips' (Deepwell 1995: 22), Meecham insists the NC 'does
nothing to undo the structures that hold in place received opinion which argues that
women's art is derivative and not innovative enough to gain a place in the canon
defined in a masculine culture . . . a hard paradigm to shift' (Meecham 1996: 74).
Although critical and contextual resources for art and design are no longer totally
dominated by references to male artists an imbalance certainly exists. Within the fields
of art production and art education, you are still confronted by entrenched hierarchies
'the overwhelming masculinities of historically privileged knowledges' (Grosz 1995:
45) which require critique and questioning rather than (in)difference. In the face of
such established elitism, attempts to insert women retrospectively into a visual canon
constructed by white males has never been sufficient.

> Janson's *History of Art*, the most widely used text book, didn't mention a
> single woman artist until Janson died. Then his son revised it, including a
> big 19 out of 2,300.
>
> (Guerilla Girls 1995: 26)

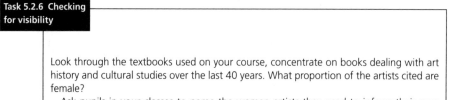

Task 5.2.6 Checking for visibility

Look through the textbooks used on your course, concentrate on books dealing with art
history and cultural studies over the last 40 years. What proportion of the artists cited are
female?

Ask pupils in your classes to name the women artists they used to inform their own
production.

(Allen (1996: 86) has analysed successive NC Orders to uncover bias.)

The feminist critique of art history (Nochlin 1991 and 1999; Pollock 1988; Chadwick 1990) has made most teachers aware of Artemisia Gentileschi, Rosa Bonheur, Berthe Morisot, Mary Cassatt, Käthe Kollwitz and Frida Kahlo. More recently the Turner Prize has introduced names such as: Helen Chadwick, Paula Rego, Rachel Whiteread, Mona Hatoum, Cornelia Parker, Angela Bulloch, Cathy de Monchaux (Plate 21), Gillian Wearing and Sam Taylor-Wood, Tomoko Takahashi, Fiona Banner, Catherine Yass, Anna Gallaccio, Gillian Carnegie, Rebecca Warren and Tomma Abts. Through newspaper journalism young British artists (yBa) Sarah Lucas and Tracey Emin have become well-known figures. However, it is often media notoriety rather than consideration of artistic acumen that has helped to develop their reputations. McRobbie (1998: 55) contends Tracey Emin's work and reputation 'owes more to the girls just wanna have fun humour of More magazine than it does to her feminist elders Cindy Sherman or Mary Kelly'. Indeed, Emin's weekly confessional diary in the Independent reinforces this view.

While most teachers are aware of the need to move beyond the limited definitions and debates promoted by the seventies feminists, many are ill-informed about more recent feminist interventions and readings. Pollock (1996a) highlights the false dichotomy between past and contemporary 'generations' and calls instead for 'constructed correspondence'. She also reiterates the importance of 'geographies' which are cultural and social as well as political. There is no homogeneous community of women artists, critics or academics. Petersen and Wilson (1976) reveal how women artists were frequently referred to as wives or lovers rather than artists in their own right. Chadwick (1990) explains how some women artists have received considerable public and critical attention but cautions against:

> . . . the dangers of confusing tokenism with equal representation, or the
> momentary embrace of selective feminist strategies with the continuing
> subordination of art by and about women to what is, in the words of
> Pollock 'the gender free Art of men'.
>
> (p. 349)

Hoorn's research reveals how the impetus for inclusion of women in the writing of art history has not necessarily altered the view of women as inferior artists. Her analysis of art texts shows the subtle ways in which women's artmaking is devalued through limited discussions and reference to women's personal characteristics rather than contributions to artistic practice (in White 1998).

This serves as a reminder that you need to scrutinise all texts for such practices and encourage pupils to do the same. All too often, the media either trivialises or pathologises the work of women artists. Journalist Hunter Davies describes Jenny Saville in the following way: 'She doesn't look like the artist, more like a lower-sixth-former, so young, so small, so conventionally dressed' (Rowley 1996).

Jenny Saville's work is a useful example. As a figure painter she is seen as part of a long tradition often compared with Freud and Bacon and a British painterly school. Comparison with artists such as Jo Spence, Cindy Sherman and Orlan confers a matriarchal lineage and invites a different interpretation, one which confronts traditional constructs of female subjectivity. Rowley (1996) compares more typical representations of 'the supine female object body,' where the female model is observed by the male artist, to the self-examination that Saville undertakes, the way 'scale' and 'gaze' and 'perspective' are used as interventions which work with, yet against, traditional modes.

This requires that you, as a teacher, ask: Why am I teaching this? Does it still apply? Is it relevant? Am I presenting pupils with fixed interpretations or am I asking them to consider different readings?

Task 5.2.7 More than tokenism?

Look through your GCSE SoW; identify how women artists have been represented in your supporting resources.

Is representation based only on materials, skills and techniques or on other issues?

Is gender important in their work?

What other constructs inform women artists' practice: race, religion, social class?

Do, or can, their artworks act as 'interventions', questioning traditional hegemonic practices?

Can they be recognised as part of a matriarchal lineage or are they grouped alongside male artists as part of the same genre?

With other members of your tutor group develop a SoW which encourages pupils to question traditional representations using the work of women artists as interventions.

Consider how pupils might be encouraged to use these findings to inform their own production in art, craft and design.

Freedman offers a useful way forward when she promotes the exploration of 'sites of contestation and frayed boundaries' (Freedman 1994: 48). Rather than women (and other marginalised groups) being injected into the curriculum and their stories adjusted to conform with the modernist model of history you should consider areas of contestation or disagreement. Introduce pupils to different critical positions and help them to understand that interpretation is open, how each one is a construct and why some have been privileged over others.

Pollock (1996a) insists that:

> It is vital to show that the present is historically shaped. Sexual difference and sexual divisions in society are not natural but historical and that is why they can be challenged and changed. The past as tradition – in Art History it becomes the Canon – is used to justify the present status quo. Validated by time, the canons of great art brook no discussion or serious consideration. Feminist interventions have to disrupt canonicity and tradition by representing the past not as a flow or development, but as conflict, political, struggles on the battlefields of representations of power in the structured relations we call class gender and race.
>
> (p. 12)

The easily accessible resources and publications dealing with classical and modernist genealogies (the canon) need no promotion, they can be found in any bookshop on any high street. Gretton (2003) suggests that they can usefully provide the starting point for raising issues of power relations, marginalisation and stereotypes. However, deconstructing the canon in this way can be seen to reinforce its hierarchical position.

Wolff (1990) reveals that just as women are more or less invisible in mainstream histories of modernism, the same pattern can be seen to be developing in postmodernism (p. 6). Various working parties set up to inform the development of NC 2000 advocated a postmodern approach to art education. However, there is a growing suspicion that postmodernism has failed to resolve issues of marginalisation. By claiming to embrace all previously marginalised groups within its wider remit and including them within its discourses, postmodernism can be seen to have subsumed and thus silenced them (Burgess and Reay 2006).

Panting (1999) points out how 'debating gender has become fraught with anxiety and is considered anachronistic and unnecessary' (1999: 20). She reveals how work that deals with gender is itself on the periphery; how it is no longer synonymous with feminism but has been subsumed into the broader picture of identity politics.

Task 5.2.8 Identity politics

Repeat Task 5.2.6 but this time examine the work of male artists and subject their work to the same scrutiny.

Consider how the representation of men has changed in art and the media over the last 40 years.

> Not only do we have to grasp that art is a part of social production, but we also have to realise that it is itself productive, that is it actively produces meanings . . . it is one of the social practices through which particular views of the world, definitions and identities for us to live are constructed, reproduced and even defined.
>
> (Pollock 1982: 9)

Resources

The Women's Art Library (WAL) (http://make.gold.ac.uk/): Until 2003 WAL was a voluntary-aided organisation constituted as an educational charity since 1982 to promote women artists working in any medium in any part of the world. In 2003 its Arts Council funding was cut and it was forced to close down. Fortunately the library at Goldsmiths College, University of London agreed to take in most of its resources which includes over 140,000 slides of artworks divided into three main sections: historical, documentary, contemporary, and 5,000 files of press cuttings, a comprehensive library of books, videos, monographs and exhibition catalogues and a growing collection of artists' books, a comprehensive 'Women of Colour Index' set up by artist Rita Keegan. In 1995 WAL produced a slide pack of 24 women artists. This provides a useful selection, from Artemisia Gentileschi's *Judith Beheading Holofernes* (1618) to Sonia Boyce's *She Ain't Holding Them Up, She's Holding On* (1986). It identifies recurrent themes such as identity, memory, motherhood and war. It highlights the range of media used by recent women artists including examples as various as Zarina Bhimji's (1992) *Installation of suspended and children's kurtas* to Tess Jaray's public art floor tiling for the Centenary Square in Birmingham.

Other key feminist, women's libraries are the Feminist Archive in Bristol, the Glasgow Women's Library and the Fawcett Library at London Guildhall University.

There are two journals dedicated to Women's Art: *MAKE* by WAL (1983–2002) and *n.paradoxa* (http://web.ukonline.co.uk/n.paradoxa/), the international feminist art journal exploring feminist theory and contemporary women's art practices.

Broude, N. and Garrard, M. D. (1994) *The Power of Feminist Art*, New York: Harry N Abrams. A chronological series of articles on feminist practice in the USA since 1960s supported by quality reproductions.

Dalton, P. (2001) *The Gendering of Art Education*, Buckingham: OU Press, looks at the way art education has always been implicated in producing gendered identities.

Hiett, C. and Orbach, S. (1996) *Venus Re-Defined*. Liverpool: Tate. A resource pack produced to accompany a display of sculptures by Matisse, Renoir and Rodin at the Tate, Liverpool. Drawing on feminist critique themes are identified. The pack aims to involve pupils in discussion about the work and surrounding discourses; it is evaluated in *JADE*, 1996, 15, 3.

Hyde, S. (1997) *Exhibiting Gender*, Manchester: Manchester University Press, examines ways in which women's art and representations of women are collected and 'represent-ed' in galleries and museums. It presents full colour reproductions of pairs of unidentified works, one by a woman and one by a man and asks, 'Can you tell which one is the woman's work?' Issues raised are further discussed in the text.

IRIS (http://www.irisphoto.org/live/index.asp) The Women's Photography Project: celebrates the contribution made by women practitioners to the development of photographic theory and practice.

MAKE, Issue 81, Nov 1998, 20 years of Women's Art. This is a special edition charting the development of women's art from 1978–98 in an accessible series of illustrated articles.

VARO (http://www.varoregistry.com/): an electronic registry of contemporary international women's art. Includes work by Audrey Flack, Barbara Kruger, Sophie Calle, Rachel Whiteread.

If you are interested in the way gender identities are constructed the following books will be of interest; they are sociological texts, therefore they make no direct reference to art and design education. Both books reinforce the notion that gender identities are multiple not singular, not necessarily fixed or constant, and that they are subject to a range of powerful influences.

International Journal of Art and Design Education, 26, 1, Spring 2007, edited by Professor Nick Stanley, Special Edition: Lesbian and Gay Issues in Art & Design Education.

Mac an Ghaill, M. (1996) 'The Making of Men', in P. Woods, *Contemporary Issues in Teaching and Learning*, London: Routledge, in which the author discusses the many different types of masculinities he observed in one comprehensive school.

Mirza, H. (1992) *Young, Female and Black*, London: Routledge, deals with the way personal agency as well as socialisation theory contribute to constructs of femininity.

Myers, K. and Taylor, H. with Adler, S. and Leonard, D. (eds) (2007) *Genderwatch: Still watching . . .*, Stoke-on-Trent, UK and Sterling, USA: Trenthers Books.

Contemporary art

> Contemporary art introduces many aspects of popular culture: photography,
> video and computers. It raises relevant issues, it uses contemporary materials
> and technologies, it erodes traditional boundaries because it does not always
> fit neatly into traditional categories of painting, sculpture or print and art,
> craft, and design.
>
> (Burgess and Holman 1993: 9)

Examples of artwork referenced in this chapter have been deliberately selected from
contemporary practitioners. It is important that you introduce pupils to artists who
share the pupils' life and times: but remember you cannot introduce pupils to exciting
installation pieces and then expect them to be prepared to respond enthusiastically
using only powder paint. You must carefully consider ways in which such work can
impact on pupils' own production. There are numerous examples of contemporary
artists who deal with natural, found, reclaimed objects and materials (Bill Woodrow,
David Mach, Andy Goldsworthy, Rona Pondick, Meret Oppenheim, Annette
Messager, Yinka Shonibare). Their practices translate into classroom practice. Others
(Helen Chadwick, Cornelia Parker, Mariko Mori, Anna Gallaccio, Bill Viola, Tony
Ousler) use ephemeral, degradable pieces, space-consuming installation works or
highly sophisticated expensive technologies which make application to the classroom
all but impossible. When using contemporary art to inform pupils' work it is important to explore the ways in which it can reinterpret the old and/or introduce new
issues and ideas rather than encouraging only transcription or pastiche. Remember,
contemporary conceptual cliches can be as limiting as modernist 'masterpieces'.

The work of contemporary artists such as Joseph Beuys, Barbara Kruger, Mary
Kelly, Jenny Holtzer, Hans Haacke, Christian Boltanski and Willie Doherty can
seem incomprehensible without reference to its political content. The work of
Robert Gober, Tracey Emin, Judy Chicago, Hannah Wilke, Andre Serrano or the
Chapman Brothers of an overtly sexual or abject nature raises interesting issues of
censorship and freedom of expression (work distanced by time is often not regarded as
so problematic). In an article entitled, 'Monsters in the Playground' I suggest:

> The deliberate avoidance of 'difficult' subject matter in art is tantamount to
> paranoia, a failure to acknowledge that it is all pervasive in a society
> dominated by the mass media and its scopophilic apparatus. Although absent
> from the curriculum the monsters lurk behind the back of the teacher in full
> view of the students, the unacknowledged 'other' that in every other respect
> pervades their lives.
>
> (Burgess 2003: 108)

However, the extent to which such challenging work can be discussed with pupils
depends on the ethos of the school and the attitudes and values of staff and parents. As
a student teacher, you must always consult your mentor first. The school environment
should be one where, within reason, young people are encouraged to look critically at
challenging images, in order to understand the ways in which they represent negative

stereotypes and call into question morals, attitudes and values. Art can be seen to have a privileged place in the curriculum, one where it is possible to explore such ideas particularly with GCSE and post-16 groups. Artists play games with intentions in order to encourage engagement with social, political, ecological or moral issues, sometimes provocatively, sometimes deliberately confrontational. The painting of Myra Hindley by Marcus Harvey (1995) was reviled and refused by many, for others it raised significant issues about the role of art in contemporary society.

Contemporary art has become a popular topic for reportage. Along with the Turner Prize, which receives annual rebuke from the media, contemporary shows exhibiting the most radical and unexpected use of materials attract the most coverage. It is through this avenue that pupils' impressions of what comprises contemporary practice are often formed and reinforced. Such articles provide a cheap and potentially informative resource for teaching and learning. Also scrutinise TV viewing guides and ask media resources staff to tape programmes relating to material and visual culture.

> **Task 5.2.9 Collecting 'critical' comments: cuttings**
>
> With an 'A' level or other post-16 group complete the following:
> Ask students to bring in any articles or reviews about contemporary art. Ask them to compare reviews in different newspapers and magazines to investigate the ways in which the media construct attitudes and values.
> Invite pupils to write reviews for their own and others' work.
> Set up cuttings files (lamination helps preserve otherwise degradable paper.)

Resources

Information on contemporary art is not difficult to obtain, although books and catalogues with good quality reproductions are unavoidably expensive, often beyond the limitations of capitations. Contemporary Galleries such as Camden Arts Centre, the Tate Modern, Ikon, Laing, Serpentine, Whitechapel and the ICA have excellent bookshops with catalogues and publications available by mail order. Making your archive from images available from the web is one way of building resources: providing you keep within the copyright laws! Pump almost any artist's name into Google or a similar search engine and you will be surprised at how easy it is to find visuals and biographical information (however, interpretative texts should not be used without critically analysing the content). A less well exploited alternative is the wide variety of magazines currently available, the list is endless but the following are recommended in addition to those already cited: *Art Monthly, Contemporary Visual Arts, Frieze.*

The Department of History of Art & Design at Manchester Metropolitan University, 0161-247 1930, produces and sells slide sets and videos. They deal with design, architecture and fine art, cultural diversity, popular culture. Many are based on exhibitions and original research. It houses the Design Council collection of slide sets which are available on loan, and sells slide sets produced by galleries and museums.

The Art Gallery Handbook: A Resource for Teachers (2006) edited by H. Charman, K. Rose and G. Wilson. A Tate Gallery Publication which, although pitched at the 5–14 age range, provides a useful introduction to learning in galleries.

The following provide good colour reproductions of contemporary art:

Archer, M. (1996) *Installation Art*, London: Thames & Hudson.

Art & Design Magazine published six times a year by Academy Group Ltd.

Blurring the Boundaries: Installation Art 1969–1996, San Diego: Museum of Contemporary Art.

Button, V. (2005) (4th Edition) *The Turner Prize*, London: Tate Publishing.

Morgan, S. and Morris, F. (1995) *Rites of Passage: Art for the End of the Century*, London: Tate Gallery.

Price, D. (1998) *The New Neurotic Realists*, London: The Saatchi Gallery.

Sensations Exhibition Catalogue (1997), London: Royal Academy of Arts.

UNIT 5.3 PARTNERSHIPS WITH OUTSIDE AGENCIES

Giroux (1994) insisted that:

> Teachers cannot locate teaching in one space, they need to engage with other educators in a variety of sites in order to expand the meaning and places where pedagogy is undertaken.
>
> (p. 166)

By the end of this unit you should be able to:

- identify some of the important national and local contributors to resource based learning;
- examine how resource packs support visits and follow-up work;
- identify the role of the professional artists in schools, LEAs and Local arts boards.

Galleries and museums

Over the last 10 years there has been a series of Government initiatives from the DCMS and DFES to research and promote galleries and museums as important learning resources for all pupils. Recently liaisons with such outside agencies have been included as an important element in a wider government agenda for education including 'Every Child Matters' (http://www.everychildmatters.gov.uk/), 'Youth

Matters' (http://www.dfes.gov.uk/publications/youth/), 'The Children Act' (http://www.dfes.gov.uk/publications/childrenactreport/) and the extended school day. Included in these government initiatives are: *'The Renaissance in the Regions Programme'* which supported regional 'hub' museums in nine regions in 2003–06 to evaluate and develop their education programmes; results of the first phase can be found on their website (www.mla.gov.uk); *Creative Partnerships* (http://www.creative-partnerships.com), the Government's creativity programme for schools and young people, managed by the Arts Council and *En-quire: inspiring learning in galleries* (2005–6), a national research project managed by Engage (the professional association for gallery educators) aimed at 'inspiring learning in galleries' which is currently in its second phase to be completed in 2008 (http://www.engage.org/projects/enquire.aspx). Detailed information on the background for all these projects can be found on www.inspiringlearningforall.gov.uk. Although many of these initiatives are predicated on the notion that creativity cuts across subject boundaries and extends to learning out of school hours, the implications for art and design departments in school are still considerable. You should find out if, and how your teaching practice school is involved and how you might participate.

In a special edition of the American journal *Art Education* entitled 'Art Museum/School Collaboration', Mayer (1998) suggests that in gallery education there is a shift away from object-centred values towards people-centred values. She claims that:

> The most significant change in art historical theory that semiotics offers educators concerns the 'receiver' – the viewer. Central to semiotics are theories of reception . . . the goal of semiotic analysis of works of art is not to produce interpretations of those works, but to investigate the processes viewers use to make sense of visual art.
>
> (p. 19)

The idea that the viewer has agency and is not just a passive recipient is significant and calls into question more traditional ways of gallery education. Mayer believes that by engaging pupils in this way, 'they become empowered as active, equal thinking members of an expanding community of inquiry' (*ibid.*) as critically literate.

Task 5.3.1 Gallery education

Most galleries and museums, whether small local galleries or large national institutions are willing to allow student teachers to observe gallery educators/artists/art historians working with school groups.

Arrange to go to a number of gallery talks for different age groups (always phone first).

Enquire whether the content/delivery has been negotiated with the class teacher.

Observe different approaches to interpretation.

Identify what determines the approach: the type of work; the age of the audience; the specialism of the leader.

What types of discussion take place: are questions open, closed, framed?

Is the talk descriptive, analytical, challenging, provocative, does it raise issues, introduce or invite different readings?

This is a useful exercise, as, although the situation is slowly improving, a significant number of galleries and museums still do not have any educators, therefore if you are to work with gallery and museum exhibits you will need to understand how working this situation can differ from the classroom.

Anderson refers to 'an expanded concept of museums' insisting that no culture can be adequately represented in museums by material culture alone (Anderson 1997: xiv). He suggests that the museum walls should not be seen as a boundary. He calls for a concept of museums that includes their natural and cultural environment: one which reflects not just the physical and economic environment of the museum but also how the institution can also depend on non-material culture: 'shared values, ideologies, oral traditions, rituals, ethical standards and beliefs that give meaning and symbolic significance to our material world' (ibid.). Given this definition no one can claim not to have a 'museum' equivalent in close proximity to the school; a local cemetery, church, mosque or synagogue, historic buildings, local festivals, football grounds, visiting and permanent funfairs, all have their cultural significance and value. Anderson claims that museum education develops 'cultural literacy', the capacity to understand, respect and interact with people from different cultural backgrounds; it develops a sense of our own identity (ibid.: 7). He suggests that museums can demonstrate ethical leadership; that their values, codes of morality and expectations of behaviour (embraced by this concept of cultural literacy) help to shape those of the public. Whereas in the past most museums claimed to provide a neutral space, excluding controversy, this is now changing and the museum can be seen as a site for different voices, a place that encourages questioning and enquiry: 'children need a gallery environment that allows open and exploratory learning and encourages them to question and challenge' (ibid.: 23). Feminist critics, whilst acknowledging some change, do not believe it was ever neutral and suggest most exhibition spaces have a long way to go before they can claim to operate as dynamic and reflexive cultural forms (Duffin 1995; Duncan 1995).

Over 30 per cent of museums have libraries and study collections; the potential for teachers and older students to use these for independent on-site and online learning has yet to be exploited to the full. The development of digital technology links between the museum and education sector has yet to realise its potential, but given the rapid nature of development in this field it is one that teachers can ill afford to ignore. The National Portrait Gallery's online activity, www.global-leap.com, makes possible videoconferencing between pupils and gallery educators. The Tate offers a free online introductory course on Contemporary Art, www.tate.org.uk/ita/, aimed at sixth formers and their teachers.

Galleries and museums are a rich resource for art and design education. Both permanent and temporary exhibitions at such venues are accompanied by catalogues and teachers packs. Many offer talks and practical workshops run by art historians, critics or artists. Some are involved in outreach projects involving artists in community projects including school placements. The benefits of establishing links with local exhibition spaces and collections should not be overlooked. They provide an opportunity to involve pupils in discussions about the type, origins, quality and significance of the work on display, to make comparisons with other exhibition spaces, and to consider issues of patronage, funding, curatorial decisions, target audience;

issues which can all too easily be overlooked when working with a large national institution where there is a temptation to accept what is written as the 'correct' interpretation.

Selwood (Selwood *et al.* 1994: 37) claims that secondary teachers equate the accumulation of cultural capital with education. She is critical of the way teachers bring char-a-bancs of pupils from the north, spending hours on the motorway, in order to briefly visit a national collection based in London just to be able to 'tick off items on the cultural agenda' rather than gain an intimate knowledge of less famous local resources. They reveal that this attitude is commonplace, citing the fact that a third of all visitors to the best known western museums never get as far as the galleries. 'Their object in visiting is to buy souvenirs, savour the ambience, have a cup of coffee and say they have been. Their visit is simply a symbolic gesture' (*ibid.*).

Resource packs: packaged pedagogies

Hughes (1998b) suggests that student teachers entering the teaching profession all too quickly forget their graduate training which becomes 'rapidly overlain with the orthodoxy of the school curriculum' (p. 44). As a result of the pressures from an overfull traditional, subject-based, mandated curriculum, art and design in all its practical and theoretical complexity has to be delivered to most pupils in less than an hour. In order to cope teachers use commercial books and resource packs: expediency is taking over from philosophy and teachers are in danger of becoming ahistorical.

You need to approach resource packs with a degree of skepticism. Although many are excellent it is all too easy to presume that because the content of such packs has been selected by reputable publishers it is sound. It is usually 'safe', selected to appeal to the widest audience possible. Most commercial packs, whether of slide packs, postcard packs, video tapes, DVDs or online teachers' notes, deal with information about established art; artists and cultures. Many deal with content, context and history but are less well developed in terms of pedagogy. Some promote what is disparagingly called a 'Blue Peter' approach to education. Bancroft (1995) examined a number of packs selected from advertising material and catalogues sent to her school. Her evaluations identified that most packs promoted the limited canon of DWEMs. Although 9 per cent of the images were by Black artists this was often accompanied by patronising (mis)information with work from other cultures placed in a 'timeless past'. The artists in the packs may be, in the main, dead, but issues such as 'death' are studiously avoided. All too often packs are a sanitised collection of favourite images chosen to cause no offence but which fit neatly into a 'safe' curriculum and are easily subjected to a traditional, formalist analysis. Textual information is in the form of biographical notes, questions are 'closed' suggesting that there is a correct response.

Unpacking Teachers' Packs (Clive and Geggie 1998) is an evaluation of teachers' packs from ten London galleries and museums by Engage – the national association for gallery education.

Task 5.3.2 Unpacking teachers' packs

Contact a number of local and/or national galleries and museums and request recent teachers' packs. (You can often download these from their website.) However, it is useful to make contact so that you can ask to be placed on their mailing lists to receive posters advertising exhibitions and invites to free teachers' evenings.

With other student teachers consider the following for each pack:

- Who is the target audience – is it teachers, pupils or both?
- Does the pack provide information and/or raise questions?
- Are fixed curatorial views presented or are different readings given?
- What is provided: text, images, historical, contextual information, further reading, etc.?
- How is it presented?
- How could it be developed to make it appropriate for use with pupils at KS3?
- Does the pack provide useful inset material or is it limited to a taster tempting you to visit the exhibition?
- Are connections made with the NC, GCSE, 'A' level and post-16 examination criteria? Is this necessary?

Artists, visits and Arts Council England: Regional Arts Boards (RAB)

Artists are often provided with subsidised studios on the understanding that they engage in community work including projects with local schools. Information about such schemes is readily available from local education authorities (LEAs), gallery education departments and libraries. However, the first port of call should be your Regional Arts Board (RAB) which is in a position to advise you about educational projects involving artists. Joint applications between artists and teachers which have a clear endorsement from the school's senior management team are most favourably received. RAB also support larger programmes involving a number of schools, including cross-phase and cross-arts initiatives or identified regional educational priorities in the visual arts. For example, environmental projects, site-specific or urban-regeneration projects bringing together professional artists from a range of specialisms to work collaboratively with schools, town planners and environmentalists. RABs can also give up-to-date information and advice about funding.

Task 5.3.3 Networking

Contact the RAB (http://www.arts.org.uk) and ask them to send you their most recent newsletter or bulletin. Find out about recent educational projects, how they were set up, funded and evaluated.

Ask to be placed on a mailing list informing you of local and national projects.

The role and value of the professional artist as a resource for teaching and learning has given rise to interesting debates, especially in secondary schools where the teacher has a degree in art or design and may practise as an artist part-time. The potential benefits (for artist, teacher and pupils) of such collaborations are enormous but so are the disadvantages if it goes wrong. The placement of an artist in education is seen as one answer to the problem of the separation of art and artist from the public, enabling young people to observe artists' working processes, work collaboratively with them, engage them in discussion; see (Pringle 2002) 'We did stir things up: the role of artists in sites for learning' at www.artscouncil.org.uk/documents/publications/ phpShHUNy.pdf) However, working with or alongside an artist must never be seen as a panacea. It is important to help pupils understand the complexity of contemporary practice; introducing them to the work of one artist can give them a limited view of contemporary art. Pupils need to be introduced to the diversity of practice and a well-planned residency in school enables them to extend their definition, understanding and skills in art, craft or design. Such a programme should also involve an element of curriculum development so that, once the artist has left, the concepts and processes can be revisited and continue to inform the curriculum.

Resources

Dahl (1990), Manser and Wilmot (1995), Sharp and Dust (1998), and Oddie and Allen (1998) provide comprehensive information on artists' placements in education including: contacts, selection of the artist, funding, planning, Inset possibilities through to evaluation of the project. Taylor (1991), Dickson (1995), Clive and Binch (1994) provide examples of exemplary practice. Burgess (1995) stresses the need to ensure that any placement involves curriculum development. Xanthoudaki, M. *et al.* (eds) (2003) comprises chapters on researching visual arts education in museums and galleries. Pringle (2006) elucidates the scope and distinctiveness of contemporary gallery education and the nature of the learning and teaching experience within galleries.

En-quire: Inspiring Learning in Galleries (Engage 2006), a report on an Engage research project funded by the DCMS and the DES as part of the Strategic Commissioning for Museums and Galleries Education programme. Sections are downloadable from www.en-quire.org.

Public art, architecture and the built environment

> The usual public art story, as we know, is of pieces plonked in forecourts and pedestrianised streets in ill-conceived bids to brighten up (or compensate for) botched bits of urban design. Couched mostly in the language of modernism, and therefore intrinsically siteless and nomadic, public art often succeeds merely in transforming places which were once public into an annex of the contemporary art gallery – that forbidding, empty space where only the initiated feel comfortable.
>
> (Usherwood 1998: 46)

Gablik (1995), like Usherwood (1998) questions the meaning of art in a postmodern age. She suggests that the focus should shift from culturally sanctioned spaces, such as galleries and museums to social, natural and community spaces. Similarly, Lacy (1995) and Neperud (1995) contend that educators need to move away from art on gallery walls to seeing the natural and built environment as spaces of social responsibility and aesthetic improvement. If these are important sites for artistic production then they should be recognised as resources for teaching and learning. Mark Fisher MP (1994) insists: 'We are not going to get good critical appreciation of the built environment without having it understood at school level'.

Avery (1994) insists that built environment education must encourage pupils to be socially responsible and consider the needs of gender, race and class in relation to the environment. She believes that the introduction of literature which focuses on women and the built environment can inspire pupils to investigate their surroundings critically and recognise how these impact on others. She provides an extensive list of relevant references including Women and Built Environment (WEB) and Matrix, a feminist architectural practice. Adams (1989) stresses the importance of fostering aesthetic responses to the environment. She expresses her concern that learning which emphasises socio-economic and political aspects too often neglects the subjective view and individual response (see Plates 5 and 6). She quotes Jonathan Raban to reinforce this perception: 'the city as we know it, the soft city of illusion, myth and nightmare is as real as the hard city one can find on maps, in statistics, monographs on urban sociology and architecture' (p. 192). Both Adams and Avery agree that in order to encourage attitudinal changes, support from outside agencies is important.

Building on her earlier research, Adams (1997) presents a convincing case for 'public art' education. She suggests that projects to change the school environment echo the function of many public art projects, that: 'it is site-specific, concerned with environmental improvement. It develops a sense of place and aims to create a sense of identity and local distinctiveness' (p. 236). Thurber (1997) asks: 'What sources and resources, what "sites to behold", do you have in nearby communities that are waiting to become rich sites for inquiry in Art for your students?' (p. 39). One can safely assert that the majority of school and college grounds are often a sadly neglected resource. They can be best described as flat, featureless places, consisting of vast areas of tarmac and close-mown grass (Ker 1997: 61). Redesigning a 'close-mown grass' lawn into a landscaped garden or the school's own sculpture park involving local artists, town planners, designers and architects is an ambitious project for student teachers. However, the use of particular locations as a stimulus for site-specific plans and maquettes is worth considering. This is excellent preparation for more ambitious projects once you have secured a teaching post.

Task 5.3.4 Redesigning the school environment

Survey the school environment and identify possible areas to redesign.

Locate plans of the school and its grounds. If the school does not have copies then the local town hall or library should.

Devise a SoW which encourages pupils to investigate the school environment, to look at it critically with a view to improving it. Geographical, environmental, health and safety issues and cost implications should be taken into consideration.

Ensure aesthetic response and social and political understanding are all included.

Resources

Your RAB and/or LEA will have information about local artists currently working on environmental projects both independently and with pupils. They can provide information about previous projects in the area that they have sponsored. Find out how these projects were set up and funded; identify opportunities for you to get involved in the future.

Building Schools for the Future, www.bsf.gov.uk/bsf/: the Government aims to ensure that by 2011 all secondary pupils will be learning in twenty-first-century facilities. Some schools will be rebuilt, others redesigned. It is important that you keep up to date with these developments, consider the implications for learning and teaching in art and design and, where possible, involve students in discussion about potential changes to their learning environment.

English Heritage promote greater understanding of the historic environment and produce a vast range of comprehensive user-friendly support for teachers; online and printed resources, image archives, information on site visits, and project case studies. Regional officers can offer professional advice on using sites and the local historic environment, www.english-heritage.org.uk.

The Royal Institute of British Architecture (RIBA) architecture gallery offers a programme of exhibitions designed to attract school parties and families alongside architecture students and design professionals, www.architecture.com. *Designs on Britain* is an online teaching pack to support schools interested in working with an architect on a real local project, www.artsinform.com/dob/teacherspack.html.

Art & Architecture Journal is a quarterly independent magazine of contemporary art and architecture focusing on art in the public context worldwide as a cross-cultural and multi-disciplinary activity, www.artandarchitecturejournal.com.

Architecture Centre Network co-ordinates and promotes the work of 22 Architecture and Built Environment Centres found in every region of the UK. The independent centres work with decision makers, schools, communities and professions, through programmes of education, exhibition, development and empowerment, with the aim of improving the quality of the built environment in their local area. See individual centre websites at www.architecturecentre.net for further information and contact details.

The Commission for Architecture & the Built Environment (CABE) aims to improve the quality of people's lives through good design. Their website, www.cabe.org, offers up-to-date research, publications, an image library and teaching resources for exploring and assessing the design and care of the built environment; buildings and public spaces. It includes a Citizenship and Art & design local safari guide and *How Places Work*, a school visits programme and guide that offers teachers advice and ideas for making visits to buildings and public spaces stimulating and informative.

Engaging Places is a new initiative by DCMS to unlock the educational potential of the historic and contemporary built environment. The aim is to create a one-stop-shop for built environment education, providing tools, resources and support for schools to engage with the fabric of their local communities.

Muf: muf is an all-women and architecture practice committed to working in public space.

Muf's practice is about making spaces work as well as about making spaces. Its focus on process involves a recognition of the political context of how projects are constructed, and reveals complex attitudes to design and creativity. As well as a website they also have a manual: muf (2001) *This is what we do: a muf manual*, London: Ellipsis Press.

Placecheck is a method of assessing the qualities of a place to define what is special, local and what improvements are needed, an accessible process for focusing people on working together to achieve them, www.placecheck.info.

Victoria & Albert Museum's *Architecture for All* is a joint project with RIBA to promote the study, understanding and appreciation of architecture. Architecture for Schools provides free tours of the architecture gallery and introduction to buildings from different cultures. Resources include online workshop case studies, images and handling collections of original artefacts, plans and drawings, www.vam.ac.uk.

UNIT 5.4 CHANGING DEFINITIONS – NEW TECHNOLOGIES ICT

> The increasing importance of 'new technology' in the development of digital imaging, multimedia and video is widely recognised but, as yet it is unclear how those of us responsible for visual education should respond. For many of us, our own understanding of the technology and its potential is limited and we are reluctant to replace tried and trusted methods and subject methods with unknown content.
>
> (Binch 1997: 6)

By the end of this unit you should be able to:

- recognise how new technologies are changing the way educational resources are accessed and used;
- appreciate the continuities and discontinuities between still and digital imagery;
- identify venues, sites and publications which enable you to develop RBL.

Photography and new technologies

It is with caution that I make photography a special case. It is perhaps, more correctly identified as one of the many media within contemporary art practice. Dewdney (1996) recognises that, although the art, photography, technology continuum provide developing historical narratives, technology will not be limited by its predecessors:

> New technology has a culture of very rapid transmission. On the one hand it is a technological development which builds upon and supercedes others before it like the magic lantern, photography, film, telegraphy, radio and television. On the other hand, it is a cluster of technologies which converge to produce a hybrid of machines which can combine visual, text and auditory information transmission, extremely large information storage with manipulation and authoring programmes. For art, media and design, the quickening pace of new information technology heralds the dissolution,

not only of previous distinct forms and practices of design, photography, animation, film and fine art but also the categories and disciplines of knowledge upon which those practices rest.

(p. 86)

Photography, once a distinct, albeit small area of art and design education, has all but been subsumed by the digital revolution. However, whether chemical or digital, photography still plays an important role in the lives of young people: family albums, newspapers, books, posters, postcards, advertisements, compact disc covers and so on. It is all too easy to 'respect' the photographic image as a factual record; often it carries no mark of its maker to remind us that it is a construct. When using photographic images in the classroom (still, digital or moving) you should always consider the way the producer has intervened with the 'real', how images are created, mediated and circulated. Stanley (1996) insists 'the very ubiquity of manipulated images may, if not studied seriously, inoculate us against a close and continual suspicious inspection of their contents, intent and manipulative strategy' (p. 4). You should help pupils to understand that photography does considerably more than record or confirm an event. Whether displayed on a gallery wall or on the back of a bus, photography should be viewed critically, especially if it is to inform practice. Buchanan (1995) promotes the notion of critical literacy; a combination of visual literacy and critical studies which he believes can counter the false separation between theory and practice. Pupils should be encouraged to be 'critically literate', to uncover the ways photographic imagery can represent and position the subject, build pathos, indicate status, aesthetise the politics of third world poverty or reinforce gender stereotypes; both male and female. Works by Richard Billingham, Adrian Piper, Ingrid Pollard, Martha Rosler, Cindy Sherman, Yinka Shonibare, Jo Spence and Sam Taylor-Wood present us with questioning, occasionally disturbing, sometimes ambiguous, occasionally playful, images which challenge typical expectations of the photographic image. An image can be interpreted in many ways. Artists often construct images that suggest unlimited and unstable meanings. In contrast, advertisers often produce images aimed at a particular audience (Morley 1992). Ferguson (1995) insists that advertisements not only sell products, they also sell politics. Jones (1996) states that at secondary level the range of work in ICT applications needs to, 'be made more responsive to pupils' backgrounds and interests . . . There are opportunities for well-planned Art/IT projects which have, for example, a social or ethical dimension' (p. 1).

Task 5.4.1 More than just a pretty picture

Any representation can be 'problematised for the ways in which it both produces and covers over forms of cultural self-representation that are constructed within dominant historical, hierarchical, and representational systems' (Giroux 1994: 103).

With other members of your tutor group, devise a Year 12 SoW which encourages pupils to consider the way photographic images 'frame' the way people represent ourselves and others.

Use work from two or more of the above artists plus images selected by students to contextualise making.

The advances in digital photography, including the improved quality and reduced cost of the computer print-out, herald the demise of the single-lens reflex camera and its environmentally unfriendly chemical processes. However, Stanley (1996) calls for caution when he asks educators to ponder the following question: 'do we, on the basis of the development of digitised imagery to date, have reason to rethink our views about the significance of photography?' (p. 98). He suggests you consider the relative merits of an interactive package with that of a family photo album. He warns, be 'wary of special pleading by those who offer the future which all too often bears an uncanny resemblance to the past' (*ibid.*). Similarly Tate (1997) insists teachers need to 'evaluate the somewhat messianic claims we keep hearing about how we ought to reconstruct the curriculum to take into account the need of the information super highway' (p. 2).

Resources

The journals: *Aperture, Creative Review; Creative Camera* and *Doubletake, 20:20* (The National Magazine for Photography and Media Education) provide up-to-date information on exhibitions and also suppliers.

Isherwood, S. and Stanley, N. (eds) (1994) *Creating Vision*, London: ACE. This ends with a useful selection of resources and contacts.

Creative Camera (March 1999) includes a Guide to UK Galleries. This identifies 154 photographic galleries throughout the UK. The majority are art galleries or museums which devote a significant proportion of their programme to photography, 40 are 'dedicated' photography spaces. My personal selection includes:

Dazed & Confused: 112–116, Old Street London EC1V 9BD Exhibits challenging contemporary photography.

Lighthouse Media Centre: Chubb Buildings, Fryer Street, Wolverhampton, www.lighthouse.co.uk. A dedicated space for lens-based and digital media.

The Pavilion: 2 Woodhouse Square, Leeds LS3 1DA. A photographic arts centre for women, www. pavilion.org.uk.

Viewpoint: Old Fire Station, The Crescent, Salford M5 4NZ (0161 737 1040). Exhibits and commissions a wide range of photo-based work.

The Photographers' Gallery: 5 & 8 Great Newport Street London, N1 [0171 831 1772] Exhibits a wide range of image based works including historical and contemporary.

National Museum of Photography, Film and Television: Pictureville, Bradford, BD1 1NQ, www.nmsi.ac.uk/nmpft (01274 727488). Exhibits both historical and contemporary works.

Victoria & Albert Museum: Holds the national collection of the art of photography in the nineteenth and twentieth century with 'Education Boxes' designed for use with visiting groups.

(Con)Fusion: the way forward?

> It is essential that we try and form a contemporary view of art and IT
> developments for ourselves, using our subject expertise and working towards
> an understanding of where art, craft, design and technology are taking each
> other.
>
> (Jones 1998: 5)

In 1998 NCET produced *Fusions: Art & IT in Practice*, a resource 'to support NC Art at KS3 and beyond'. It suggested that teachers needed to form a 'fusion' between traditional practice and new opportunities. However, when subject to closer scrutiny it can be seen to be just reproducing the dominant orthodoxy electronically. It failed to raise issues or refer to artists who are currently developing work which explores how ICT can be productive rather than just reproductive. Reference to the work of 'others' was restricted to the established canon with one notable exception: Barbara Kruger. As Meecham (1999) points out: 'documents like Fusion and the AVP Art Computer [1998], software CD-roms and videos – modernism's technologically validated publishing arm – still reproduce the old hierarchies but faster' (p. 81).

Fusion made no claims towards providing exemplary materials; its authors suggested that it provided a stimulus for discussion about how ICT can enhance the creative process. However, the document was supported by HMI, funded by DfEE and distributed free to all state schools; therefore it was perceived as the validated approach. Although OFSTED (2005) suggest that 'creative use of information and communication (ICT) is emerging' in secondary schools, you will find that a limited 'reproductive' approach is still dominant in many schools. If you accept this easy 'solution' you will have missed an opportunity to use new technologies to move beyond the existing school art orthodoxy:

> The fragmented, often contradictory, multidisciplinary and intercultural
> references to knowledge that students interact with through visual
> technology may have more to do with student understanding of the subject
> than does curriculum based on the structure of a discipline. Such
> postmodern visual experience should not be made to fit into modernist
> curriculum frameworks. Instead, the interpretative, didactic, even seductive
> power of the imagery should be given attention in school.
>
> (Freedman 1997: 7)

Freedman reminds you that computer 'images speak to us'. Unlike other resources which can be carefully selected to reinforce learning, the images accessed by pupils via the computer are not so easy for teachers to control. Pupil access to unsuitable materials can be denied (material deemed unsuitable because of its racist, pornographic or violent subject matter). There are other important issues in relation to new technologies which should not be ignored. It is your responsibility to ensure that you promote a 'screen culture' that refuses social isolation: recognise your responsibility of the teacher to devise projects which extend communication not isolation (see Unit

9.4). Also consider the gendered use of ICT (Spender 1995) research to date shows usage in the classroom is dominated by males.

Internet and WWW

There is an endless supply of digital art on the Net; from digitised images of cave paintings from Vallon-Pont-d'Arc in France to the latest techno art at the Pixel Pushers Gallery. The website http://www.world-arts-resources.com provides information on current exhibitions and access to high-quality images. Downloading images from the Net is time consuming and the quality is dependent on numerous variables from original site to the colour adjustment on your printer. Clearly it is a poor substitute for the real thing. Many artists are now producing work directly for the Web or adapting work to the new media. The Arts Ed Net Getty Foundation, http://www.getty.edu/education/, provides a 'gateway' to art-related sites on the Web, an extensive list of museums around the world. This should be visited and useful sites bookmarked for use by pupils. The interactivity of ICT makes possible new relationships between HE-Schools; Schools-schools; schools-galleries and museums. The National Grid for Learning closed on 13 April 2006 to be replaced by Becta (British Educational Communications and Technology Agency) which provides a 'content search service' for learning and teaching (http://contentsearCh. becta.org.uk/search/index.jsp?clear=y).

The value of the Internet and www as a resource for art and design education is indisputable. However, trawling through cyberspace for resources is time-consuming and distractions from the initial task almost inevitable. While it is important that you develop the skills necessary to surf the Net and utilise search engines to locate interesting and sometimes unexpected resources there needs to be a valid reason before you use valuable time during lessons teaching pupils how to do the same. All you need to do is to identify relevant sites and store them under bookmarks, thus providing pupils with ready access. Pupils should be encouraged to note down any relevant website addresses they come across at home or in ICT and other lessons and include these in an art directory (bookmarked) and store pertinent images in folders (virtual sketchbooks). It is important that you realise that some pupils have highly developed skills in ICT. They will have acquired these at home, in cybercafes and in other areas of the curriculum. Sefton-Green (1998) believes that young people who regularly 'surf the Net' at their own pace may well find the regimented structure of a teacher-led curriculum tedious (p. 12). Indeed pupils may know much more than you do about new technologies. Respect this expertise and recognise it as a valuable resource.

Resources

There is a never-ending supply of useful websites. My selection to date includes:

Ubuweb: www.ubu.com – an archive particularly of sound artists but also film and text, downloadable onto MP3 for free.

www.jodi.org: an artists' site showing how the Internet can be a creative medium in its own right. These artists have mischievously used the medium to produce disturbing encounters with technology which use and misuse programming coding and/or reference computer games.

ADAM: www.adam.ac.uk/ – Art, Design, Architecture & Media information gateway: a searchable catalogue of hundreds of Internet resources. ADAM gateway is due to be replaced by the forthcoming Arts and Creative Industries (ACI) Hub and until this is up and running only the ADAM gateway's search engine will be available.

ART on FILE: www.artonfile.com/ – Images of Art, Architecture & Design. ART on FILE's mission is to document important new developments in the built environment. Many projects combine art, architecture, urban planning and landscape design.

ARTEC: www.artec.org.uk/ – Digital arts and multimedia, 'laboratory for artists and creative professionals to explore this new territory'.

Computer ARTWORKS: www.artworks.co.uk – contains Latham's virtual sculpture.

ACTLAB: www.actlab.utexas.edu/ – at the boundaries where technology, art and culture collide. Worried about computers as the sewing machines of the twenty-first century, ACTLAB calls instead for 'room to stretch and do risky things'.

Dia center for the arts: www.diacenter.org/ – contains arts projects for the Web, long-term installations, information re art education in the USA.

Franklin Furnace Gallery: www.franklinfurnace.org – avant-garde performance and intervention, they are: 'committed to serving emerging artist and their ideas; and to assuming an aggressive pedagogical stance with regard to the value of avant-garde art to cultural life'.

Institute of Contemporary Art (ICA): www.ica.org.uk.

Lisson Gallery: www.lisson.co.uk.

LUX Centre: www.lux.org.uk – video and digital artwork by contemporary artists using electronic technology.

The Metropolitan Museum of Worldwide Arts Resource: www.wwar.com/.

MOMA: www.moma.org – Museum of Modern Art, New York.

TEST: www.test.org.uk – a digital research facility providing public access for anyone wanting to explore the creative potential of digital media.

Twenty-four Hour Museum: www.24hourmuseum.org.uk.

The Walker Art Centre: www.walkerart.org/.

White Cube: www.whitecube.com – a project room for contemporary art.

RBL: the way forward?

In the foreword to the Green Paper, *Teachers Meeting the Challenge of Change*, Blair (1998) suggests that teachers have suffered 'decades of drift'. He insists that the profession needs to 'engender a strong culture of professional development'. It calls for a contractual duty for all teachers to keep their skills up to date. Your CPD requirements as a newly qualified teacher will be identified in your Career Entry Development Profile (CEDP); a statutory 'Induction' programme enabling you to achieve these is now an entitlement for all those entering the teaching profession. The CEDP serves to remind you that initial teacher education is a launching pad, that teaching needs constant refuelling and refreshing. During your ITE course you need to identify and establish contacts with resources in the field of art and design that will enable you to identify your CPD.

Task 5.4.2 Identifying CPD resources

- Identify the art and design Inset/CPD provision available in your placement schools.
- Discuss with your mentors resources available locally and nationally; make a record of these.
- Contact AAH, ENGAGE and NSEAD (visit website) to find out what is currently being offered.
- Find out about proposed conferences and publications via the above or through your LEA or RAB.
- Get your name on as many mailing lists as possible.
- Identify provision for CPD accreditation, links with modular MA in Art & Design Education, Museums & Galleries, Critical Studies and Art History.

It is no simple task to develop the type of RBL that engages with the issues raised in this chapter. It demands that you appreciate difference, understand context, recognise agency (your own pupils and others), make critical comparative judgments on the basis of evidence and empathy, not received opinion. Giroux (1994) insists that teachers must be researchers and work with each other in producing curricula. Lippard (1995) identifies teachers as agents of exchange as well as change agents. She believes that educators have to shift the emphasis away from bringing 'great' art to people for their veneration, towards working with people to create meanings both in and through art:

> We are laying out the ingredients but still looking for the recipe. Once there are more cooks, everyone will use the ingredients differently . . . Critical consciousness is a process of recognising both limitations and possibilities. We need to collaborate with small and large, social, political, specialised groups of people already informed on and immersed in the issues . . . At the same time we have to collaborate with those whose backgrounds and maybe foregrounds are unfamiliar to us, rejecting insidious notions of 'diversity' that simply neutralise difference.
>
> (p. 114)

Resources

AAH is the professional association for art historians. Its annual conference engages with current debates in the discipline, and across disciplines. The schools sub-committee organises events for students and teachers (www.aah.org).

EnGAGE is a professional body which promotes greater understanding and enjoyment of the visual arts by engaging with the public, artists, galleries and educators (www.engage.org). Address: Suite AG, City Cloisters, 196 Old Street, London EC1V 9FR.

The *NSEAD* must be one of the most significant resources for art and design. Many of the articles cited in this book come from iJADE, the society's journal which provides an international forum for the dissemination of ideas, practical development and research findings. The important role that the NSEAD play in producing and selling publications for art and design education should not be overlooked. In addition it provides Inset and organises conferences. Detailed information about its activities can be found on its website (www.nsead.org).

Standards Site includes a discussion forum, allowing teachers to share with each other their experiences in implementing strategies for raising standards in school – including video clips. Not art and design specific, it nevertheless has the potential to promote curriculum development projects to a wide audience both in and beyond the subject field. It is important that as student teachers you recognise the importance of advocacy and ensure that visual and aesthetic literacy are promoted alongside numeracy and literacy. It includes best practice, schemes of work and lesson plans for teachers to access and customise, and encourages teachers to share teaching materials (www.standards.dfee.gov.uk).

6 Practice in PGCE Art and Design

Nicholas Addison and Lesley Burgess with Alison Hermon

INTRODUCTION

This chapter is divided into two sections: the first examining your Practical Teaching Experience (PTE), the second your Higher Education (HE) course studies. Some units evidently belong to both; indeed the two components are indivisibly related. It is only the opportunities and resources offered by these respective sites that determine differences: their common concern is pedagogy. The way the two components are sequenced varies from course to course, but the HE component is often front- and end-loaded to allow for a sustained period in school. The way these components have been divided in this chapter does not therefore run in strict chronological order but looks at different issues and practices.

OBJECTIVES

By the end of this chapter you should be able to understand:

- the trajectory of your PTE, its different components and cumulative effect;
- the ways in which course studies enable you to develop and transform subject knowledge into effective pedagogical practice with specific reference to SEN;
- the means by which the partnership model enables you to plan, develop, evaluate and record:

  1 your own studies and practice of art, craft and design;
  2 a broad theoretical and philosophical base on which to build educational practice;
  3 specific school experiences for developing the TDA Professional Standards.

UNIT 6.1 PRACTICAL TEACHING EXPERIENCE: A PARTNERSHIP MODEL

By the end of this unit you should be able to:

- understand how to use focused observations to gather knowledge of learning and teaching;
- understand the importance of planning and reflection for effective teaching;
- consider ways of developing competence in teaching in relation to the Standards.

The partnership between schools and HE in the education of teachers provides an ideal situation in which to relate theory to practice and vice versa. The interrelated elements of the partnership are made coherent and dynamic through a process of personal reflection in combination with theoretical study and collaborative action (Schon 1987; Prentice 2003). This provides a basis for the development of your teaching and enables you to contribute across a wide range of art educational practice. The partnership model is complex; responsibility lies with you, as a postgraduate student, to negotiate your position and role within the partnership.

6.1.1 Induction to practical teaching experience

Before taking responsibility for teaching you are required to undertake a number of activities and tasks in your placement schools directly related to subject application, including classroom observation and shadowing, collaborative teaching and lesson analysis. Documents (proforma and observation schedules) supporting these activities should be provided by your placement schools, but the following tasks indicate something of the range you should cover:

- collect information about the school's catchment area and pupils' social and cultural backgrounds;
- identify and discuss the department's educational philosophy and its relationship to the whole school ethos, the National Curriculum (NC) and external examination syllabuses;
- investigate the perceptions of the department by looking at OFSTED reports, league tables, the local press, etc.;
- collect information about pupils' prior knowledge including previous schemes of work (SoW);
- shadow pupils and teachers to observe different pedagogies in action;
- assist with teaching through team-teaching, advising individuals and working with small groups of pupils;
- finalise PTE timetable;
- plan and resource lessons;
- identify appropriate assessment procedures.

Task 6.1.1 Induction exercise

During the induction periods at your PTE schools you should find out and record the following information:

- the names of colleagues;
- lesson times;
- sises of classes/groups;
- the composition of groups; mixed ability, etc.;
- available materials and equipment (including Health & Safety requirements);
- available resources; visual material, videos, books, DVDs, etc.;
- the types of courses on offer in art and design;
- departmental and school policies; Equal Opportunities, discipline, differentiation, assessment, homework, etc.;
- examples of Programmes of Study (PoS), Schemes of Work (SoW), etc.;
- your teaching timetable;
- groups you are expected to observe/assist/teach;
- provision for, and use of, Information Communication Technology (ICT) in the school and your department;
- Special Educational Needs (SEN) provision within the school and your department.

6.1.2 PTE, school 1

Progressively during the PTE you are introduced to teaching through systematic observation, planning, practice and reflection. As you gain confidence you work with individuals and small groups and participate in team-teaching with experienced colleagues. Throughout your PTE you are supported by school tutors, your subject mentor and a teacher responsible for professional studies, all of whom monitor and review your progress in relation to the Professional Standards (TDA 2006a). Your HE tutor visits you and, with school colleagues, observes and assesses your teaching, evaluating your development and setting targets.

During your PTE you may be preoccupied with surviving the mechanics of teaching, however, it is essential that you are ready to apply theory to practice and use both to develop a personal educational philosophy. It is also important for you to establish yourself as a professional colleague across the partnership, someone who enters into and contributes to the life of the whole school. To assist you in planning and evaluating this experience you are required to keep a teaching file.

6.1.3 The PTE file: guidelines and criteria for assessment

The PTE file is your primary means of recording your developing practice and is a vital resource for your teaching. Your file should contain records of your schemes of work, lesson plans and evaluations. It is important that you adopt a systematic approach, filing each teaching group or SoW in a logical order. An A4 loose-leaf file with dividers is ideal for this purpose. From your file, your tutors and colleagues should be able to gain insights into the nature of your teaching and the quality of the learning experiences you provide.

SoW and lesson plans should contain four sections:

a) Intentions: aims and objectives

You are advised to produce a cover sheet clearly explaining the following:

- why the SoW has been chosen (aims);
- the learning objectives;
- how learning is to be assessed;
- how inclusion is to be addressed;
- how the SoW work relates to cross-curricular/interdisciplinary perspectives.

b) Preparation

Identify:

- materials and equipment: the things pupils work with; paper, printing rollers, inks, computers, etc.;
- resources: the things you use to support your teaching; DVDs, reproductions, objects, visits, texts, etc.;
- Health & Safety Regulations.

c) Organisation

- time: structure, sequence, pace;
- space: arrangement of learning environment and resources;
- people: differentiation, interactions and collaborations.

d) Reflections

Following each lesson you are required to record an evaluation of your:

- effectiveness;
- pupils' responses and the extent to which they have realised the learning objectives;
- implications for the following lesson and differentiating need.

Your PTE reflections and notes should clearly show the ways in which you have made connections between your HE workshop experiences and your teaching, and the requirements of the NC for Art and design, GCSE and post-16 courses. They should reflect your resourcefulness and include relevant visual material: photographs, illustrations, drawings, diagrams, samples, reproductions, catalogues, trigger sheets, etc.

Planning is the key to effective teaching. You are more likely to feel confident if you have:

- understood the educational aims and learning objectives of a SoW;
- researched and resourced all relevant aspects of subject knowledge and its application to classroom activities;
- related knowledge to the particular pupils you are to teach, including their prior knowledge and special educational needs;
- fully prepared yourself with questions, key words, visual aids and materials;
- considered the arrangement of the room and the different activities of particular lessons, their pace and the consequence of each on the following lesson;
- considered what methods of assessment to employ.

Once this is done, and only then, are you ready to face the question of how to 'manage' classes, how to interact with and motivate pupils. These are elements of teaching that you can only learn through experience in the classroom.

6.1.4 Guidelines for teaching

General conduct

1 Remember to learn the names of your mentors and department colleagues but also those of the Head teacher, SENCO and support staff, ICT co-ordinators, librarians, etc..

2 Be punctual. If you are unwell and unable to attend school, inform the secretary by telephone before morning school and leave a message for your HE tutor.

3 Remain in school all day (unless special prior arrangements have been made with your tutors).

4 Observe school regulations, e.g. about smoking, use of common rooms, dress code, health and safety, fire drill, etc.

Subject knowledge

5 Develop and promote an understanding of the intrinsic and extrinsic values of art and design in the school curriculum and the wider community (Unit 12.2), advocating the importance of visual and aesthetic literacy across the curriculum.

6 Support your lessons with systematic plans and carefully selected resources, ensuring coverage of the NC Art and design PoS and examination syllabuses. Always take into consideration the variety of needs of pupils, relating work to their age and conceptual understanding by using appropriate language and activities.

7 Think of yourself not only as a teacher concerned with practical activities but as someone who can promote understanding and appreciation of art and design through critical, contextual and historical studies. Stimulate intellectual curiosity and communicate your interest in art, craft and design as social practices.

Planning

8   Prepare thoroughly: prepare yourself to meet and to work with your pupils. Find out about their skills, attitudes and previous experience. Attend to practical details, e.g. health and safety, distribution of materials and equipment, storage of work, organisation of space and time.

9   Have constantly in your mind the quality of your pupils' experience of art, craft and design in your lessons. Identify clear learning objectives; develop SoW that challenge pupils and ensure high levels of pupil interest and motivation.

10   There are important decisions to make (some made by you, others by pupils) for example decisions about: timing/pacing, starting points, media and scale of work, the degree of teacher direction or pupil autonomy.

11   Recognise the need for continuity and progression. Find out about pupils' previous learning in art and design. Plan sequences rather than isolated lessons. Conceive every lesson as making another lesson possible.

12   It is important to anticipate what your pupils should be able to gain from your presence. Try to imagine what the teaching/learning event might be like before it happens. Record your anticipations and compare them with your post-lesson evaluations.

Teaching and class management

13   Reinforce the school/department's Code of Conduct. Recognise the importance of establishing and maintaining a good standard of discipline through well-focused teaching and positive reinforcement.

14   Learn the names of your pupils as efficiently as possible (a diagram of seating arrangements may be helpful). Keep a register.

15   Maintain classrooms, studios and workshops in a clean and safe condition to ensure they operate successfully. You are responsible for ensuring that pupils clean and clear all work surfaces, tools and equipment at the end of every lesson. If you are not conversant with a particular tool, material, or item of equipment, it is your responsibility to ask for assistance and try them out before introducing them to pupils. At no time should pupils use electrical tools or other potentially dangerous equipment without the support of a qualified teacher: remember you are not 'loco-parentis'.

16   Identify pupils who have SEN and/or English as an Additional Language (EAL) and ensure that they are given positive and targeted support. Remember also to identify pupils who are very able, they too need support through differentiated, challenging work and extension activities.

Monitoring assessment, recording, reporting and accountability

17   After each lesson you should reflect upon your teaching and the pupils' learning. To what extent have your aims and objectives been met? Become conscious of yourself as a cause in what happened; make a written record of your evaluation; assess and record each pupil's progress systematically providing targets for learning.

Subsequent PTE

During subsequent PTEs you can build on the experience and expertise acquired in your first school to develop high levels of teaching skills and classroom management. Placement in other schools also provides you with the opportunity to develop an understanding of how different contexts affect learning.

Regular feedback and evaluation of your progress, co-ordinated by your subject mentor, takes place in each PTE. This leads to a report assessing your teaching and identifying targets for further developing your practice in relation to the Standards. You should endeavour to achieve these targets while consolidating and developing your existing strengths. Experienced teachers continue to observe and support you in your teaching. In addition your HE tutor visits you to monitor your development. You are given notice if you are chosen as one of the representative sample of students to be observed by external examiners at the end of the PTE. A final report measures your achievements against the Professional Standards for the Award of Qualified Teacher Status.

6.1.5 Special educational needs (SEN)

Alison Hermon

> All children have the right to a good education and the opportunity to fulfil their potential. All teachers should expect to teach children with special educational needs (SEN) and all schools should play their part in educating children from the local community, whatever their background and ability'.
> (DfES 2004b)

By the end of this section you should be able to:

- understand what constitutes a special educational need (SEN);
- begin to identify specific arrangements and procedures for SEN in the art and design department.

Introduction: What is a special educational need?

One of the requirements for meeting the Standards for QTS is for you to address the wide range of educational needs of all the pupils you teach in the mainstream classroom in order to support and enable them to reach their potential. This can be a significant challenge even for an experienced teacher, as pupils' needs can be very diverse and complex, especially for those identified as having a special educational need (SEN).

The SEN Code of Practice (2001) gives the following definition:

> Children have special educational needs if they have a *learning difficulty* which calls for special educational provision to be made for them. They have a *learning difficulty* if they:

- have a significantly greater difficulty in learning than the majority of children of the same age; or
- have a disability which prevents or hinders them from making use of educational facilities of a kind generally provided for children of the same age in schools within the area of the LEA . . .

(p. 6)

The SEN Code (2001) recognises that children with SEN will have needs and requirements falling into at least one of the following four broad areas. These are:

- communication and interaction;
- cognition and learning;
- behaviour, emotional and social development;
- sensory and physical development.

(See The Four Areas of SEN table in Earle and Curry 2005: 14.)

Many children fall into more than one of these areas, as there is a wide spectrum of special educational needs that are frequently interrelated. There are specific needs that are directly associated with particular conditions. At the same time, there are groups of pupils with special needs who do not have SEN. For instance, the National Curriculum (1999) states that 'not all pupils with disabilities have a SEN. Many pupils with disabilities learn alongside their peers with little need for additional resources beyond the aids which they use as part of their daily life' (p. 29). Therefore, pupils may exhibit special needs if they come from a social group whose circumstances or background are different from most of the school population. However, these are distinct from the individual experience of learning difficulty that is the foundation of SEN. Needless to say, mounting evidence suggests that certain groups in society are overrepresented in particular forms of SEN provision (Frederickson and Cline 2002).

Current arrangements in mainstream schools for those with SEN

Arrangements for SEN pupils vary between schools and LEAs. However, your placement school will have a policy for SEN provision based on the code (2001), which identifies a number of fundamental principles:

- A child with SEN should have their needs met.
- The SEN of children will normally be met in mainstream schools or settings.
- The views of the child should be sought and taken into account.
- Parents have a vital role to play in supporting their child's education.
- Children with SEN should be offered full access to a broad, balanced and relevant education including . . . the National Curriculum.

You need to liaise with your subject mentor to ascertain how this is interpreted in subject-specific ways in the art and design department in relation to documentation and practice. Indeed you need to be fully aware of your responsibilities as well as learning about the most effective strategies for implementation. This includes familiarising yourself with the SEN register, which identifies specific pupils, their needs and provision. This provision is based on a 'graduated response' to needs identified on three levels or 'stages of action'. These are:

- '*School Action*' when the requirement for extra provision is minimal. This is initiated by the SENCO (the Special Educational Needs Co-ordinator) acting in response to concerns about pupil progress and performance. The action may involve the deployment of extra staff, resources or specialist equipment.
- '*School Action Plus*' when a pupil continues to experience difficulties under these terms, necessitating further support including external services.
- '*Statutory Assessment*' leading to a '*Statement of SEN*' if a number of intervention programmes have been implemented without success and a pupil's needs are identified as more severe.

Within the school, there will be other corresponding documentation noting necessary procedures and strategies to support individuals, such as the IEP (Individual Education Plan) identifying a small number of short-term manageable targets for action to help SEN pupils to progress. These include details of provision, teaching strategies, review dates and expected outcomes relating to targets drawn up with regard to specific areas for development, mainly focusing on generic and core subject needs. For this reason, do not expect support in art and design as this is usually in limited supply.

Through observation and interaction with the class you might find that some pupils who you think warrant extra attention have not been noted. This could be because learning difficulties have arisen through issues related to art and design activities which may not have been formally recognised or which pupils have learned to hide. Under these circumstances, you need to liaise with your subject mentor, class teacher and/or SENCO to know how best to proceed. In the last OFSTED report on SEN (2004), it was recognised that nationally one in five children have difficulty in learning requiring some form of extra help and one in 30 have a statement drawn up. It is possible that mainstream provision will be deemed unsuitable for a pupil with a statement. When this is the case, they are usually placed in some form of alternative provision such as a special school, pupil referral unit or a combination of these or other options, depending on the nature of the learning difficulty. However, the process of statementing and placing pupils is not a straightforward one with difficulties facing all parties involved. In 2002, the Audit Commission's report identified the following problems:

- too many children are waiting to have their needs met;
- parents lacking confidence in the system, leading to pressure for statements; and
- some children who could be taught in mainstream are being turned away.

One of the key difficulties is that there is 'an inbuilt conflict of interest in that it is the duty of the LEA to both assess the needs of the child and to arrange provision to meet those needs, all within a limited resource' (HC 2006: 32). However, in the last 20 years there has been a significant increase in the number of pupils with SEN attending mainstream schools influenced by radical policy changes initiated in the early 1980s. Therefore, you need to question how and why SEN was originally conceived, what are the key factors underpinning the legislation and their impact on current policy and practice.

Task 6.1.2 Special educational needs: policy into practice

Look up departmental policy with regard to SEN. What are the arrangements and how do they relate to the subject?

Through classroom observation begin to identify pupil differences, in terms of learning and behaviour. In so doing, check particular strengths and abilities that fall outside the criteria through which art and design is usually assessed, particularly in terms of pupils' social skills and pre-cognitive skills.

The impact of Warnock: evolving themes and issues within special education

By the end of this section you should be able to:

- understand the main principles underpinning the advice given by the Warnock report and how it has influenced SEN policy;
- have a greater awareness of evolving themes and issues within special education;
- recognise some of the overarching benefits and pitfalls of inclusion for pupils with SEN in mainstream schools.

In the UK, special educational needs (SEN) was introduced as a legally defined term by the Education Act (1981) following the advice of the Warnock report (1978). This recommended that the existing statutory categories of disability defined in medical terms should be abolished and replaced with a 'statement' of SEN, a detailed profile of pupils' needs following assessment. In contrast to the previous system, special educational needs were conceived as lying on a continuum with ordinary needs, with common goals achieved by individuals through differentiated routes depending on the level and severity of need. Some of the reasons for change were pragmatic, for instance the difficulty of children fitting neatly into the different categories of disability. Other reasons reflected a growing body of opinion upholding the view that disability was the direct result of discriminatory barriers created by society, the 'sociological response' rather than emerging from deficiencies in the individual, the 'medical or psychological model' (Clough and Corbett 2000). It was no longer considered appropriate to perpetuate a sharp distinction between the disabled and non-disabled. Furthermore, 'when separated from their peers or singled out for special treatment, children are less likely to reach their potential' (Richardson *et al.*

1995: 332). Indeed, a negative label associated with a disability or learning difficulty could be further compounded by the stigma of attending a special school. Therefore, addressing issues of social exclusion in terms of changing the location of pupils' education was seen by many to be a very positive step and during the 1980s and 1990s there was a considerable decline in the number of children attending special schools. In 1994 the Salamanca World Conference declared that 'Regular schools with this inclusive orientation are the most effective means of combating discriminatory attitudes, creating welcoming communities and achieving education for all . . .' (UNESCO).

It would appear that government policy for SEN is promoting this approach to inclusion. Indeed, revisions to the statutory framework for inclusion in the 2001 SEN and Disability Act strengthened the rights of SEN pupils to attend mainstream schools. In 2004 the SEN Strategy 'Removing Barriers to Achievement' guidance stated that 'the proportion of children educated in special schools should fall over time' (p. 37). Although good practice is evident, there are those who are critical, including Baroness Warnock who expressed a concern that some children are being forced into mainstream schools against their best interests (2005). A year earlier she had highlighted the fact that 'children with learning difficulties were taught mainly by classroom assistants or removed into units isolated from their contemporaries' (2004: 15).

From this perspective, the question needs to be raised whether this model of inclusion is indeed 'inclusive' in practice as well as appropriate for all. Corbett (1997) has suggested that inclusive rights activists only promote inclusive values and ignore individual difficulties with some pupils needing a separate setting in order to thrive and be fully supported.

A recent report from the House of Commons Select Committee on SEN (2006) has cited failings in the system as it stands today noting that '. . . there is considerable inequality of provision – both in terms of quality and access to a broad range' (p. 6). The report believes that the problems have been caused by government policy, which is inconsistent and 'confused'. For example, some of the institutions accepting SEN pupils are struggling to cope with rising numbers, as well as with insufficient resources, training and guidance. Equally, it is believed that 'the standards agenda remains the much greater priority of the government' and increasing competition between institutions is acting as a disincentive for the most successful mainstream schools to take their fair share of SEN pupils (p. 47).

However, rather than feeling dismayed over inefficiency and ill-judged expenditure, there are those who feel relieved and pleased that SEN is finally being included in debates within general educational developments (Corbett 2001). As Barton has recognised when developing more inclusive approaches to education '. . . existing school systems – in terms of physical factors, curriculum aspects, teaching expectations and styles, leadership roles – will have to change' (1998: 85).

Task 6.1.3 Inclusion

If indeed, 'the notion of inclusion . . . is about a philosophy of acceptance . . . about providing a framework within which all children can be valued equally' (Thomas *et al.* 1998: 15) you need to ask the following questions:

- Is inclusion for all pupils with SEN in mainstream art and design departments an unattainable ideal?
- To what extent are the needs of individuals or the whole class compromised by an inclusive approach? Therefore, are differentiation and inclusion a contradiction in terms?
- Will pupils with SEN feel a greater sense of belonging in a special school where everyone has a SEN? From this perspective, is this more of an inclusive setting than mainstream school?

Valuing differences within existing frameworks for learning in art and design for pupils with SEN

By the end of this section you should be able to:

- acknowledge statutory provision for pupils with learning difficulties in art and design;
- recognise possible benefits of art and design education for pupils with learning difficulties;
- understand some of the difficulties children with SEN experience in art and design and how through practice, these might be perpetuated or overcome.

The 'Index for inclusion' (Booth *et al.* 2002) promotes the view that 'inclusion starts from a recognition of the differences between students.' And '. . . the development of inclusive approaches to teaching and learning respect and build on such differences' (p. 7). You can begin to achieve this through differentiation within your planning and teaching in art and design, acknowledging routes to achieving common goals can be different for individuals and groups, including those with SEN. At the same time, it is widely accepted that effective teaching for pupils with SEN is often effective for all pupils, however, '. . . as schools become more inclusive, teachers need to be able to respond to a wider range of needs' (*ibid.*). For this reason, accommodating those with more extreme needs is given particular emphasis within statutory provision. As part of a common framework across subjects, the NC Inclusion Statement determines that learning needs are met by art and design teachers through:

- using visual and written materials in different formats as well as non-sighted methods of reading, such as Braille for the purpose of communication;
- ensuring access to sensory experiences through visual and tactile materials/resources in Art and design demonstrations and activities;

- investigating the environment, (including visits to art galleries) in order to make up for the lack of first-hand encounters;
- providing access to specialist aids, equipment, adapted and/or alternative tasks as well as extra support to promote participation in practical work and to aid response;
- identifying achievable short-term goals in activities, (a 'small steps' approach) noting the importance of positive reinforcement and the amounts of time involved, to foster a supportive learning environment and to build pupil self-esteem.

(QCA 2002)

Other recent materials, such as the QCA's *Planning, Teaching and Assessing the Curriculum for Pupils with Learning Difficulties: Art and design* (2001), the DfES's, *SEN: Training Materials for the Foundation Subjects* (2003), and Earle and Curry's *Meeting SEN in the Curriculum: Art* (2005) provide you with greater detail in terms of how these fairly broad statements translate into classroom practice and methods of assessment. As you read the guidance you will become aware of the emphasis placed on modification to the curriculum, such as adapting work from an earlier or later key stage to enable pupils to progress and demonstrate achievement. It needs to be noted that differentiation for high-achieving pupils requires a different type of support in terms of tasks in order to stretch their abilities. This group tend to excel in many subjects, as well as in art and design. (For further information see QCA *Gifted and Talented* guidance at www.qca.org.uk/8774.html). Simultaneously, there are exceptional cases where pupils with literacy and numeracy needs display highly developed mimetic skills or advanced motor skills with which they can transform materials. It is noted that one of the benefits of art and design for pupils with SEN is in the promotion and encouragement of alternative ways of working and to appreciate this as part of a personal response to experience in which individual difference is valued. 'For many people the artist is perceived as an outsider and different, challenging the concept of normality, the conventional and the acceptable' (Hermon and Prentice 2003: 270).

Despite this, the difference associated with SEN can have negative connotations arising from the label of a 'learning difficulty'. Booth *et al.* (2002: 4–5) suggest that this has occurred because of a lack of emphasis on external factors 'creating barriers to learning and participation'. In contrast, Warnock (2004) claims that SEN has been misrepresented as a single category rather than constituting a broad continuum of needs. From either perspective, the label of 'difference' in terms of SEN can give rise to feelings of failure, low esteem and limited expectations for those involved. Consequently, many pupils with SEN want to be seen as the same as their peers rather than different from them in order to feel accepted and valued. It is imperative, therefore, that art and design departments acknowledge issues of conformity and difference to challenge negative identities for those with SEN rather than perpetuate them. For instance, it has been suggested that prevailing norms of 'school art' place undue emphasis on specific ways of working, restricting recognition of other approaches which are likely to be adopted by pupils with learning difficulties (see Chapter 3).

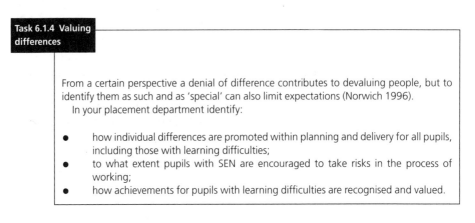

Empowering learning through challenging perspectives of difference: integrating contemporary art practice in art and design education for pupils with SEN

By the end of this section you should be able to:

- recognise the importance of personal identity and contemporary art practice within art education and how they can be used to address, challenge and overturn negative views of difference for pupils with SEN;
- understand how selected ideas, materials and processes can promote innovation, risk-taking and autonomous learning for pupils with learning difficulties;
- acknowledge teaching and learning strategies which are enabling and empowering for pupils with SEN.

Personal identity for pupils with SEN

A focus on 'personalised learning' and the individual needs of children and young people, especially the most vulnerable, has been central to recent government reform within education as well as external services. These concerns are reflected in the *Every Child Matters (ECM)* (2003) and *Youth Matters* (2005) policy papers. Although educators have acknowledged some of the conformity issues facing those with learning difficulties, the fostering of creative growth through focusing on personal qualities is seen to be an essential part of developing a positive sense of identity for all pupils, including those with learning difficulties (Corbett 1999). Therefore, it is important for you to address these areas despite the potential risks and drawbacks involved.

Contemporary art practice for pupils with SEN

It has been claimed that one of the most compelling reasons for integrating contemporary art practice in the secondary art and design curriculum is that it engages with current cultures and times and can be seen by pupils as familiar territory and relevant to their experience (Dawe Lane 1996; Burgess and Addison 2004). This applies to all pupils including those with learning difficulties. Likewise, critics of current arrangements emphasise the importance of '. . . a non acceptance of givens which are outside the influence of the learner, the society in which they live' in proposals to reconceptualise the curriculum within the framework of a postmodern agenda. From this perspective cultural difference is valued and traditional distinctions and boundaries of practice challenged (Swift and Steers 1999: 3).

Unfortunately, for those with learning difficulties, it is possible that moving outside known frameworks will foster feelings of insecurity. Also, recognising personal differences can risk foregrounding negative aspects of self and achievement. Nevertheless, these were the key areas addressed in a boundary-breaking project in a small special school in Brighton in which pupils with SEN, physical and neurological disabilities were inspired to create a 'different' identity through focusing on contemporary art and fashion (Hermon and Prentice 2003).

'Re-fash trash: fashion with a difference': an art and design project for pupils with SEN in a special school

The purpose of the project, to create a different or alternative identity, provided a group of secondary pupils with visible and hidden disabilities to address positive and negative views of difference. The reason for implementation was to promote a more positive view of themselves through working in the context of art and design, allowing them to challenge convention and celebrate non-conformity. From this perspective, a different appearance could be 'worn' in order to increase social status by defying convention. The project involved school-based work, a gallery visit and final performance. By working at the boundaries of contemporary art and fashion the concept of difference was promoted. Pupils were challenged to create a positive but non-conforming body image, reconceptualising art forms and making imaginative connections between subject matter, materials and process, using resources within and outside tradition.

Planning a different identity

A postmodern stance acknowledges that different points of view are recognised as well as experimentation with media outside conventional distinctions. Therefore, although it was important that pupils were involved in expressing their own preferences for clothing, their ideas were extended through visiting an exhibition at the Hayward Gallery showing 'deconstruction fashion' garments. In terms of social theory and the visual arts it is believed that '. . . deconstruction involves the unpacking of social constructs that have become so embedded in society as to appear natural' (Efland *et al.* 1996: 28). Within this context, fashion is recognised as the embodiment of non-functionality, something akin to an anti-fashion statement or as 'a late twentieth century emblem for change and risky transformation' (Gill 1998: 26–28). Reflecting these ideas, pupils saw that makers had used media from different contexts outside conventional approaches to dressmaking, including recycled materials in order to promote difference in terms of identity, making, material and idea. Each deconstructed garment exposed how it had been transformed and how its status had been increased. In this way postmodernism blurs the boundaries between traditional distinctions, fact and fiction, fantasy and reality.

Making a different identity

Using these as models for their own ideas, pupils were liberated from having to conform to make garments in a conventional way. Instead pupils developed their own ways of working using readily accessible materials. Therefore, the making promoted innovation as well as learner autonomy, with the teacher in an enabling role that nurtured rather than controlled to support individual responses and challenge notions of assumed dependency. At the same time, recycling media and ideas empowered students to transform thinking and materials through changing contexts, realising the importance of their own solutions as well as alternatives presented by others. Therefore, it was important to realise that one individual could present many, even conflicting identities in one or more garments. Also, these could be exchanged and related to each other in individual and collective ways with peers.

Presenting a different identity

In terms of presentation, pupils decided to model their different identities on a 'cat-walk', organised in the school hall to an audience of parents and friends. This was a risky decision as a public display could still perpetuate rather than challenge a negative identity. Foucault (1988) has described accounts which attempt to challenge the kind of identities that are constructed for special needs pupils as 'technologies of self'. These are 'transgressive', challenging the limits imposed by others. In this respect, individuals are allowed 'to shape their own identities by subverting the norms that compel them to repeatedly perform as disabled subjects' (Allan 1999: 69). The focus on difference was therefore essential as pupils were seen and valued *because of* their differences, not despite them. As one pupil remarked 'I felt very glamorous, I wouldn't have worn them normally. It gave me a bit of a buzz that I could go on the catwalk . . .' and another 'it's about what you can do and not what you can't. We all have different things we can make and do . . .' (Hermon and Prentice 2003: 227).

Through the medium of the performance they were able to give voice to otherwise marginalised views. Furthermore, it was recognised that by challenging the hierarchy of art in the artworld, instruction could change social relationships and transform attitudes so that pupils were able to see themselves in a more positive way. By focusing on contexts which challenge conformity and promote identity, issues can be addressed which encourage more positive interaction between art and society for social reconstruction, addressing the inclusion of teachers, students, staff and the wider community.

Conclusion

It is clear that the crossing of boundaries in art and design encourages pupils to go beyond the limits of orthodoxies and expectations. Also, it is important that in the process, they engage self-critically with personal values, self-identity and issues of empowerment. Therefore, to include any young person you have to be concerned with the whole person. This can be neglected if you focus on only one aspect of a student such as SEN. Also, in a mainstream classroom, it can have a negative impact in diverting attention away from the difficulties experienced by other pupils without the label (Booth *et al.* 2002). According to Corbett, from this perspective 'it is important not to see inclusion as distinct from education generally. The valuing of differences is a facet of certain educational values, not just of an inclusive ideology' (2001: 30).

Further reading

Corbett, J. (2001) *Supporting Inclusive Education: A Connective Pedagogy*, Routledge/Falmer: London.

Hermon, A. and Prentice, R. (2003) 'Positively different: art and design in Special Education', *International Journal of Art and Design Education*, 22, 3.

HMSO (2001) *The Special Needs and Disability Act (SENDA)*. London: HMSO.

OFSTED (2004a) *Special Educational Needs and Disability: Towards Inclusive Schools*. London: OFSTED.

QCA (2001) *Planning, Teaching and Assessing the Curriculum for Pupils with Learning Difficulties: Art and design*.

Warnock, M. (2005) 'SEN: A New Look', *Philosophy of Education Society of Great Britain*, June issue.

6.1.6 Health and safety

It is particularly important that you are aware of your responsibilities regarding health and safety, and ensure that pupils develop safe working practices. The DfEE (1995) *A Guide to Safe Practice in Art and Design* has been updated by the NSEAD in 2004 (see http://www.nsead.org/hsg/index.aspx). It is essential that you are familiar with the content of this publication. It offers advice on all aspects of safe practice in art and design, including the legal framework regarding health and safety responsibilities and good practice guidance.

This publication identifies that safe working practices are dependent upon:

- commitment and a sound health and safety policy;
- common sense, good management and organisation;
- general awareness of requirements and shared responsibility;
- properly planned and maintained accommodation;
- appropriate techniques, use of tools and materials;
- use of adequate safety devices;
- a knowledge and awareness of potential hazards.

Usual classroom practice in art and design need not be unduly restricted because of fears for health and safety. Most practices are safe within the accepted bounds of risk-taking. Pupils should be trained to work sensibly and safely, and to acquire positive attitudes towards safe practice. Teachers (including student teachers) give a clear lead by their own planning and personal example.

As a student teacher in your PTE schools, you should have knowledge of the Health and Safety legislation as it relates to schools. You need to ensure that you are reliably safe in the practical activities that you do yourself and in those that you ask pupils to follow. See a copy of the Safety Handbook in your school and be sure to be aware of and understand its contents. Your schools have a Health and Safety Representative who you can consult.

Health and safety and special educational needs

Account should be taken of:

- the pupils' ability to understand instructions, follow them and understand any dangers involved;
- the pupils' ability to communicate any difficulty or discomfort;
- the physical ability which might affect the pupils' ability to perform a task safely;
- any medical conditions which may be adversely affected by exposure to equipment or material.

It is important for teachers to know about the medical conditions of any pupils which may give rise to risk in certain activities, e.g. respiratory problems like asthma: clay dust, aerosol sprays, screen wash; skin allergies such as dermatitis: dust, glues and pastes; epilepsy: rapid optical sensations in screen-based technologies.

Ask for this type of information from each class teacher and seek further information from the SENCO.

Remember to count out and in, equipment which has the potential to be abused, e.g. knives and scissors, adhesives and solvents.

6.1.7 Ways in which the school mentor can support you

It is the subject mentor's responsibility to organise, supervise and assess your PTE in conjunction with your HE tutor. The following list indicates the ways in which these responsibilities manifest themselves throughout the year and provides issues to discuss in your debriefing meetings.

During Induction your subject mentor:

1   introduces you to the other members of the department and support staff;
2   arranges working and storage areas, pigeon-holes, etc. to enable you to plan, evaluate and record your teaching effectively;
3   provides you with copies of the department's policies, systems and handbooks;
4   prepares a timetable of classes for your PTE covering a range of age, ability, motivation, etc.;
5   provides class lists for the groups you teach and liaises with form tutors, year heads, etc.;
6   arranges a regular time for a weekly debriefing tutorial;
7   supports your planning by providing suggestions about SoW and their relationship to PoS, individual lessons and their sequence within a SoW;

advice on resources, approaches, methods and materials, addressing the Attainment Target levels within the NC Art and design, providing occasions for joint marking, assessing and responding to pupils' work: this should take place between you and the relevant class teacher.

During Practical Teaching Experience your subject mentor:

1   ensures you receive your entitlement as indicated in the course handbook;
2   discusses your PTE file: preparation of SoW, record of lessons observed, reflections on and evaluation of lessons, and monitoring of assessment of pupils' progress;
3   observes your teaching and afterwards discusses the lesson in detail and on a number of occasions provides written comments on the lessons observed, including setting specific targets;
4   gives you increasing responsibility for devising your own SoW (in line with the department's PoS) and adding to, plus developing, teaching resources;
5   takes an active interest in reading and commenting on your HE tasks, enquiries and assignments;
6   takes an active interest in the ongoing development of your Career Entry Development Profile (CEDP) and assessment record file;
7   helps you to develop your teaching Standards (TDA 2006a);
8   completes reports in time for you to read and respond to them before the final date for submission.

Feedback to you could take the following forms:

●   a weekly meeting in a suitable place for discussion between you, your mentor and/or other teachers involved;
●   a discussion between you, your school and HE tutors about your overall progress to set targets for further development with reference to your PTE file and notes about the lessons observed.

Good practice occurs when:

●   you take an active role in the feedback sessions;
●   you are able to evaluate your own teaching;
●   your lessons' aims or means of assessing pupil progress and achievement are used to structure feedback;
●   feedback is thorough, comprehensive and, where appropriate, diagnostic;
●   there is a balance of praise, criticism and suggestions for alternative strategies;
●   points made in feedback are given an order of priority, targeting three for immediate action;
●   you receive written comments on the lessons observed.

Further reading

Capel, S. *et al.* (2005) *Learning to Teach in the Secondary School*, London: Routledge.

Hermon, A. and Prentice, R. (2003) 'Positively different: art and design in Special Education', *International Journal of Art and Design Education*, 22, 3.

Robinson, K. (1982) *The Arts in Schools: Principles, Practices and Provisions*, London: Calouste Gulbenkian.

UNIT 6.2 DEVELOPING PEDAGOGY IN PARTNERSHIPS

By the end of this unit you should be able to:

- identify strategies to develop your subject knowledge and application;
- develop ways to record your Practical Teaching Experience and curriculum studies.

6.2.1 Workshops in art education (also see Chapter 2)

a) Developing subject knowledge and application through workshops

Teachers of art and design are expected to have a wide range of conceptual, critical and technical skills in order to address the broad-based curriculum required by the NC Order for Art and design and the examination boards. During the PGCE year you should participate in as many workshops as possible, developing existing skills and learning new ones. Your course may not differentiate between technical and concept-based workshops. However, occasionally it is instructive to consider them discretely so that particular needs can be addressed. All workshops should have a critical dimension. These workshops can be sited at a variety of venues, including: colleges, schools, studios, galleries, museums, the natural and built environment.

It is an effective use of human resources for you to organise workshops with members of your tutor group. The content of these is negotiable, building on the experience, knowledge and expertise of the group members. You can, in teams of three to four, plan, organise and present workshops of this kind. These teams can usefully produce a supporting teaching/resource pack including some or all of the following:

- aims and objectives of workshop;
- a visualisation of processes, diagrams, photographs, etc.;
- critical, historical and contextual support material;
- relationship to the NC for Art and design and public examination syllabuses;
- implications for classroom management and organisation;
- names, addresses and contact numbers for resources including suppliers and galleries and museums;
- health and safety checklists, etc.;

- useful books, websites, DVDs, etc.;
- evaluation.

The resulting packs can be shared with your tutor group or collated into a central resource.

Task 6.2.1 Audit of technical skills and needs

Identify the technical skills and processes with which you:

- are familiar;
- feel confident to teach;
- are required to teach;

and those for which you:

- require basic training;
- require a revision workshop.

Prioritise your needs from this audit and devise a programme to meet your requirements. Remember that in addition to timetabled workshops you should find ways to learn from the expertise of fellow students and colleagues and utilise the facilities and INSET at your PTE schools.

Concept-based workshops

Through concept-based workshops you are encouraged to investigate issues and ideas using a full range of skills, critical, conceptual and technical, shifting the emphasis away from a limited technical and formalist training towards a more reflexive education. You engage with concepts which help you to define what art is, how and why it is a significant area of social production and how you can contribute to debates on pedagogy by reflecting on your observations and practice in schools. They provide a structure within which you can engage in practical activities to draw upon and extend your experience and expertise. You are asked to identify and investigate different ways of recording experience using the processes of art, craft and design, encouraging you to respond by moving from familiar towards less familiar ways of working. You can examine and exploit materials and processes in new combinations as you question traditional approaches. You are invited to locate practice in its historical, social and cultural contexts. Central to this activity is your awareness of the function and potency of visual metaphor and analogy, the equivalents by which you are able to embody and communicate your experiences in visual and other forms. Subsequently, you can develop strategies to make connections between concepts, skills, processes, products, visual and verbal modes of communication, the NC and external examinations. In addition you are encouraged to consider how to overcome constraints such as limited time, space and materials so that you can plan effectively for learning in the classroom. Fundamental to the concept-based workshop is the way it promotes praxis, the linking of theory and practice.

Criteria for evaluating workshops

Engagement in concept-based workshops enables you to:

- acquire, develop, refine and apply a wide range of skills, knowledge and understanding;
- demonstrate a broad base of practical skills across art, craft and design (including ICT);
- acquire and use skills of reflection and critical analysis;
- develop a personal response;
- locate ideas in wider contexts: cultural, historical, political, social;
- show evidence of reading to inform theoretical understanding;
- make connections between professional practice and ways of teaching and learning in art, craft and design;
- recognise the implication for lesson planning – content, methods, resources;
- make explicit links with the National Curriculum and/or examination syllabuses.

Task 6.2.2 Concept-based workshop self-assessment

Informed by these criteria, devise a self-assessment proforma to help you identify your strengths and targets for further development.

6.2.2 Approaches to ICT capability in art and design

The NC for Art and design states: 'Pupils should be given opportunities to apply and develop their ICT capability through the use of ICT tools to support their learning . . . (DfEE 1999: 34)

The new digital technologies are determining the development of ICT, whether lens based, audio or multimedia. It is therefore essential that you develop your computer literacy in line with required standards detailed in the Professional Standards for the Award of Qualified Teacher Status (TDA 2006a).

ICT forms a fundamental part of professional practice: it is central to design and increasingly an area for exploration in contemporary art. Many of you already possess highly specialised skills in this area; transferring them to the classroom is therefore your primary task (see Ash 2004).

Task 6.2.3 Developing practice through ICT

Consider ways in which digital technologies can create new forms and modes of investigation (see existing advice at www.becta.org.uk and http://www.teachernet.gov.uk/wholeschool/sen/asds/asdgoodpractice/ict/.

Develop a SoW for a KS3 class that incorporates these forms/methods.

Throughout the course, both at your HE institution and in the partnership schools, you have the opportunity to develop your ICT skills, some of which may have to take place in twilight sessions, depending upon your level of capability. By the end of the course you have to provide evidence of your ability to use ICT to support subject pedagogy.

Task 6.2.4 ICT: beginner, user or expert?

Define your present ability in relation to the following categories:

- *'Beginners'* – little or no ICT experience at all;
- *'Users'* – a fair ICT experience, but no insight into pedagogical uses;
- *'Expert users'* – much ICT experience, but no insight into pedagogical uses;
- *'Leaders'* – much ICT experience and significant insight into pedagogical uses.

(*'Leaders'* are invited to contribute to the training of *'Beginners'* and *'Users'*.)
Monitor your development in relation to these categories as the year progresses.
Devise a training programme to develop your skills and identify opportunities on your timetable to use ICT with pupils.

How ICT capability can be developed in art and design

Communicating and handling information:

- by using painting and drawing software, enabling pupils to electronically acquire their own and others' images by scanning, digitising, exploring, investigating and developing them through desktop publishing and multimedia;
- by providing access to a wide variety of visual resources through Internet and CD/DVDs, including multimedia art galleries, multimedia presentations, search engines and visual databases; helping pupils develop their knowledge and understanding of the methods and techniques used to produce artefacts, their function and their historical and cultural contexts.

Modelling:

- by using painting and drawing software and 3-D modelling packages, enabling pupils to explore situations and visualise the outcomes of different approaches, methods and techniques.

It is a requirement of the Standards (TDA 2006a) that all students provide evidence that they can use ICT effectively in their subject teaching.

Table 6.2.1 Some routes for developing your ICT capability

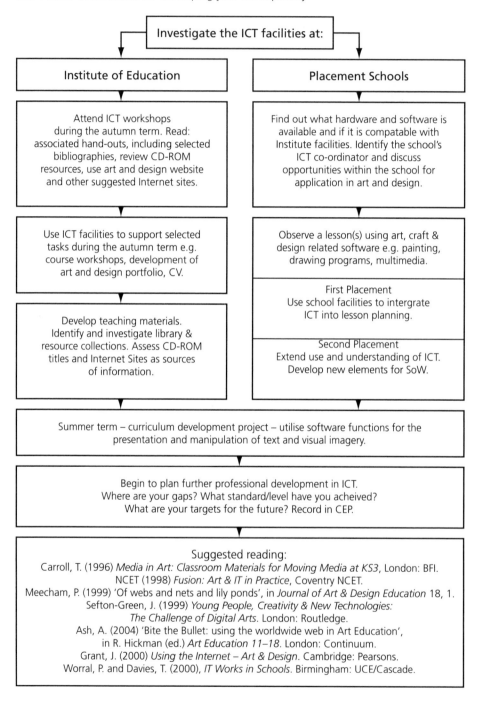

Investigate the ICT facilities at:

Institute of Education

Attend ICT workshops
during the autumn term. Read:
associated hand-outs, including selected
bibliographies, review CD-ROM
resources, use art and design website
and other suggested Internet sites.

Use ICT facilities to support selected
tasks during the autumn term e.g.
course workshops, development of
art and design portfolio, CV.

Develop teaching materials.
Identify and investigate library &
resource collections. Assess CD-ROM
titles and Internet Sites as sources
of information.

Placement Schools

Find out what hardware and software is
available and if it is compatable with
Institute facilities. Identify the school's
ICT co-ordinator and discuss
opportunities within the school for
application in art and design.

Observe a lesson(s) using art, craft &
design related software e.g. painting,
drawing programs, multimedia.

First Placement
Use school facilities to intergrate
ICT into lesson planning.

Second Placement
Extend use and understanding of ICT.
Develop new elements for SoW.

Summer term – curriculum development project – utilise software functions for the
presentation and manipulation of text and visual imagery.

Begin to plan further professional development in ICT.
Where are your gaps? What standard/level have you acheived?
What are your targets for the future? Record in CEP.

Suggested reading:
Carroll, T. (1996) *Media in Art: Classroom Materials for Moving Media at KS3*, London: BFI.
NCET (1998) *Fusion: Art & IT in Practice*, Coventry NCET.
Meecham, P. (1999) 'Of webs and nets and lily ponds', in *Journal of Art & Design Education* 18, 1.
Sefton-Green, J. (1999) *Young People, Creativity & New Technologies:
The Challenge of Digital Arts*. London: Routledge.
Ash, A. (2004) 'Bite the Bullet: using the worldwide web in Art Education',
in R. Hickman (ed.) *Art Education 11–18*. London: Continuum.
Grant, J. (2000) *Using the Internet – Art & Design*. Cambridge: Pearsons.
Worral, P. and Davies, T. (2000), *IT Works in Schools*. Birmingham: UCE/Cascade.

6.2.3 The art and design course portfolio, recording your PTE

The development of an art and design course *portfolio* should be seen as a continuous activity which provides a valuable record of the PGCE course and resources for your teaching. The contents should clearly communicate your understanding of art education indicating both the breadth of your experience and your personal philosophy. It should also reflect the highest professional standard of presentation: great importance is attached to it by interviewing panels for teaching posts (see Plates 5, 6, 7). It is essential that during your teaching experience you collect records and examples of work using ICT, photocopying, photography or by selecting actual samples of pupils' work. These form the core component of your course portfolio. Remember always to obtain permission when borrowing pupils' work and ensure that you respect arrangements to return it by an agreed date.

The following list gives some indication of the content of a portfolio:

- links with primary schools;
- working in partnership with outside agencies, e.g. galleries/museums;
- HE course workshops;
- examples of inclusive practices re: cultural diversity, disabilities, gender, race, sexualities, social class;
- resources: text-based, visual, multimodal;
- SoW undertaken on PTE including samples of lesson plans;
- pupils' work in progress;
- a range of pupils' responses and outcomes;
- evidence that you have assessed and recorded pupils' progress effectively;
- evaluations of your SoW;
- involvement in ICT;
- areas of interest or concerns that extend across SoW, e.g. sketchbooks, visual literacy, differentiation, SEN, equal opportunities, language;
- curriculum development projects;
- classroom organisation and display;
- assessment proforma;
- resource sheets: trigger, reference, technical;
- school-based research and aspects of professional studies;
- examples of your own work.

Some sheets will cover a single issue in depth; others will include related themes and issues. When designing portfolio sheets you should:

- clearly present ideas using text and image;
- be concise and selective;
- use terms consistently;
- ensure that texts and images complement rather than duplicate one another;
- show a balance between information, descriptive records, analyses and critical evaluation;

- consider different audiences, e.g. interviewing panels, non-specialists, pupils.

The portfolio should provide evidence that you have successfully covered the Professional Standards (TDA 2006a).

7 Assessment and Examinations in Art and Design

Andy Ash, Kate Schofield and John Steers

Unit 7.1 provides a general overview looking at terminology and methods.

Unit 7.2 focuses more specifically on classroom-based practice and the requirements of the National Curriculum (NC). This should help you to make judgements in relation to the attainment target levels.

Unit 7.3 discusses external assessment and examinations in art and design, with reference to GCSE, A/AS Level and GNVQ.

Unit 7.4 reconsiders the purposes of assessment for learning in art and design

UNIT 7.1 OVERVIEW OF ASSESSMENT: PRINCIPLES AND PRACTICE

Andy Ash and Kate Schofield

Introduction

Assessment and reporting is an area of contention in art and design. Finding a common policy is a troublesome issue for teachers, with clashes between philosophical positions and pragmatic practices fuelling the debate. The intervention of the 1988 Education Reform Act (ERA) in England and Wales provided statutory assessment procedures for the first time. This chapter discusses these developments and considers ways in which teachers can understand and make use of assessment procedures to develop ways of recording and reporting.

Since the introduction of the GCSE in 1986, consolidated by the 1988 ERA, fundamental changes have occurred in the secondary art and design curriculum. The introduction of the NC Art and Design Order (DES 1992) encouraged teachers to make explicit decisions about the assessment of pupils and to formulate comprehensive systems. The ERA established guidelines which still influence the way pupils are monitored, assessed and examined.

The purposes of assessment are:

- for the differentiation of individual pupils' learning needs;
- to provide information about pupils' progress;
- to enable pupils to understand their own learning and progression;
- to enable pupils to appreciate and relate their work to the work of others (artists, designers and craftspeople);
- for teachers to evaluate and thus develop their schemes of work (SoW);
- for teachers to make judgements about pupils' achievement in relation to attainment target levels;
- to provide information for parents/guardians and pupils;
- to enable comparisons to be made about all pupils and schools across the country.

Objectives

By the end of this unit you should be able to:

- understand the different terms used in art and design assessment;
- consider the various methods which can be applied within the art and design curriculum;
- look at a range of systems and procedures for assessment at KS3.

Some definitions

Assessment is a general term embracing all methods customarily used to appraise the performance of an individual or a group. It may refer to a broad appraisal including many sources of evidence and aspects of pupils' knowledge, understanding, skills and attitudes, or a particular occasion or instrument. An assessment instrument may be any method or procedure, formal or informal, for providing information about pupils' learning: individual discussion, homework, project outcomes, mock examination work, group critiques. You should consider existing assessment policies operating in schools alongside your own developing understanding of its purposes and value.

> The notion that one programme of assessment could fulfil four functions (formative, diagnostic, summative and evaluative) has been shown to be false: different purposes require different models of assessment, and different relationships between pupils and teacher. It may be possible to design one assessment system which measures performance for accountability and selection purposes, whilst at the same time supporting the teaching and learning process but no one has yet done so.
>
> (Gipps and Stobart 1993: 3)

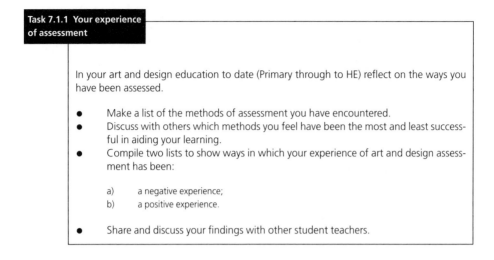

Task 7.1.1 Your experience of assessment

In your art and design education to date (Primary through to HE) reflect on the ways you have been assessed.

- Make a list of the methods of assessment you have encountered.
- Discuss with others which methods you feel have been the most and least successful in aiding your learning.
- Compile two lists to show ways in which your experience of art and design assessment has been:

  a)   a negative experience;
  b)   a positive experience.

- Share and discuss your findings with other student teachers.

Types of assessment

Formative assessment is usually continuous throughout the process of a particular learning activity. It has been recognised by most art and design educators that formative assessment is particularly constructive. It is concerned with providing on-going advice and feedback during the process of making rather than assessing and grading an isolated finished product. Formative assessment is an integral part in the development of schemes of work (SoW) which support learning; it encourages and guides the pupils' work forward. If pupils do not receive regular feedback on their work they quickly lose motivation and become unsure of their own success or failure.

Summative assessment occurs at the end of a SoW or Programme of Study (PoS) and therefore focuses on examination outcomes or more holistically on a portfolio of coursework. Summative assessment can be a culmination of assessments made over a period of time, for example GCSE coursework.

Ipsative assessment gauges the development of an individual pupil from one moment in time to another (usually the present). It is concerned with the evaluation of personal achievement rather than an individual's relationship to national or local norms. It often takes the form of pupil self-assessment, providing an opportunity for pupils to appraise themselves in a non-competitive climate. Additionally pupils' self-assessment provides teachers with insights into pupils' understanding of their progress (see Figure 7.1.1). This can then be used for diagnostic assessment.

Diagnostic assessment is used to provide evidence of pupils' ability in a given subject (through it, learning needs can be identified). It should be used constructively as a vehicle for discussion between teacher and pupil where both parties consider progress and set targets for future development. This type of negotiated assessment, in the form of a dialogue, promotes learning and a degree of pupil ownership in the assessment process. When diagnostic assessment is negotiated it becomes a powerful motivating factor.

At the end of every project you need to think about what you have learnt and achieved, and write an evaluation. This is a very important part of your project so you need to take time to do it thoroughly. Here are some questions which might help when writing your evaluation.

What have you learnt during this project and does this correspond with the learning objectives?
What have you produced?
How did you produce this work?
What materials did you use?
Which artists/craftspeople/designers have you referred to?
In what way did they inform your work?
Do you think your final work communicates your ideas effectively?
How could you develop it?

Primary colours, secondary colours, complementary colours, tessellation, star, hexagon, arabesque, curved line, leaves, flowers, calligraphy, repeat pattern, symmetry, compass, ruler, tissue paper, Islamic art, paint, felt tips.

(PGCE student 1998)

Figure 7.1.1 Self-assessment for Year 8 pupils
(from a PGCE student's PTE file)

Task 7.1.2 Evaluating forms of assessment

Collect examples of formative, summative and ipsative assessments. Evaluate their strengths and weaknesses.
To what extent do you think the self-assessment form (Figure 7.1.1) is ipsative?
What subtractions and additions are required to make it so?

Elements of assessment

Monitoring and tracking

It is important that as a teacher you monitor pupils' progress on a regular basis. In order to facilitate effective planning it is imperative that pupils are constantly assessed to establish achievement and learning: this provides evidence to evaluate the success of your teaching (gradually a system of monitoring teachers' performance is being introduced and the school records will inform this process). A proven method of monitoring pupils' progress is to use a tracking sheet. The following example shows how, lesson by lesson, you can monitor each pupil by making a quick assessment during or at the end of a session. At stages during a SoW this cumulative assessment provides patterns of learning and behaviour which indicate pupils' responses and can help you to differentiate needs in relation to different types of activity. For example,

DESIGN AND MAKE – frame project – using medieval patterns and designs

	Name Orange/Purple	1	H/W	2	H/W	3	H/W	4	H/W	5	H/W	Comments
1	Pupil A	⊠	O	□	A	□⊘	A	□	O	□	O	Producing a successful piece of work.
2	Pupil B	O	O	□	B	⊘	B	⊘	B	⊘	A	Report: last 2 weeks has made a good start.
3	Pupil C	□	A	□	A	□	B	□⊘	A	O	A	Making good progress.
4	Pupil D	□⊘	A	□⊘	A+	~	A	~	A	~	A	Works well/good idea development.
5	Pupil E	O	O	~	O	O	O	~	O	O	O	Missed a lot of lessons sp/studies.
6	Pupil F	⊠	C	⌐	C	⊠	O	⊠	C+	⊠	C+	Confidence but work is not improving.
7	Pupil G	⌐	C	⌐	B+	⌐	B	□	B+	□	A	Works very slowly but carefully on task.
8	Pupil H	⊠	C	⊠	C	□	B	~	C	⌐	C	Needs to concentrate. Keep to task set.
9	Pupil I	~	C	□⊘	C	□	B	□	O	□	B	Started project well. Keeps on task.
10	Pupil J	~⊠	B	□	A	□⊘	O	□	B	□⊘	A	Worked very well this term/motivated.
11	Pupil K	□D⊘	A+	□D⊘	A	□D⊘	A+	□⊘	A+	□D⊘	A	Moved away from other pupils. Good start.
12	Pupil L	□	B	□	B	⊘	B+	⊘	B	□	B	Is progressing well. Developing design.
13	Pupil M	⌐	B+	⌐	O	□⊘	B+	□	A	□⊘	A	Slow, but the work is of a high quality.
14	Pupil N	□	C	⌐	C	⌐	C	□	B	□	O	Refining and developing work. Slow start.
15	Pupil O	~	A	⌐	A	⌐	B	□	B+	□⊘	B+	Progressing well from a slow start.
16	Pupil P	□	A	□⊘	B	□⊘	O	O	B	□	B	Growing in confidence and ability.
17	Pupil Q	⌐	O	⊠	O	O	O	□	O	⌐	O	Tends to copy but is gaining confidence.
18	Pupil R	⊠	B	□	O	□⊘	O	O	B	□⊘	B	Very motivated and is kept on task.
19	Pupil S	⌐D⊠	C	⊠	B	D□	B	⌐D	C	D⊠	O	Limited attention span. Easily distracted.
20	Pupil T	□	B	□⊘	A	□	B	□	B	□⊘	B	Technical ability improving every lesson.
21	Pupil U	⊘	A	⊘	B+	⊘	B+	□⊘	A	□⊘	B+	Working extremely well. Keen on project.

AIMS OF LESSONS/HOMEWORK SET

1. Intro. to medieval patterns. Start designing individual patterns.
   H/W – Observational drawing in response to patterns and texture.
2. Finish design. Enlarge pattern and repeat on to frame shape.
   H/W – View from a window – observation and recording. Tone/shade.
3. Start to decorate frame layering papier mâché.
   H/W – Collage using newspaper or magazines. Fruit or vegetables.
4. Continue to add layers, building up the surface of the frame.
   H/W – Referring to skills learnt last term, draw hard and soft objects.
5. Complete project.
   H/W – evaluation worksheet.

KEY

- □ On Task
- ⊠ Off Task
- ⊘ Excellent
- ⊠ Unsettled
- D Disruptive
- ⌐ Limited
- ~ Coasting
- O Absent

Figure 7.1.2 Tracking sheet

pupil Q you will note, has not completed any homework. Your action might be to speak to the pupil and discover what the issues are and why this is the case. You may want to contact the form tutor and/or the parents.

Task 7.1.3 Tracking sheet

Examine the tracking sheet example given, look at pupil S. What action should you take? Now look at pupil E. What action should you take in this instance? Finally look at pupil K. What action should you take here?

What sort of additional aspects would you want to track/identify? Devise your own criteria and symbols relevant to your teaching.

Use this tracking sheet to monitor a selected KS3 class for six weeks. At the end of this time you will be able to see clearly how your pupils have been performing, learning, progressing, behaving.

Achievement refers to the overall accomplishment of a pupil including personal factors.

Case study 1: Refer to pupil F on the tracking sheet (Figure 7.1.2). You can see that the symbol 'off task' occurs frequently. This pupil is not achieving the SoW objectives. Differentiation in terms of tasks and teacher input needs to be considered. The implications suggest that actions needed for this case may be differentiated resources, more appropriate worksheets or additional support from you. You will need to identify the pupils' needs as there may be a combination of reasons. It may be that a particular pupil is an 'Invisible Child' as described by Pye (1988), one who goes unnoticed because they do not register as either difficult or able.

Case study 2: Very able pupils (pupil D) who, from the tracking sheet appear to be 'coasting' are shown by the frequently occurring symbol. You need to consider whether they require more challenging or differentiated tasks. Providing extension work in your initial lesson planning is a way of ensuring that they are continuously challenged from the outset. Alternatively you might wish to engage in discussion to enable these pupils to take more responsibility for their work and become involved in setting their own learning targets.

You need to select appropriate times to identify the issues raised by the tracking sheet and to decide on appropriate action. By constantly building this process into your practice, this reflective approach supports and strengthens your preparation and planning, leading ultimately to more effective teaching. It may be useful to collect visual evidence on a weekly basis in tandem with the tracking sheet. This can provide a useful focus for a meeting with your mentor.

Grading and marking

Up to the end of KS3 the responsibility for marking and grading pupils' art and design work largely rests with teachers, although they must use the levels provided by the NC. It is only once pupils reach KS4 (GCSE) that external criteria are imposed by the examination boards.

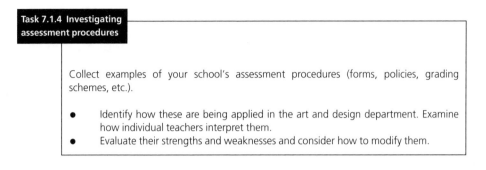

Task 7.1.4 Investigating assessment procedures

Collect examples of your school's assessment procedures (forms, policies, grading schemes, etc.).

- Identify how these are being applied in the art and design department. Examine how individual teachers interpret them.
- Evaluate their strengths and weaknesses and consider how to modify them.

Further reading

Atkinson, D. (2002) *Art in Education: Identity and Practice*, Dordrecht, Boston and London Kluwer Academic Publishers.

Boughton, D., Eisner, E. and Ligtvoet (eds) (1996) *Evaluating and Assessing the Visual Arts*, New York: Teachers College Press.

Gipps, C. (1994) *Beyond Testing: Towards a Theory of Educational Assessment*, London: Falmer Press.

Hulks, D. (2003) 'Measuring artistic performance: the assessment debate and art education', in N. Addison and L. Burgess (eds) *Issues in Art and Design Teaching*, London: RoutledgeFalmer.

Kennedy, M. (1995b) 'Approaching assessment', in R. Prentice (ed.) *Teaching Art and Design: Addressing Issues and Identifying Directions*, London: Cassell.

UNIT 7.2 ASSESSMENT IN ART AND DESIGN

Andy Ash and Kate Schofield

Objectives

By the end of this unit you should be:

- informed about assessment within the NC;
- able to consider different methods of using recording systems for assessment.

Good practice

With the growing emphasis on assessment and accountability, the collecting of evidence demands that art and design teachers pay more attention to the organisation of their time. Successful art and design teaching is based on observations and knowledge and understanding of the way pupils learn as individuals. This means providing activities and an environment where pupils:

- relate practical making with theoretical understanding: make connections between knowledge, skills and understanding; investigate the work of artists, craftspeople and designers and relate it to their own;

- work collaboratively: negotiate roles and responsibilities in the group; are open and receptive to issues and ideas within the group; are able to understand and make choices about tools, techniques and materials appropriate to their needs;
- engage in dialogue: discuss ideas with the teacher and other pupils to aid progress and understanding; develop an art-specific vocabulary; speak confidently about the work of other artists, craftspeople and designers;
- reflect on their learning in feedback and critique sessions: encourage explanation of work processes, skills and concepts; challenge and respond to each others' work.

Assessment in the NC

The NC PoS provide the basis for planning, teaching and learning and lead on to assessment during each key stage determining the knowledge, understanding and skills that should be taught to pupils as their minimum statutory entitlement. Art and design departments use the PoS to provide a basis from which to develop schemes/units/projects of work for teaching and everyday assessment. These schemes specify objectives for teaching and pupil learning and ensure that over the period of the KS pupils meet the legal requirements.

Decisions about how to mark and record progress in relation to these objectives are matters for schools to consider in the context of the needs and achievements of their pupils (SCAA 1996: 2).

Attainment Target for NC Art

Art and design is a foundation subject in KS 1, 2, 3, but is optional at KS 4 (GCSE) and post-16. Art and design PoS have one Attainment Target (AT): knowledge, skills and understanding, with four strands which outline the process of working in the subject:

1   Exploring and developing ideas;
2   Investigating and making art, craft and design;
3   Evaluating and developing work;
4   Developing knowledge and understanding.

It is recommended that you adopt an integrated approach to teaching within the four strands. Although SoW do not necessarily address all four, the PoS must.

The Attainment Target Levels are designed to indicate clearly the expectations for art and design for the majority of pupils by the end of KS3. You are expected to make summative judgements about pupils' attainment and the way they relate to the standards of performance at KS3. To make a rounded judgement you use your knowledge of:

- pupils' overall performance, based on the information gained from the continuous assessment of work across a range of contexts;
- the expectations set out in the NC Order;
- the relationship between pupils' achievement and learning objectives.

Task 7.2.1 The function of assessment

Important questions to consider during this activity are:

- How is evidence recorded and stored?
- Who is it for?
- What is it for?

Task 7.2.2 Making judgements

Using samples of Year 9 pupils' work discuss with colleagues where you would place each work in relation to NC level 5 and then 6 (5/6 is the expected attainment for the majority of pupils at the end of KS3).

A   Working towards;
B   Achieving;
C   Working beyond;
D   Demonstrating exceptional performance.

Continuous assessment: practical ways to record and assess pupils' progress

In assessing their own work pupils have to be reflective. They need to develop and use an art vocabulary as part of the requirements of the NC. Self-assessment is a vital vehicle for this process. This involvement actively promotes a sense of ownership, empowerment and responsibility. If you remain distant, aloof and appear judgemental you reinforce the traditional barriers between pupils and teachers.

Openness and accountability are increasingly important and we advise that you adopt a model of dialogue. You should engage in discussion with pupils so that together you can develop, evaluate projects and set targets for learning. In order for this to happen you should enable pupils to:

- develop a specialist art vocabulary;
- move from description to interpretation;
- apply their analytical skills;
- extend their range of concepts/experiences.

Recording systems

Recording systems enable pupils to document progress, set targets and take an active role in their learning. Some of the tried and tested methods are:

Diaries: these can be sequential and chronological, recording the process of work in progress, reviewing work completed, setting deadlines, tasks, and identifying needs. A diary may also record achievement and progress.

Sketchbooks: a compilation and development of ideas, combining visual material, texts, objects, plans for future work; a process-based book (see Unit 9.1).

Portfolios: a formal compilation of work with a more self-conscious arrangement, generally with a higher level of presentation. They may include preparatory studies and outcomes rather than process-based work. They are mainly used for work which demonstrates evidence of achieving learning objectives. Often the portfolio is presented to others as a filtered collection of art and design work representing range and depth of study.

Log book: a way of monitoring achievement, keeping records and dating tasks. It is also a way of documenting units of work completed.

Scrap book: an informal collection of useful source material, showing ideas and the way in which pupils have engaged in initial levels of research.

Photographs: these can be seen as an independent element to record moments and events (e.g. installations), individual pieces or objects or can be used as part of any of the above.

ICT: including CD/DVDs, video and audio tapes, CAD and computers. Through using various software packages, machines and processes in collaboration, material can be stored, retrieved, manipulated and used to generate art, craft and design work.

Task 7.2.3 Recording

In your practice, which of the above recording systems do you employ? Which ones do you under-use and why?
  Consider additional methods of recording: add to the list.

UNIT 7.3 ART AND DESIGN EXAMINATIONS

Andy Ash and Kate Schofield

Introduction

Units 7.1 and 7.2 examined the principles and definitions of assessment, focusing on the day-to-day methods employed in the classroom. In this unit the focus is on the external examination systems, how they have developed and relate to the NC.

Compared to other subjects in the curriculum there are noticeable differences in the approach to art and design assessment and examination of pupils' work. External syllabuses have been designed to provide an extension for pupils' work at KS3. The syllabus requirements for GCSE (KS4) are intended to build upon the philosophy and practice of the NC. This is particularly highlighted in the construction of the assessment objectives against which all pupils are marked.

Art and design 'A' Levels have been radically restructured following the Dearing Report (1997). The review of post-16 qualifications attempted to broaden the curriculum for 17 and 18 year olds which had historically been considered too narrow. To an extent this has signalled the end of the restricted specialisation which used to be the basis of the system. With new 'A' Levels and an increasing number of applied qualifications at 14–19 and post-16, pupils have a more flexible learning path preparing them for the demands of work in the twenty-first century (see www.qca.gov.org).

Objectives

By the end of this unit you should:

- be aware of external assessment and examinations;
- understand the role of the teacher in external examinations;
- be confident to apply written criteria to visual work;
- have been introduced to the assessment of GCSE and 'A' Level examinations.

The process involved in external assessment

Task 7.3.1 GCSE: continuity and progression

Reflect upon your experience of taking the GCE/GCSE examination in art and design. Consider the following issues:

- Was there an emphasis on process or product?
- How well and in what ways did it prepare you for your future art and design education career?

The GCSE examination

The GCSE in art and design comprises both an assessment of pupils' coursework done over a period of time, usually two years, along with an externally prescribed controlled or timed test (ten hours). These two elements of coursework and test make up the GCSE examination and all three examination boards adhere to the same principles. The GCSE is assessed holistically; the marks from both the coursework and test are added together to form the overall mark. This is then accorded a grade by the examination board. This is the same for all pupils and for all art and design syllabuses offered by the examination boards. They all have to comply with a Code of Practice which in turn is monitored by a government body Qualifications and Curriculum Authority (QCA). In theory, whichever GCSE examination syllabus is taken by candidates, parity of the marking is ensured through a system of monitoring by QCA.

The examination boards vary the way their syllabuses are designed and laid out. In practice, for example, some require candidates to submit four pieces of coursework while others demand three. Examination question papers for the timed test also vary in their length and presentation. The number of questions set and the amount of written stimulus or starting points depend on which examination board is selected by schools. Generally teachers investigate all syllabuses and then make a choice based on the one most closely allied to the type of work they already do in their particular department. The way in which teachers are required to mark their pupils' work also varies between one board and another. This said, all candidates are assessed according to written criteria which are common to all three.

The GCSE criteria

During the 1990s much work was done to develop and provide written criteria for assessing GCSE work and, from 1998, these were put in place and made statutory. From one viewpoint the criteria are intended to encapsulate the essence and philosophy of the NC Art and design, the idea being that all secondary pupils who follow it through the first three Key Stages are able to develop these ways of working to meet the criteria for GCSE (KS4). All candidates entering for the GCSE examination have to address the following criteria in both their coursework and test units.

Assessment objectives

The following assessment objectives are not in any order of priority, and are approximately equally weighted.

In each of the syllabuses candidates are required to demonstrate their ability to:

(i)   record observations, experiences and ideas in forms that are appropriate to intentions;
(ii)   analyse and evaluate images, objects and artefacts showing understanding of context;
(iii) develop and explore ideas using media, processes and resources, reviewing, modifying and refining work as it progresses;

(iv)   present a personal response, realising intentions and making informed connections with the work of others.

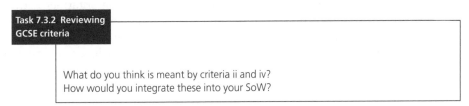

Task 7.3.2 Reviewing GCSE criteria

What do you think is meant by criteria ii and iv?
How would you integrate these into your SoW?

Past practice has meant that a number of pupils may not be familiar with some of these criteria. Although pupils may be good at working within the broad aegis of say criteria i and iii they may come unstuck (flounder) when trying to score marks in ii and iv. If pupils are taught to engage with the way their work relates to each criterion they are provided with a strong foundation for future practice, learning and development.

Endorsements

An endorsement is a specialism within a broad art and design syllabus. All three exam boards offer either an unendorsed syllabus where candidates follow a broad course of work, or an endorsed syllabus, which has a specialist approach such as textiles, graphics, 3-D, photography, drawing and painting.

Marking GCSE

Since the inception of GCSE the responsibility for the assessment and marking of pupils' course and test work has been with the teacher. The art and design teacher is therefore seen to be in the best position to judge pupils' level of attainment because they are aware of the formative, preparatory and developmental work at all stages. In the past, if a candidate for some reason performed badly in an examination there was no means of awarding a fair mark.

Post-16 education

This book is published at a time when provision for 14–19 and post-16 education has seen major changes. Both the Dearing Report (1997) and the Tomlinson Report (2004) argued for similar changes to the system. Dearing recommended three distinct pathways for post-16: academic ('A' Level), general vocational (GNVQ) and work based (NVQ). It was considered that a broad-based approach was the way forward. Students could opt to follow any particular route. Central to this idea was a new argument for breadth. Dearing recommended that post-16 should largely be in the form of academic education for those who wished to continue with school subjects and for the rest it should be restricted to key skills. The debate continued by proposing closer links between the academic and vocational educational routes. A multi-dimensional approach has been suggested which emphasises modularity. Thus, the flexibility offered to students in selecting pathways allows for sideways movement (from academic to vocational and vice versa) enabling a broader and more adaptable programme of study to suit individual needs for future employment or further study. This was particularly highlighted by Tomlinson who was suspicious of the A-level qualification and suggested a less specialised curriculum for all. However, as we write, a range of fourteen new diplomas designed to transform 14–19 education is being ratified. Each diploma offers a specialised 'pathway to the economy' constituting an 'applied' course in, for example, the 14–19 Creative and Media Diploma.

Advanced Subsidiary and Advanced Level examinations

During the first year of post-16 students take a number of AS Levels (up to five). In the second year they select from these, two to four subjects to take through to 'A' Level. In art and design the aims and criteria for AS and 'A' Level are the same. The subject criteria set out the knowledge, understanding, skills and assessment objectives common to all AS and 'A' Level specifications. Specifications in art and design should promote the development of:

- intellectual, imaginative, creative, and intuitive powers;
- investigative, analytical, experimental, practical, technical and expressive skills, aesthetic understanding and critical judgement;
- an understanding of the interrelationships between art and design and an awareness of the contexts in which they operate;
- knowledge and understanding of art and design in contemporary society and in other times and cultures.

(Edexcel 2006: 8)

The content of the proposed syllabuses integrates critical, practical and theoretical work in art and design.

Task 7.3.3 GCSE and 'A' Level: identifying difference

Examine the AS and 'A' Level examination syllabuses for art and design used by your schools.
What differences can you find between GCSE and 'A' Level?
Do they indicate continuity and progression?

Built into the AS, 'A' Level are Key Skills. They are:

- application of number;
- communication;
- improving own learning and performance;
- ICT;
- problem-solving;
- working with others.

Task 7.3.4 Key skills

Devise, plan and write a SoW, unit of work or project in which all the Key Skills are covered. Make a list of the different ways in which you can collect evidence and assess students' understanding of these skills.

Further reading

Hodgson A. and Spours, K. (eds) (1997) *Dearing and Beyond: 14–19 Qualifications, Frameworks and Systems*, London: Kogan Page.

Mason, R. and Steers, J. (2006) 'The impact of formal assessment procedures on teaching and learning in art & design in Secondary Schools', *International Journal of Art & Design Education*, 25, 2: 119–133, Oxford, Blackwell.

UNIT 7.4 RECONSIDERING ASSESSMENT FOR LEARNING IN ART AND DESIGN

John Steers

Introduction

This section poses, not necessarily answers, a number of questions for you and raises issues about assessment and examinations in art and design.

Is a national structure for assessment in art and design necessary?

Can a wholly objective, standardised system be designed for art and design?

Do attempts to achieve this affect course content for better or for worse?

Who is qualified to make judgements about standards in art and design?

What should be assessed: student capability, potential, attainment, achievement or 'effective' teaching?

Do detailed assessment criteria tend to fragment knowledge, skills and understanding?

Can criteria successfully accommodate all attainment levels and diverse outcomes?

Does the language of criteria adequately encompass the possible range of visual and tactile outcomes?

Should assessment be more holistic, more 'authentic'?

Should assessment be 'negotiated'?

Is a terminal examination necessary?

Do examinations assess qualities that cannot be found in a coursework portfolio?

7.4.1 Is there a need for assessment in art and design?

When, in the 1960s, the examination system was extended to include 80 percent of the 16+ age group there were some art teachers who declared that this would lead to the inevitable death of creativity and imagination in pupils' art and design work. Ross (1993) summarised this view when he wrote: 'The status system embodied in the mundane curriculum stresses impartiality, inequality, secularity, acquisitiveness, discrimination, epicureanism. All these are anathema to the arts and to the true artist'. He also identified a key dilemma:

> If teachers do participate in public examinations they run the risk of allowing their work to be wrestled from its legitimate roots, yet if they do not they seem to push the arts further out along the educational limb, accepting the more the arts become exceptions to the rules of schooling the less relevant they are likely to appear.
>
> (Ross 1993: 92)

However, most art and design teachers concede that assessment, including public examinations, is an integral part of a coherent teaching and learning continuum of planning, recording individual and group progress, course evaluation and reporting: it is not a peripheral, 'add-on' component. The National Curriculum (NC) introduced in England and Wales as a consequence of the 1988 Education Reform Act defines the key skills, concepts, knowledge and understanding for each subject. The key strands form the basis for level descriptions designed to identify in broad terms the differences in pupils' performance at the end of each key stage, at ages 7, 11, and 14. Optional examinations continue to be available for 16- and 18-year-old students, in England principally the General Certificate of Secondary Education (GCSE) and General Certificate of Education 'Advanced' level (GCE 'A' level) respectively.

The reality for you is that assessment in art and design is a simple fact of teaching.

> . . . from their earliest years students are subject to assessment: simple gestures, affirmative or negative reactions are all measures of performance. Teachers, as a matter of course, are continually gathering information about the effectiveness of teaching and learning, accumulating records to monitor students' progress and to assist prognosis . . .
>
> (SEC 1986b)

Perhaps the key question you should ask is not whether to assess or not, but rather: how can educationalists and teachers design valid, reliable, and effective assessment instruments that are fit for purpose?

7.4.2 Some functions, key principles and terms

The functions of assessment are varied, producing a set of different purposes that can often appear contradictory:

- providing feedback for the pupil or teacher (formative);
- diagnosing the pupil's future needs or direction (diagnostic);
- providing motivation for the pupil or teacher (formative);
- providing a means of licensing or qualification (summative);
- providing a basis for selection (e.g. for higher education or employment) (summative) (see 7.2.1).

Assessment can also be used for:

- ensuring accountability by monitoring national, local or individual teachers' standards;
- controlling the curriculum.

This is complicated by institutional misunderstandings and tensions where government and teachers tend to assign different priorities to each assessment function. The General Teaching Council for England (GTC 2004) has argued for a better balance between internal and external assessment, suggesting that the government '. . . should develop a teacher assessment model that focuses solely on formative assessment to avoid the distortion of function' (p. 1). The GTC also recommends that 'the summative assessment of individual pupils is separated from the collection of summative data to be used for national monitoring' and that a rolling programme of sampling pupil cohorts could accomplish the latter.

It is axiomatic that any system of assessment should satisfy certain *general principles*. Perhaps first and foremost of these is that well-designed assessment procedures should be valid, reliable and useful. The *validity* of an assessment is the extent to which it serves its purpose, employing methods that reflect the aims and objectives of courses and educational programmes. Also, as far as possible assessments should be designed and operated so that they give a similar result when taken by pupils of similar ability under similar conditions – this is *reliability*. A perfectly reliable assessment probably is impossible to achieve because unreliability may be caused by differences in the conditions for, or the context of, the assessment in the judgements of examiners and moderators, or bias of one kind or another. The *utility* of an assessment instrument is its convenience, flexibility and cost-effectiveness; all assessment procedures have to operate within the time and resources available. These three qualities are inevitably in tension in most assessment systems and, consciously or unconsciously, you may find yourself giving different weightings to each.

You should use assessment results to provide direct information about pupils' achievement in relation to specific objectives: therefore they should be *criterion-referenced* rather than *norm-referenced*. (Norm-referencing refers to an assessment system in which pupils are placed in rank order and pre-determined proportions are placed in the various grades. Therefore the grade given to a particular pupil depends upon a comparison between that pupil's performance and those of all the other individuals or groups, rather than upon the absolute quality of the performance. This approach may be best suited to monitoring overall standards.)

Assessment procedures should be *formative* – 'Assessment for Learning is the process of seeking and interpreting evidence for use by learners and their teachers to decide where the learners are in their learning, where they need to go and how best to get there' (ARG 2002: http://arg.educ.cam.ac.uk/CIE3.pdf). However, as well as being formative, assessment often needs to be *summative*, that is where results are intended to provide a final or cumulative assessment of achievement at the conclusion of a particular course or phase of education.

Scales or grades should be capable of comparison across classes and schools if teachers, pupils and parents are to share common and secure standards: thus your assessments should be calibrated or *moderated*. (Moderation is the means by which the marks of different teachers in different centres are equated with one another and through which the validity and reliability of assessment is confirmed – it is the process of checking comparability of different assessors' judgements of different groups of pupils.)

The ways in which examination specifications, assessment objectives, criteria and mark schemes are set up and used should relate to expected routes of educational development, providing continuity at different ages: therefore you should relate assessments to *progression*. There is a consensus that assessment procedures should reflect the aims of courses and should enhance the quality of teaching and learning without dominating or distorting these; you should therefore ensure that the curriculum you teach is not assessment-led.

The US educationalist Eisner (1972) has drawn important distinctions between *evaluation, testing* and *grading*. He describes evaluation as a process through which value judgements are made about educationally relevant phenomena. In the UK there seems to be a distinction between *evaluating* curriculum or lessons and *assessing* pupil ability or attainment. (In reality most assessment is concerned with assessing 'performance' or 'attainment' rather than underlying ability.) Testing, according to Eisner, is a procedure used to obtain data for purposes of forming descriptions or judgements about one or more human behaviours, while grading is a process of assigning a symbol standing for some judgement of quality relative to some criterion.

Strictly speaking, a *test* is any assessment conducted within formal and specified procedures, designed to ensure comparability of results between different test administrators and between different test occasions. Tests can be composed of externally prescribed multiple-choice, true or false questions, practical task, short-answer or essay writing components that are assigned a score by independent assessors according to defined rules. More loosely, a test may be one of a broad range of assessment instruments with rules of administration and marking which ensure comparability of results. In contrast, *Standardised tests* are assessment or measuring instruments that have been subjected to rigorous development and testing, that are administered and scored in a predetermined, standard manner, and for which there is published data on the reliability and validity of the instrument. Normative data (data collected using the instrument, usually from a variety of different samples of the relevant population) are also available.

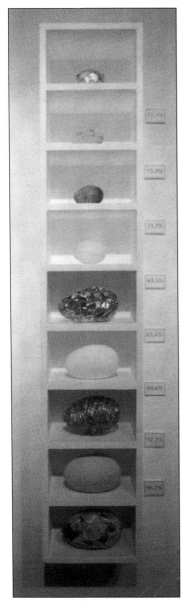

1 *The semiotics of materials*

2 *The semiotics of materials*

3 *The semiotics of materials*

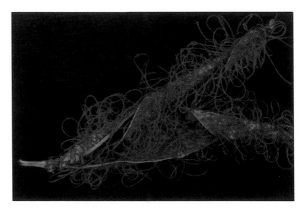

4 *The semiotics of materials*

5 *Recording a 'Sense of Place': St Pancras Chambers*

6 *Recording a 'Sense of Place': St Pancras Chambers*

7 *A Question of Context*

8 *Cabinet of Curiosities: Sketchbook?*

5, 6, 7, 8: PGCE students (Institute of Education, University of London)

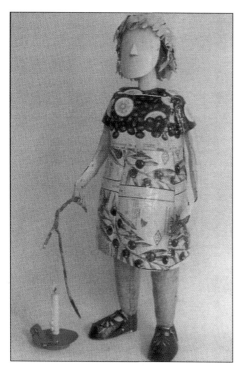

9 *Playing with Fire* (1998), Lucy Casson

10 *Spice Boxes* (1998), Natasha Hobson

11 *Gazebo* (1995), Thomas Heatherwick

12 *Heart Development Hat* (1998),
Helen Storey

13 *Between Words and Materials*

14 *Untitled* (1998), Meera Chauda

15 *Seat of Learning*

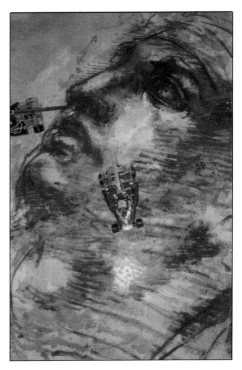

16 *'I can't draw either'*

13, 15, 16: PGCE students (Institute of Education, University of London)

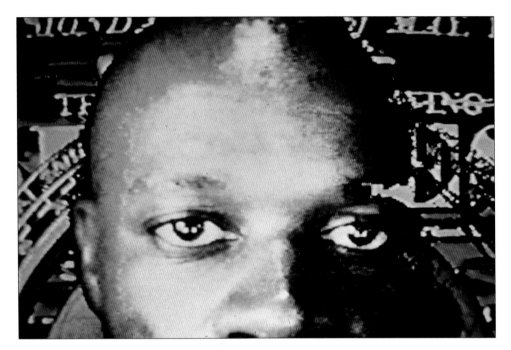

17 *Tradewinds* (1992), Keith Piper

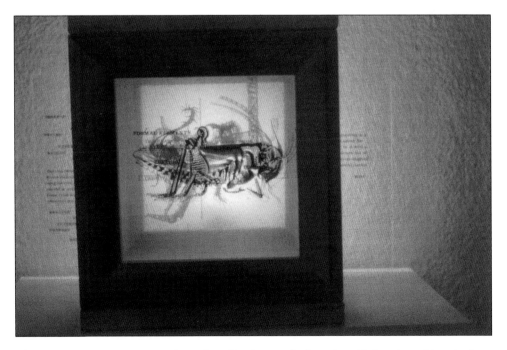

18 *Investigating Insects*, PGCE student (Institute of Education, University of London)

19 *Walking in Space* (1998), Michael Brennand-Wood

20 *Steppenwolf* (1997), Caroline Broadhead

21 *'Evidently Not'* (1998), Cathy de Monchaux

22 *Diasporic Pyramid: Monument to Asylum Seekers,* PGCE student (Institute of Education, University of London)

23 *Sculpture: colour workshop*

24 *Layering Cultures*

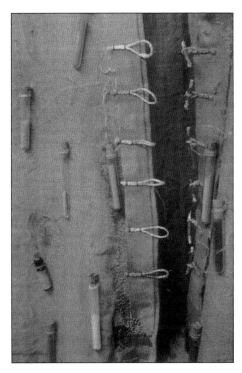

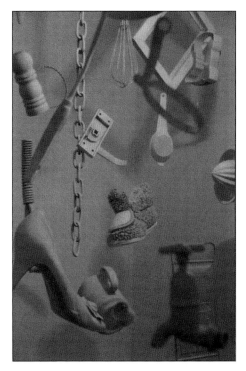

25 *Shaman's Coat*

26 *Installation: Domestic Turn*

23, 24, 25, 26: PGCE students (Institute of Education, University of London)

7.4.3 A short history of art and design assessment

As you enter the profession you will discover that there are a large number of nation-wide formal and less formal art and design assessment systems operating from primary schools to postgraduate courses which are subject to almost constant change. The combined efforts of all those involved in these systems, from classroom teachers, examiners, officers of the awarding or validating bodies and government agencies such as the Qualifications and Curriculum Authority (QCA) are almost permanently employed in trying to refine and improve procedures, often to comply with and accommodate the latest government strictures. Consequently any attempt to describe the current assessment regime is likely to be out-of-date before an account such as this can be published and you are advised to access the websites of the various regulatory (e.g. QCA) and awarding bodies (e.g. Edexcel, OCR, AQA, WJEC and CCEA). Nevertheless, a short account of the recent history of art and design examinations may provide you with a useful context for understanding the current assessment regimes.

Art and design examinations, particularly drawing examinations, date back at least to the nineteenth century but, for the purposes of this account, the introduction in the mid–1980s of the General Certificate of Secondary Education (GCSE) is taken as the starting point for a series of 'reforms' that have continued unabated to the present time. The introduction of the GCSE was not a rushed affair. The initial proposals to combine the former General Certificate of Education (GCE) Ordinary level and the Certificate of Secondary Education (CSE) dated back at least to the days of the Schools Council and Prime Minister James Callaghan's seminal lecture on education reform at Ruskin College Oxford in 1976. The new GCSE examination was aimed originally at the top 60 percentile of the 16+ ability range although from the start, in the case of art and design, the range was often much wider than this. Today virtually all school students are expected to take a broad range of GCSE examinations at 16+.

For many art and design teachers the advent of the GCSE was welcome and overdue, marking the end of the need to 'double enter' more able students for both previous examinations. Moreover, the rationale for the introduction of the GCSE was clear enough: the GCE/CSE system was incompatible with comprehensive education. According to Her Majesty's Inspectors of Schools (HMI), work in secondary schools was dominated by examinations which tended to govern the type and length of classroom activities (a criticism that, ironically, may be equally if not even more valid today). The reliability of inter-examination board standards then, as now, was considered questionable, so the new GCSE was administered by just four regional examining groups in England, one examination board in Northern Ireland and another in Wales (Scotland has its own arrangements for curriculum and assessment currently administered by the Scottish Qualifications Authority).

At the core of the GCSE proposals was a requirement that common national assessment criteria should be established for all examination syllabuses (now called 'specifications') and assessment procedures, to ensure that all syllabuses with the same subject title had sufficient content in common and so that all boards applied the same performance standards to the award of grades. To this end, grade criteria working parties were established for some subjects as early as 1981 but the working party for art and design was not convened until 1984 when the template for such assessment criteria had been firmly established.

The key to the exercise was how to define pupils' achievements through explicit criteria while not overly restricting the methods by which those criteria might be achieved. The art and design working party accepted that this approach involved many compromises and these included tacit agreement that it simply may not be possible to assess adequately or equally all curriculum objectives because, in practice, the evidence of some objectives may be too ephemeral to be valid. Lengthy consideration led to the identification of three equally weighted, closely interdependent and interrelated domains:

- a *Conceptual Domain* concerned with the formation and development of ideas and concepts;
- a *Productive Domain* concerned with the abilities to select, control and use the formal and technical aspects of art and design in the realisation of ideas, feelings and intentions; and,
- a *Critical and Contextual Domain* concerned with those aspects of art and design which enable candidates to express ideas and insights which reflect a developing awareness of their own work and that of others.

(SEC 1986c)

While it is clear that this model is only one among many that could have been adopted, it did mark a significant step forward and influenced both the development of the National Curriculum in England and later examination criteria. In developing the detailed criteria, a number of important issues emerged and significant lessons were learned. For example, the dangers of using such adjectives as 'simple' or 'sophisticated' in criterion statements because of the way these words were open to wide variations of interpretation, a lesson that has been overlooked in the interim as such terms are increasingly finding their way back into the assessment criteria lexicon.

In retrospect the introduction of grade criteria was seminal in more ways than one. It could also be seen as the thin end of a wedge leading towards a state where, as Eisner (1985) warns:

> . . . infatuation with performance objectives, criterion referenced testing,
> competency based education, and the so-called basics lends itself to
> standardization, operationalism, and behaviorism, as the virtually exclusive

concern of schooling. Such a focus is . . . far too narrow and not in the best interests of students, teachers, or the society within which students live.

(p. 367)

From the mid-1980s, through successive agencies such as the School Curriculum Development Committee (SCDC), the National Curriculum Council (NCC), the School Examinations Council (SEC), the School Examinations and Assessment Council (SEAC), the School Curriculum and Assessment Authority (SCAA), to the present QCA, governments have sought once and for all to 'nail the jelly to the wall' through repeated attempts to define the content, aims, objectives and assessment parameters and criteria for school subjects, including art and design. But there has been relatively little new thinking and in reality much of this relentless process has consisted of 're-packaging' the subject by means of repeated editing and précis of assessment documents to make them fit the common template currently favoured for all curriculum subjects.

The publication of Curriculum 2000 and the latest specifications for qualifications gave the impression that all questions of what constitutes good practice in schools had been resolved at last. Of course this was just an illusion: it was inevitable that before long another perceived change of circumstances or belated admission of inherent problems would require another round of 'reform'. Another comprehensive review of GCSE and GCE Advanced level, as well as the introduction of new national diplomas, is underway at the time of writing in 2006.

7.4.4 The washback of art and design examinations

The idea of *washback* or *impact* is common in examination and testing literature. It is suggested that assessment procedures have much to do with what actually happens in classrooms with claims of 'teaching to the test' and 'assessment leading the curriculum' being too common to be ignored. Claims can be made for both negative and positive washback, but there is very little hard evidence to support the argument that examinations and other assessment procedures directly influence teaching, and most evidence appears to be anecdotal rather than based on empirical evidence (Mason and Steers 2006).

Nevertheless, for many years concerns have been expressed about an increasing orthodoxy of approach and lack of experimentation in art and design education. In the early 1980s, when the GCSE was still a proposal, one experienced chief examiner and the then chair of the Schools Council 16+ art sub-committee, expressed concerns about the washback of the examination system on classroom practice and prophesised problems with standardised assessment criteria:

> The existing relationship between curriculum and examination syllabuses is a 'dog and tail' affair. The influence of external examinations has, to some extent, bred a species within the genus of 'School Art'. . . . The question of whether the 'tail wagging the dog' is a satisfactory state of affairs must be linked with the possibility that the existing dog is a mongrel that defies

simple definition. This is not to say that some mongrels are not more healthy than some more easily categorised pedigrees, but it does make the establishment of national criteria guidelines more difficult – more difficult in the sense that criteria will necessarily be based upon generalisation of a plethora of objectives and practices – generalisation which will undoubtedly influence the future of art education.

(Price 1982: 399)

Binch (1994), at one time chair of the art and design panel of the largest examination board, reflected on how the examination system introduced into England and Wales in the mid-1980s, with its strong emphasis on process, influenced the style of work throughout secondary schools. He claimed that frequently this led to a single, linear classroom methodology where:

> . . . the starting point is usually investigation and research, followed by the development of ideas and some experimental activities, and the completion of a 'finished' piece of work . . . The model reinforces the insular nature of 'school-art' and, even when reference is made to external sources, it is usually based on the same methodology of objective drawing and visual analysis.

(p. 124)

The approach described by Binch has proved very reliable over the years, often producing 'safe' work that, of its kind, is often undeniably of high quality and on which teachers can depend for the award of good grades by the examination boards. Today the examination pressures in secondary schools are overwhelming and influence classroom practice not only at KS4 but throughout KS3. In a high-stakes education system it should be no surprise to you that teachers are adept at finding effective prescriptions for their pupils to follow that satisfy the various demands of the awarding bodies and, in turn, league tables, inspection and threshold payments (a form of payment by results). But whether such a dominant approach is in the best interests of pupils is another question.

How has this come about? Since the early 1980s examination syllabuses have increased considerably in size and the detailed guidance they provide. Typically in the earlier generic GCE or CSE syllabuses a few paragraphs sufficed to outline art and design, but today separate art and design syllabuses (now termed 'specifications') of 20–40 pages are the norm. Now it is hard to believe that 25 years ago syllabuses rarely contained specific aims, objectives, subject content or mark schemes, principally because the GCE 'O' level examinations generally were externally marked and it was not thought necessary to provide such information for teachers or candidates. Differences between syllabuses offered by the 20 or more examination boards were marked, but over time there has developed a near total conformity between the three English awarding bodies, no doubt as a consequence of increasing requirements to use the prevailing examination templates of the QCA and its various antecedents.

The earlier syllabuses invariably included a wide range of optional papers with a focus in particular on fine art, design and craft skills such as, for example, lettering, photography, print-making, pottery, theatre design and mural design. The opportunities to pursue such specialisms in depth appears to have diminished partly as a consequence of the decision to introduce the 'unendorsed' art and design examination and partly by the 'rationalisation' of recognised titles for endorsed papers. By the beginning of the twenty-first century the overwhelming majority of art and design candidates were entered for unendorsed papers and much of the work submitted took the form of drawing and painting. It is clear that one consequence of the changes that have taken place has been that some specialisms have been lost in many schools along with any real opportunity to choose between studying a particular aspect of art and design in depth or to following a more general course of study.

Not all changes have been negative in their impact. For example, there has been a move away from question papers that offered little more than one-word 'starting points' or instructions to the examination supervisor about how to set up a specified still life group or pose a model. Instead the most recent question papers favour formats that provide more support for candidates, for example, by providing a detailed design brief. Another recent approach is to provide a 'question paper' with a single common theme to cover a wide range of art, craft and design activities offering a long and discursive discussion of ideas that candidates might wish to develop. The idea clearly is to motivate candidates to respond as creatively as possible to the required 'terminal test'. Twenty years ago the time for examination was often short, typically 2–3 hours in which to produce a drawing or composition. The time for the GCSE terminal examination is now 10 hours for all examining groups and there is a common pattern of question papers being issued to candidates some weeks in advance of the terminal examination in order to give them opportunities for research and to plan their work.

Of particular significance has been a marked trend away from an optional, formal art and design history element towards a general requirement for a critical and contextual studies component. This shift has been largely uncontentious but it is not unproblematic. Pragmatism has played a part in the widespread acceptance of critical studies in secondary schools because the 'old' art history demanded considerable teaching time and was only suited to reasonably academically minded students. In contrast, critical studies at examination level are often dependent on pupils researching 'personal studies' in their own time. The outcomes of this approach are varied with some exceptional projects in evidence. More generally, however, 'descriptive and non-contextual studies' might be a more accurate title and questions need to be asked about what coherent knowledge of art and design practice pupils gain from often insufficiently directed 'research' with a narrow focus that often encourages internet-assisted plagiarism accompanied by problematic pastiches of style (see Chapters 3 and 10). There is a need for you to develop a clear rationale for the inclusion of critical studies based on some coherent view of content and cultural transmission, real critical thought and reaction, and articulate debate.

The past 20 years has seen a trend away from holistic assessment with no published criteria or mark schemes to a process of aggregation of component marks, based on mark schemes closely related to the published assessment criteria. It is becoming evident that this tends to lead to fragmentation, with teachers teaching to specific assessment objectives in the knowledge that, if students provide evidence of some obvious engagement with the ideas and practices embedded in the criteria, they will be rewarded almost regardless of the actual quality of their work. Of necessity, assessment criteria are drafted to be generally applicable to a wide range of specialisms and activities. As a consequence you will find they are not always easy to apply to some of the more unusual outcomes and may tend to inhibit some of the more creative responses that reflect, for example, contemporary practice in art and design. In effect, when there is over-reliance on criteria they act as regulatory devices through which both teaching and learning practices are normalised.

Although coursework was not generally a component of the old GCE 'O' level examination, it was a valued element of the GCSE from the outset. However, from the early 1990s onwards governments have sought to reduce the importance of coursework in all subjects in favour of the terminal test (in art and design the permitted ratio of coursework to terminal examination is presently 60:40). What constitutes a 'unit' of coursework is not always well defined although the best and most committed candidates continue to select from a considerable body of work.

Most of the change that has taken place has ostensibly been in the interests of increasing examination reliability. However, art and design examination reliability remains an issue, particularly given the much-increased size of the entry for GCSE art and design examinations for each awarding body now that these are fewer in number, in total over 200,000 candidates are entered for GCSE art and design each year. In the past, smaller examination boards with relatively small entries employed small teams of examiners. Although not dependent on formal mark schemes or assessment criteria, it seems reasonable to assume that a good level of consensus could be reached on standards, especially as one examiner often assessed all the work for a particular component. At the present time at least one awarding body employs over 200 moderators and, in these circumstances, it is clear that reliability must be dependent entirely on effective standardisation procedures.

Achieving accurate standardisation presents a considerable challenge to the awarding bodies and ever-tighter drafting of assessment criteria and their rigid application is not the answer. Boughton (1995) succinctly identifies the inherent problem:

> . . . any attempt to use written statements intended to describe the range of complex and subtle characteristics of visual expressive work at any level of schooling will be less than adequate . . . The qualitative nature of the arts . . . cannot be effectively captured in words alone. Linguistic representation of the arts is at best reductionist, and at worst misleading.
>
> (p. 146)

Common criteria have value in helping to focus the assessors on particular issues but they do not provide absolute measurement standards, assessment of the arts in schools still requires aesthetic judgement and connoisseurship based on experience of what students of a particular age *can* achieve. Moreover, criteria are written at a level of generality intended to accommodate any possible response. In practice this presents real difficulties for you and may only work when like is compared with like: there is a perception that anything more challenging that does not conform to the particular conceptual framework of the examination is less likely to be fairly rewarded, thus encouraging further orthodoxy.

In September 2000, the government introduced reformed GCE Advanced level examinations, together with new vocational 'A' levels and a key skills qualification. The new 'A' level was divided into six units, each assessed by an external exam or coursework. The first three units of the 'A' level make up Advanced Subsidiary, or AS level and the examinations can be taken at intervals during the course, or taken at the end of the two-year course. The introduction of modularity into the examination system created new problems for art and design. One immediate concern was a greatly increased workload for teachers while another more philosophical concern was with the 'washback' of the examination on teaching and learning. In art and design the first year of a GCE 'A' level course was often seen as a time for experimentation, for trying out new ideas, for supporting students' individual creativity by giving them the confidence to explore their own ideas and making them less dependent on teacher-led projects. This was threatened by the new exam because, in effect, every project, every piece of work from the start of the course needed to be fully realised and of the required standard. The doubtless unintentional message appeared to be don't bother being creative, avoid risks, play safe, do what is expected, a message equally damaging to students and to the discipline.

But nothing is forever in the world of examinations and in 2006 the examination was once again under review and, at the time of writing, it seems likely that the number of modules will be reduced in the new specifications.

Task 7.4.2 The washback of art and design examinations

In relation to your placement schools, would you say that the curriculum is led by the examination board's assessment criteria?

Do you consider that the National Curriculum attainment targets adequately describe what pupils might achieve at KS3? If not, why not?

7.4.5 What kind of assessment for art and design?

It is evident that you need to reappraise assessment procedures continually, try to improve the values implicit in evaluation, assessment and examination practices, question for whom the results of such practices are addressed and for what purpose, and evaluate their general and relative usefulness. To this end it seems necessary for you to develop a range and variety of assessment instruments and select from them for clearly identified, different purposes (Swift and Steers 1999: 14; GTC 2004).

Although you may base individual assessment for some purposes on comparison with other individuals or groups (i.e. norm-referenced), you should not depend on comparative means alone because such an approach inhibits the ability to respond to innovation or difference. For example, you need to monitor carefully the way standardisation is increased through the development of ever-tighter subject core specifications for GCSE and GCE 'A' level, especially for washback effects. There is a continuing danger of a further decrease in the variety of examination options and an increasing tendency to standardised pupil responses.

The learner's intention should be crucial to the way you or an examiner 'reads' the work. *Negotiated assessment* places considerable emphasis on the importance of the teacher sharing assessment criteria with the pupils and taking into account the pupils' intentions as well as the practical outcomes when assessing projects that might be either pupil motivated or externally set. Discussions at various stages of a project, aimed at understanding the pupils' intentions and approach to a particular brief, and various forms of pupil self-assessment can all be taken into consideration for both formative and summative assessments.

Swift and Steers (1999) and Atkinson (2001) also warn of the importance of avoiding 'hidden' criteria such as 'accuracy', 'likeness', 'in perspective', 'expressive', unless they are clearly stated and explained in their various meanings. It is important that you acknowledge hidden agendas and make them explicit and open to debate if they are to be used at all. You should reject spurious notions of objective judgement, since in reality any claim to objectivity in art judgements is based on the amount of comparative, experiential knowledge the teacher or examiner possesses and can apply. The reality is that differences of opinion and interpretation are axiomatic: for example, consider the concept of 'originality' as applied to children's art.

Authentic assessment strategies reject testing in favour of procedures which require pupils to engage in long-term, complex and challenging projects reflecting real-life situations. For art and design a structured *portfolio* may prove to be a particularly authentic assessment instrument. Torres Pereira de Eça (2005) proposes that such a portfolio could take the form of a folder, exhibition, work journal, CD-ROM, web page, etc., and that the data for assessment might include:

- Reports or notes, either visual or written, of previous experiences, interests, etc.
- Final visual products such as paintings, drawings, sculptures, prints, graphic design, product design, multimedia, photographs, films, video records of performances, installations, exhibitions, etc.
- Visual or written preliminary studies, developmental records
- Investigation reports and data, critical inquiry (written and visual)
- A self-assessment report that might include interviews – tape, video, digital records of the students' intentions, progress, investigations, achievements, presentations, self-assessment and 'crits'.

(p. 211)

In conclusion, there is a need for you to draw on and design a range of assessment instruments, affording a sensible balance between internal and external assessment. With them you should be able to reward experiment, challenge and encourage independent thinking and making, while discouraging playing safe and formulaic responses. This process enables you to place more emphasis on formative evaluation of work during the learning process and summative and authentic portfolios.

Further reading

(ARG) Assessment Reform Group (2002) 'Assessment for Learning: 10 principles', accessed at http://arg.educ.cam.ac.uk/CIE3.pdf on 24 August 2006.

8 Issues in Craft and Design Education

Lesley Burgess and Kate Schofield

The units in this chapter encourage you to engage with issues of craft and design and consider their status in secondary art education. You are invited to question traditional definitions, attitudes and values, discuss the roles of 'practical thinking' and 'intelligent making' and explore your developing understanding of these concepts in relation to their pedagogical application.

Unit 8.1   Focuses on craft and looks briefly at its history in relation to learning and teaching in secondary schools and challenges the limited definition perpetuated in many departments. The role of intelligent making is investigated and different models of delivery are explored. Examples of work by contemporary craftspeople are used to show how crafts are not static but constantly being updated and redefined. The important role played by the Crafts Council in developing an awareness of the above is acknowledged. Reference to their research projects is highlighted. Gender, class and ecological issues are referred to and artefacts and products from different cultures are briefly considered in order to raise awareness.

Unit 8.2   Focuses on design, and subjects it to a similar analysis, this time using contemporary designers to define new directions and their potential application in the secondary classroom. Initiatives developed by the Design Museum and Design Council provide examples of good practice.

UNIT 8.1 CRAFT

By the end of this unit you should be able to:

- challenge preconceptions and question existing orthodoxies both in school and in the wider community;
- formulate your own working definition of craft;
- understand different ways of promoting craft in the curriculum.

What is craft?

> Craft is a fluid, technological activity that cannot easily be categorised into one set of attitudes or life styles.
>
> (La Trobe-Bateman 1997: i)

> . . . an object designed and made by one person, possibly with the use of assistants . . . craft is the workmanship of risk.
>
> (Crafts Council 1998)

> Craft is like a four-letter word; it has such grungy connotations.
>
> (Blackburn 1998a: 27)

> . . . craft is a shaky business.
>
> (Perry 1998)

> . . . the decorative and applied arts.
>
> (*Crafts Magazine*)

As you can see there are many conflicting and partially formed, even ill-formed, definitions of 'craft'. This can create confusion and ambiguity.

Craft does not appear as common currency in relation to the arts until the mid-nineteenth century with the beginning of the Arts and Crafts Movements. Prior to this there were many partially formed definitions of craft. Greenhalgh (1997) charts the evolving meaning of the term. In the eighteenth century the term was used to describe political acumen and shrewdness: 'to be crafty'; the term also had connotations of power and secret knowledge, e.g. Freemasons, witchcraft. Greenhalgh asserts that craft was gradually divorced from design in the 1920s when it was, 'intellectually isolated from both the pursuits of beauty (art) and purpose (design)' (Greenhalgh 1997: 40).

Practical skills have been taught under a variety of labels and categories, handicrafts were later separated into light crafts and heavy crafts. As terminology changed previous labels became unpopular and even disparaging.

> The term 'handicraft' has developed connotations implying elementary practical activities which require a minimum of skill and thought and little knowledge to achieve success.
>
> (Glenister 1968: 6)

Berkeley (1987) suggests that an old ethos exists where practical and creative school subjects remain on the fringe. Art and design and music are exceptions as they are considered 'culturally credible'.

Brief history of craft in schools: the split between fine art and craft

As Dormer pointed out, 'A hundred years ago the link between craft knowledge and fine arts was taken for granted' (Dormer 1994: 7). Today the situation is different.

In the 1970s and early 1980s art and design teachers perceived craft as outside their remit. It was firmly located in departments of Craft, Design and Technology (CDT). In the main the work undertaken by CDT teachers was seen as belonging to a separate discipline as their training route was often radically different to that of art and design specialists. Although CDT encompassed a wide variety of activities and teaching styles it was felt that a pupil's level of understanding and overall intellectual grasp could only be comprehensively manifested through designing and making. This was managed in a linear fashion, in response to a perceived need and with a prescribed range of materials and tools. In contrast teachers of art and design emphasised a less prescriptive, more experimental and open-ended approach. Importance was placed upon process and individual responses rather than problem-solving and working towards a predetermined outcome. This concurs with Dormer's view that, 'an emphasis upon individuality in art has aided the decline of craft' (*ibid.*: 26).

During the 1980s with the growing importance of critical studies in art education, the division between art and craft became increasingly pronounced. Art was seen to promote the creative and aesthetic dimension. In contrast, craft still emphasised skills within a given tradition, the efficient use of different materials, tools and techniques. This concept of craft was not limited to schools but was reinforced by the general public.

Dormer sums up this position:

> There is a view that craft knowledge, because it is communal (it has been created by many people), conflicts with originality. The prejudice against craft tends to be crude: craft, it is thought, is bound by rules, and it is assumed that rules necessarily conflict with freedom of thought, imagination and expression.
>
> (*ibid.*: 7)

At the beginning of the twentieth century craft and art were easy to tell apart in schools. They were defined by both materials and gendered practices (Dalton 2001). Art used materials, such as paint, canvas, marble, bronze, and tended to be the preserve of boys whereas crafts used materials such as fibres, clay, wood, paper and glass and tended to be a domain for girls. Art had 'serious' content and no utility function while craft objects were made to be used. This suggests that applied art is not serious art, an entirely prejudiced view. Craft has evolved and old distinctions have begun to dissolve. At the beginning of the twenty-first century craft is seen, and respected, as a component part of the creative industries, as defined in DCMS Creative Industries Mapping Documents 1998–2004. Craft is now seen as a creative industry, alongside advertising, antiques, film, video and photography, software, computer games and electronic publishing. 'Making It in the 21st Century', a socio-economic survey of crafts activity in England and Wales 2003 published by Arts Council England and the Crafts Council, identifies 32,000 makers within the crafts industry, generating a turn-over of £826 million: 'Combining employment satisfaction and vocational stability with artistic enquiry and business risk, makers provide an illustrative model of the new creative entrepreneurs of the 21st century' (p. 2).

Craft and the National Curriculum

The NC Art Working Group (DES 1991b) were clearly of the opinion that it was, 'unhelpful to dwell on the divisions between art and craft that seem to exist at a pro-fessional level' (p. 13, 3.28) suggesting that art and craft are so intimately linked that they are best treated as one. However, as a student teacher it is important that you have an informed understanding of these issues. It could be argued that in order to under-stand how they relate to each other you must first understand their individual and par-ticular qualities. Indeed, they have different histories which overlap and interweave.

The NC Art 5–14 (DES 1991a) insisted that, 'a craft approach to materials and tools should not be subordinated to an art directed one' (p. 13). However, many schools took this as an opportunity to neglect craft, craftspeople, or craft history, as a valid component of the art curriculum. They assumed that through the realisation of ideas and intentions art and design provides insight into the properties of materials and their sensitive handling, thus craft is subsumed. The Crafts Council has been keen to dispel this idea. Teachers have been criticised for adopting a populist position in which craft is secondary to school sculpture; they have failed to afford it a place within the curriculum. The NC states: 'Art and design includes craft' (DfEE 1999: 14).

Task 8.1.1 The status of craft	Identify ways in which your PTE school interprets the NC Order. Does it see craft as subsumed within art or as a discrete element? Where is it placed?

The Crafts Council

An important contributor to the debate is the Crafts Council which has played an important role in raising the profile, status and understanding of the crafts. Following the implementation of the NC they identified the need to revisit, redefine and raise the profile of the crafts in education. This was a timely debate as OFSTED (1995a) identified that many schools give insufficient attention to this area of the curriculum. The Crafts Council has worked closely with centres of Initial Teacher Education initiating a dialogue to question preconceptions and to investigate issues and concerns in craft education, particularly contemporary craft practice. They have given a high priority to the support of critical and contextual resources in school.

The findings from the survey 'Pupils as Makers' (Crafts Council Survey Part 1: 1995; Part 2: 1998) reinforces concerns about the inward-looking nature of school practice and highlights the need for teachers to support pupils' practice through contact with professional makers and their work. It is important that you question preconceived ideas and question the boundaries through your teaching. With the support of Crafts Council publications you can move away from the definition of the crafts as merely traditional, decorative and unchanging to a recognition that contemporary crafts are dynamic and diverse akin to the work of contemporary fine art. You need to consider not only the qualities that differentiate the disciplines but also the blurred boundaries where some of the most exciting practice exists. Pupils as Makers prompted NSEAD and Arts Council England to initiate Making it Work, an organisation working between the cultural and education sectors, making the case for support for the contemporary crafts in education through projects in schools and CPD for teachers.

Between 1997 and 2007 a tranche of conferences have provided dialogue and challenged orthodox practice within the crafts. Pixel Raiders 1, 2002, and Pixel Raiders 2, 2003 looked at the impact of new and digital media on the contemporary crafts. The Challenging Crafts Conferences, 2004 and 2007 challenge traditional crafts practice and consider the implication of new technologies on craft.

The term craft brings with it connotations which art and design teachers have found problematic. At a conference of the Crafts Council in October 1997 writer and critic Pam Johnson suggested that most people are limited in their definition of the term. She insists that it is important that educators establish a shared understanding of ways of defining intelligent making (her definition of crafts) and the role and the importance of practical, creative learning. Contributing to this debate can increase your knowledge and understanding of the scope and range of professional practice in the crafts and consider the relationship between developing practical skills and knowledge and understanding of craft's historical, technological and cultural contexts. Too many teachers still define the crafts in terms of traditional practice, e.g. corn dollies, thatching or a representation of craft through reductive commercialised kits such as tapestry packs, model aeroplanes, kites.

Task 8.1.2 Defining craft

a)   In no more than 12 words give your own definition of craft.

Discuss your definitions in tutor groups.

b)   Identify from the following those terms which fit your definition of craft:

aesthetics, system, hand thrown, manufactured, concepts, product, process, mass produced, personal, unique, pattern, decoration, traditional, contemporary, precious, three-dimensional, artefact, object, model, icon, symbol, pictures, two-dimensional, realistic abstract, private, multimedia, ICT, cottage, primitive, industrial, sculpture, imagination, identity, self-expression, ideas, creative, materials, skills, heritage, non-European, reproduction, restoration, applied, hand-built.

Johnson (1997a) suggests that craft can fall into four different cultural positions. These are:

- *Utilitarian* – Made for use, distinctive, batch production, mindful of unit cost, profitable: the sort of objects and artefacts found on sale in Habitat or at a crafts fair;
- *Decorative* – Embellishment of a place or person, celebratory, sensuous, sublime, lyrical, affirming rather than disturbing;
- *Interface* – Collaboration with manufacturers, architects and designers, part of a whole, emphasis on skill and knowledge of materials;
- *Expressive* – Production of meaning, transmission of experience, narrative, challenging perceptions, defamiliarised, exploration of cultural values, associative potential of materials and forms, one-off, posing of questions.

However, Johnson acknowledges that there is a danger in fixing these categories which she claims are constantly shifting and open to reinterpretation. She recognises the crafts as a field with uncertain boundaries which encourages debates. This reflects the statement by Buck:

> Borders are never comfortable places to be but they are often where the most interesting things happen. Nowhere are the boundaries more volatile and complex than those that lie between art and craft. It is a particular feature of Western culture that such a distinction between these two branches of artistic practice exists at all; and even the most cursory examination of the art–craft dialectic throws up some disconcerting insights into our current cultural climate.
>
> (Buck 1993: 8)

Because many contemporary craftspeople are beginning to imbue their work with art concepts, they are demanding more recognition; and because many artists are combining craft materials and processes into their work, the boundaries are being blurred and constantly redefined. As student teachers you need to establish working definitions and communicate them to your pupils, recognising that you need to redefine them as they in turn become old categories and ideas.

Task 8.1.3 The crafts and visibility

In Chapter 5 you were asked to name six artists. Can you now name six craftspeople?
Now name six *contemporary* craftspeople.
Discuss in tutor groups:
    To what extent are craftspeople studied in your placement schools?
    How are they used and referenced within SoW?
    In what ways have you included craftspeople in your own SoW?

Statements by craftspeople

Michael Brennand-Wood (Plate 19)

Patterning is a visual language; a genealogical study of patterns reveals a rich history of association and meaning. I am interested in understanding the connections between the information I collate, the reasons I associate specific ideas together. The adventure is to discern *why* and make visual sense of the clues amassed. The examination of materials, both conceptual and physical, is an important aspect of my work, layered shapes and patterns can act as mnemonic devices, triggering memories of places, experiences and alternative realities. Currently I am investigating the relationship between the bio-morphic and the cellular aspect of textile constructions. Fabrics are a skin, structural elements unite to form a flexible surface, a process that parallels the evolution, growth and mutation of cells within the body.

Caroline Broadhead (Plate 20)

Over the last 25 years the emphasis of my work has evolved from jewellery, and the relationship of objects to the person, through to work that expresses ideas about the person and their relationship to the world. I have worked on ideas around the changing nature of an object when worn or not; the ability to manipulate and change the identity of an object; the potential for movement and/or restriction when being worn. I have worked with the idea of garments being metaphors for people carrying information that reveals the personality. The continuing theme is the body, the way we present, package and decorate our bodies, and how that wrapping can reveal underlying tensions and contradictions. More recently I have become interested in shadows – the negative double of the body – as an indication of a presence, or absence; the exaggeration or distortion of a person; the transparency of shadows and the way they affect the surfaces on which they fall; shadows that trace the pathway of the sun. The shadow as an image with no substance, the alter ego, the dark side, the inescapable other, gives me the opportunity to focus on the non-material. By giving 'body' to a shadow I am giving more importance to the things that are not immediately apparent.

Thomas Heatherwick (Plate 11)

For me, the process of making – the hands on experience of creating things – is the most important way of learning. Making helps us to better understand and appreciate the world in which we live. Everything that surrounds us is made, either by man or nature. We can all learn through making. It informs the link between abstract concepts and physical reality. Through making we can learn to solve problems, adopt more practical ways of thinking, understand the way things work, the nature of materials and the relevance of theories and mathematics, as well as witness the realisation of our ideas. I believe that making should be an integral part of the curriculum for all schools and colleges, not just as a means of teaching art and design, but as a source of learning and inspiration about the world around us.

PGCE Student (Plate 10)

My sculptural ceramics were inspired by a five-month placement at the Bezalel Academy of Art and Design in Jerusalem. The 'cocktail of nationalities' I discovered on my placement enabled me to study different cultures and ceremonies. I grew especially interested in the Jewish 'Havdallah' Spice Boxes. After the Sabbath, Jews light the spices contained in the boxes as a sign of renewal for the coming week. These ceramic sculptures are made from porcelain clay; thrown, hand built and press moulded from plaster casts of fruit and bubble wrap. They are decorated with oxides and coloured clay slips, then salt and lustre fired to 1280°C and 780°C.

Helen Storey (Plate 12)

In June 1997 I embarked on a project with my sister (a developmental biologist at the University of Oxford). Our work together was instigated by The Wellcome Trust's Sci/Art initiative – a concept which sought to bring artists and scientists together to produce bodies of work that could explore each others' worlds and capture the public's imagination at the same time. Our collection entitled 'Primitive Streak' portrayed the first 1000 hours of human life in textiles and clothing. It has been through the experience of this project that I have come to understand, and fully appreciate, first hand the power of visual communication. This work has succeeded in a way none of my previous work has. It underlines the vital need for an education system that encourages and harnesses the skill of the eyes; not just as a means to copy or take in information but as a gateway to engage an individual's imagination and the ability to make personal sense of the world around us.

Task 8.1.4 Responding to craft practice

Consider the statements and work (Plates 10, 11, 12, 19 and 20) of the above craftspeople.

1   How would you develop a SoW which has as its focus one of the above?
2   How would you integrate one or more of these craftspeople into a SoW?
3   Can the methodologies used for critical and contextual studies for fine art be applied directly to craft?

Recycling, scrap, reinvention

> Craft finds us the diamonds in the landfill.
>
> (Press 1996: 13)

The idea of recycling materials is not new. People from all ages and in all cultures have recycled as a necessity for their existence. Denise Mucci Furnish, a contemporary quilt maker, compares her work, which includes lint from electric clothes dryers, with the earliest examples of quilting found in ancient China. The Chinese believed that the wearer's spirit lived on their clothes: worn out fragments were stitched onto new fabrics and the cycle began again. Today this idea of reusing has become an aesthetic in its own right.

Recycling is currently one of the important issues in contemporary crafts. An awareness of environmental issues has led to a reassessment of materials and an investment in recycling technology to which craftspeople have responded with both interest and ingenuity. It is important to establish a positive attitude to the reuse of materials by presenting examples of craftspeople who have continued to develop a recycled aesthetic: Jane Atfield, Lucy Casson, Michael Marriott, Tejo Remy, Lois Walpole. The use of recycled materials should always be carefully considered and selections made on the basis of appropriateness to your projects. They should not be regarded as an easy, cheap alternative in which financial and resourcing implications may outweigh pedagogical concerns.

In the Catalogue 'Recycling' Press (1996) states that the value of the stuff people throw away draws attention, in a way only the crafts can, to the properties, uses and abuses of materials to create models of 'sustainable enterprise in an un-sustainable wasteful industrial system' (Press 1996: 14). The following statements provide differing rationales for recycling:

> I use recycled and found materials for environmental and aesthetic reasons . . . I have always found it liberating to use these materials because it takes the angst out of experimenting.
>
> (Lois Walpole in *ibid*.: 52)

> I utilise found objects and do not exclusively recycle but start by making shapes from existing objects. They come from a wide variety of sources including plumbing shops and cookery stores. The biggest challenge would be trying to get a client to pay a commercial price for a piece of scrap.
>
> (Tom Dixon in *ibid*.: 80)

> My primary reason for using reclaimed materials is aesthetic. I like the quality of old wood and printed time. If it's been washed by the sea too then so much the better. Obviously the economic factor is a bonus and it can allow more experimentation without fear of expensive failure . . . working with something that's already had a life and that past history can sometimes work against you as well as with you.
>
> (Kirsty Wyatt-Smith in *ibid*.: 56)

> I am sculpting with recycled tin plate, wire, wood, plastics and found objects. My way of working has been to discover as I work, evolving different methods of using the materials to obtain a certain effect, my techniques are discovered through messing about with metal. I like to use non-precious materials and simple techniques. The fact that it is a non-precious material enables a freedom that a similar new and therefore expensive material would not allow.
>
> (Lucy Casson (Plate 9) in *ibid*.: 48)

Task 8.1.5 Using scrap materials

In tutor groups discuss the above statements and their application to the classroom. Devise a SoW based on one of the concepts about 'scrap' offered by these three craftspeople. Support your SoW with the work of others that relates to these categories.

Comments: attitudes and values

Consider the following statements:

> Outside the insular world of contemporary crafts, the word 'craft' carries all the associations which are guaranteed to hasten its educational decline. Craft in our culture is dumb. It is associated with manual skills, domesticity, occupational therapy and our less than smart school children. In other words, crafts is working class, female, mentally unstable or just plain thick – and very usually all four. It succeeds in scoring highly on all the prejudices of a dominant culture that is anti-working class, patriarchical and believes that intelligence is only demonstrated, and thus can only be measured and valued, through the use of words and numbers – most especially when the latter is preceded by pound signs.
>
> (Press 1997: 47)

> Craft today is seen as an accessible quasi-art form. It is essentially a middle class concept. Crafts are purchased by the educated elite – they buy craft objects because they are 'one-off' or limited editions and therefore expensive – Liberty's posh!
>
> (Johnson 1997b: 42)

> Craft practice promotes a set of values that recognise 'making' as a basic human need – values that will become increasingly important as we progress through an information age; Crafts, education and practice integrates creativity, problem solving and manufacture, a process which facilitates innovation – and therefore has an important contribution to make to the needs of industry and the business world.
>
> (Ball 1998: 4)

Task 8.1.6 Self-positioning in relation to the crafts

In tutor groups discuss the above statements as a starting point to debate your own position.

As you have discovered, personal narratives, informed and developed by personal experience and prior learning, including class, gender, race and sexuality, condition your position to and within art, craft and design. It is essential that you question your own preoccupations and identify existing school orthodoxies and their current craft practice.

Gombrich (1992) insists that people's judgment of what they see is affected by the mental framework they use when looking. He suggests that if people are told that something falls into the craft category their attitudes towards it are going to be different than if they had assumed it to be 'fine art' (p. 27). Blackburn (1998b), organiser of two selling exhibitions of modern crafts for Sotheby's, admits that getting people to take crafts seriously as an art form is an 'uphill struggle'. She states: 'I've known wealthy collectors who admire something and ask the price, you say £1,500 and they almost pass out with shock. But they look at a painting for £20,000 and can't get their hands on it quickly enough' (p. 24).

It is worthy of note that the persistence of this hierarchy is a particularly British phenomenon. Michael Brennan-Wood (Plate 19) who has been at the forefront of British textiles for over 20 years, is better known abroad, particularly in Japan.

Politics

Attitudes and values are in fact perpetuated, albeit unintentionally, by government institutions and organisations such as the Crafts Council, the Arts Council and the Design Council, and, to an extent, the Engineering Council. They are responsible for establishing boundaries in relation to their funding. The Crafts Council is structured around materials: ceramics, glass, wood, metal, textiles, etc. The subsuming of the Crafts Council under the auspices of the Arts Council is likely to have a profound effect on attitudes. A new cultural body is called for in a paper published in 1998 *The Comprehensive Spending Review: A new approach to investment in culture*, where it is posited that there is a need to break down artificial barriers within art forms. Johnson states:

> The categories of art, craft and design now strain to contain the diversity of hybrid objects circulating within visual culture. The hybrid textile object tests the limits of our current exhibition funding structures . . . why not a Council of Visual and Material Culture and with it a more fluid use of exhibition space?
>
> (Johnson 1998a: 26)

Practitioners have for many years endured feelings of inferiority and are unable to articulate and theorise about their work, 'their craft is a wallflower at a cultural ball' (*ibid.*: 34).

Social, class divisions

Class divisions are also significant. Parker and Pollock (1981) state that traditionally, history of art constructs hierarchical classifications for art and craft separating artist from artisan. Fine art becomes the possession of the privileged classes while the applied art or craft is associated with the working class and the unpaid work of women.

> The art/craft hierarchy suggests that art made with thread and art made with paint are intrinsically unequal: that the former is artistically significant. But the real differences between the two are in terms of *where* they are made and *who* makes them.
>
> (Parker 1984: 5)

In the middle ages women worked alongside men in embroiderers' guild workshops. By the eighteenth century such work was perceived as a female occupation and by the nineteenth century it was recognised as the pursuit of women of wealth, fine stitchery, or the skilled work of working class-women for which they were poorly paid.

During the Industrial Revolution mass-made factory products separated out skill from knowledge and understanding. The workers on the production line, unlike their predecessors, had no real understanding of the objects they were making. Each worker was restricted to the knowledge required to produce a small part of the production. The worker was no longer a craftsperson but 'an animated tool' (Frayling 1990: 163).

Gender

It is impossible to write a unit on crafts without addressing gender. In the past craft has displayed a gender divide, which was based on historical circumstances. Right up until the 1980s schools' curricula dictated that activities such as wood and metalwork should be followed by boys, while girls were assigned needlework and cookery classes. Traces of this gender divide continue to linger in many schools today. It is your responsibility to challenge these traditional attitudes towards gender and examine any reoccurrence of stereotypical attitudes. It might be useful to look back at the list of craftspeople you provided earlier in the unit. (How) does your list reflect these issues? Have you provided an equal balance between men and women craftspeople?

In a short succinct paper, Dalton explains the position of craft in schools in the 1980s. She claims: 'In schooling the conditioning process continues to train young women into habits and skills that make association between women and textiles a natural one' (Dalton 1987: 31).

Attempts to include boys in textiles and to afford equal status and value to metal, wood, plastic, fabric, food and clay, have failed to change the divide which equates boys' craft activity with the workplace and girls' craft activity with domesticity. Dalton reveals how fashion and textile manufacturers further perpetuate this divide by recognising the potential to exploit a young female audience. They provide textile teachers with ready-made packs, including wall charts, visual aids and cheap patterns. Girls are taught garment and soft-toy making through a systemic use of pre-designed, precut paper patterns.

Shreeve (1998) suggests another reason for this gendered divide. She claims that characteristics of tacit knowledge, such as intuition, hunches, knowhow, are ascribed to the feminine and given lower value. Society in general places more emphasis on explicit knowledge and craft, for its sins, falls into a less-valued category.

Parker (1984) cites the way advertising and the media constantly reinforce a gendered view of the crafts. The changes in the secondary examination system in the 1980s, from GCE to GCSE, provided an opportunity for educators to reconsider stereotypical gender divisions. The NC Art proposals (DES 1991a) reiterate the need to remedy earlier divisions insisting that it is important for teachers to select methodologies, topics and materials which attract the interest of both sexes. The document claims that it is important to present pupils with examples of the work of artists, craftworkers and designers of both sexes insisting that this way, 'both girls and boys can grow up knowing that the full spectrum of media techniques and skills is open to them' (p. 60).

You must avoid the notion that gender stereotypes in the crafts are unusual. They are not. Roles are often reversed across history, place and time, e.g. in sixteenth-century England knitting was primarily a male occupation and in West Africa the Fori men of Dahomey produce woven cotton cloth which is used for appliqués. Feminism in the arts during the 1970s and 1980s protested against the distinctions between 'art' and 'craft' grounded in different materials, technical training and education. The way activities such as sewing, piecing, cutting, applique have been assigned in history to women, despite the fact that men also practised them, can be seen, in part, as responsible for the shift in attitudes in education. Chadwick claims that during the 1970s distinctions between 'art' and 'craft' began to 'fray around the edges' (Chadwick 1990: 332). The movement known as Pattern and Decoration, which attracted both men and women, used fabric and surface elaboration partly as a reaction to the gender-biased use of the term 'decorative'. Judy Chicago's *Dinner Party* (1979) blurs distinctions. It celebrates women's historical and cultural contributions in a large installation which incorporates sculpture, ceramics, china, painting and needlework (Chicago 1996). Similarly, Faith Ringgold and Miriam Schapiro explore issues of exploitation, race and gender in their work (Chadwick 1990). Jefferies (1995) insists textiles is a hybrid in the expanded field. She quotes Sarat Maharaj's essay, 'Arachne's Genre: Towards Intercultural Studies in Textiles', which argues that avant-garde textiles practice, 'maps out an in inside/outside place – "edginess" – it cites established genres and their edges as it cuts across them', to throw out of joint, 'handed-down notions of art/practice/genre/gender' (*ibid.*: 168).

Learning through intelligent making

Theories about learning can be called upon when discussing the crafts. Perhaps the most widely used excuse for not linking thinking with making in craft is the belief that learning in the crafts is tacit. The dictionary defines tacit as: silent; passed over in silence; unspoken, implied or indicated, but not actually expressed; arising without express contract or agreement. So, within the context of making in craft, tacit learning could be thought of as the ability to do something, to weave willow, to stitch fabric, to weld silver, as having an inherent ability to work with one's hands and eye. It was

Polanyi (1964) who insisted that people know more than they can tell, that tacit learning includes: implicit thinking, intuition, hunches, skilled performances.

Polanyi's explanation of tacit knowledge, as the kind of knowledge that cannot be fully articulated, is commonly used by makers to account for their difficulty in explaining how they have worked their materials and what skills they have employed in making. As Dormer reminds you: 'Craft relies on tacit knowledge' (Dormer 1997: 10). Tacit knowledge of craft making is acquired through experience and it is this knowledge that enables pupils to do things as distinct from their talking or writing about them. Closely allied to this theory is that of implicit learning or the inherent ability to learn from a task without rational or conscious deduction. In education, explicit knowledge is the given. The unseen and the unspoken aspects are under-valued or ignored within art and design. Often you find that pupils have, what appears to be, an innate ability to work with a material and realise a successful outcome without being able to articulate how they have done so.

Haptic knowledge has close links with both tacit and implicit learning. Haptic knowledge is that which you acquire through tactile and kinaesthetic physical sensa-tions. It depends on feeling and touch as a means of communication. This sense of touch can be undervalued in art and design. Shreeve (1998) provides a powerful poetic example of a student working on a piece of embroidery. She describes the way different parts of the body respond to the machine and production of the work:

> Without conscious effort, the eyes, hands and feet all respond and adjust minutely to changes in the total production of the piece. The body gently moves in response to the work, the performance of the machine and to the desired outcome. Fingers walk along the fabric to gently guide and obtain feedback from the embroidery process. At no point has the student needed to articulate exactly what is required; this has been processed at a level below the conscious threshold.
>
> (Shreeve 1998: 43)

This example highlights the importance of Kinaesthesia, the faculty of being aware of the position and movement of parts of the body. In addition, the synaesthetic response should not be overlooked. Sounds of tacky ink sticking to the printing roller, the purring of the sewing machine or the smell when flux melts letting in the silver solder, are part of and inform intelligent making. Shreeve suggests that sometimes language has no function in craft learning, she identifies simultaneous doing and speaking as difficult.

Task 8.1.7 Tacit and haptic learning

In your experience of art and design what other examples can you give that rely on tacit, implicit and haptic learning?

Watch a teacher demonstrating a making activity, e.g. throwing a pot, carving a piece of wood, using a sewing machine. Identify the extent to which s/he uses language to explain the activity. Do you think that language is always necessary?

Crafts in context

> In many cultures the clear distinction between the major or fine arts
> (painting and sculpture) and the minor or applied arts (craft), is an
> irrelevance. The European Renaissance established this division by raising
> the fine arts to a liberal, and thus cerebral, art, fabricating its intellectual
> credibility through a supporting body of theory, and debasing its sibling by
> designating it mechanical . . .
>
> (Addison 1997: 4)

Traditionally in many cultures the hierarchy of crafts or the western representation of a divide between art, craft and design does not exist. Crafts are not special, pigeonholed as particular representations or with particular domains. Teachers have for many years recognised that the visual and material cultures other than from Europe and north America provide a rich resource for teaching and learning in art and design. The NC (DfEE 1999) states that pupils at KS3 should 'be taught about: . . . artists, craftspeople and designers from Western Europe and the wider world . . . in a variety of genres, styles and traditions, and from a range of historical, social and cultural contexts' (pp. 20–21). However, you must recognise the need to locate artefacts from different cultures in context, avoiding decontextualised comparisons and western categories of art and craft. You should realise the danger of projecting a western aesthetic.

You need to move away from formal analysis as a way of perception and think about the following questions: What materials are used? How are they made? Why are they made? And by whom?

Criticism

> They say: 'Learning a craft is easy, it's thinking that's difficult'. Yet in learning
> a complex craft it is the head that hurts, not the hands.
>
> (Dormer 1991a: 38)

The identification of ways in which the crafts contribute to an understanding of contemporary, visual and material culture has preoccupied makers and crafts critics throughout the last decade (e.g. Harrod 1997; Johnson 1998b; Greenhalgh 1997; Dormer 1997; Rowley 1999; Jefferies 2001).

It is rare to find an art and design department that differentiates between art and craft in their approach to interpretation and evaluation. One of the main anxieties around craft criticism is the attempt to talk about a wide range of practices as if they were one thing. Johnson (1995) asks us to consider models of craft criticism and questions whether it is appropriate to use art historical models, 'Let us hope that the critical debate will go forward by acknowledging and outlining the differences and contradictions rather than trying to resolve them prematurely or, indeed, to hide them away' (p. 44).

Task 8.1.8 Critical evaluation of art and craft

In tutor groups discuss:
Do you need to adopt different critical approaches to art and craft?
What possibilities or problems does a differentiated approach produce?
Does such differentiation serve to reinforce or question hierarchies?

Secondary school pupils are fascinated by, and surprisingly well informed about, contemporary technology. However, they often fail to make the connection between innovative practice and the crafts they experience at school. As student teachers you need to ensure that you don't only perpetuate the traditional definition of craft. Stungo (1998) suggests that you:

> . . . envisage a world where fabrics are not inert materials to be cut and stitched but live, intelligent substances; a world where dyeing is redundant because natural fibres can be grown in any colour, a world where fibres carry vitamins and sunscreens in their weave to protect your skin and clothing comes with built in pollution monitors, mobile phones or just any other kind of technology that you care to think of.
>
> (p. 26)

Ways forward for craft

In an article written in 1998 we tentatively suggest two alternative ways forward:

- to foster craft as dependent on a distinctive form of intelligence, with a unique but alternative way of knowing with its own history and traditions, secure within its own boundaries;
- to promote craft as an important and integral facet of contemporary culture; a field that is not static or fixed but is sufficiently confident in its distinctive features to reject limiting boundaries and transcend the restrictions of conventional categorisations.

(Burgess and Schofield 1998: 129)

The first promotes learning *in* the crafts while the second emphasises learning *through* and *in* the crafts. Our preference is for the latter: which do you choose?

Crafts Council

In addition to local and national museums, resources and publications concerning the crafts are available from the Crafts Council. It has an extensive reference library, a reference desk and photostore (www.craftscouncil.org.uk).

- Reference Library: contains over 4,000 books, 100 journals and 40 videos;
- Reference Desk: provides listings and contact details of craftspeople in any crafts discipline throughout the country;
- Photostore: is an interactive, user-friendly picture library system which enables visitors to search quickly for images and information by maker, object, material and technique. Containing over 35,000 images of contemporary British craft from architectural glass to toys and spanning 20 years.

Further reading

Burgess, L. and Schofield, K. (1998) 'Shorting the circuit', in P. Johnson (ed.) *Ideas in the Making*, London: Crafts Council.

Crafts: The Decorative and Applied Arts Magazine.

Dormer, P. (ed.) (1997) *The Culture of Craft*, London: Crafts Council.

Jefferies, J. (ed.) (2001) *Reinventing Textiles: Gender and Identity*, Winchester: Telos.

UNIT 8.2 DESIGN

By the end of this unit you should be able to:

- understand why design remains an area of neglect, contention and confusion;
- develop different ways of promoting design in the curriculum.

James Dyson, a trustee of the Design Museum, claims that he is so concerned about what he sees as a national failure to inspire upcoming generations to develop design that the rest of the world wants to buy that he intends to spend £12.5 million establishing the Dyson School of Design Innovation in Bath with its first cohort of 900 16–18 year olds moving into the purpose built premises in 2008.

Design Education: Now You See it Now You Don't, the report of a seminar organised by the National Society for Education in Art and Design (NSEAD) and the Design and Technology Association (DATA) at the Royal College of Art in July 2006, points out that design education in mainstream education continues to be much neglected. During the seminar delegates, educators from both Art and design and Design Technology, voiced their concern about the status of design in the secondary curriculum and its failure to keep up to date with the rapid development of design in the real world. This neglect is repeated in the subject section in the Annual Report for 2004–05 of Her Majesty's Chief Inspector of Schools which makes no specific mention of design education despite it supposedly being a key element of 'art and design'. However, it does note that the art and design curriculum has narrowed, and suggests '. . .more needs to be done to transform the relationship between art and design in schools, the creative industries and the cultural sector' (OFSTED 2005).

Established narratives of design, like those of craft, are being challenged and deconstructed. Myerson (1993) suggests that design as an object fixed in time and space, has been replaced by design as a process, fluid, changing, perplexing, and increasingly unable to be contained by traditional disciplines or methodologies.

Design is one of the most contested areas of the curriculum. Its position and status have shifted regularly over the last three decades.

Task 8.2.1 The status of design

In tutor groups: think back to when you were at secondary school.
  What did design mean to you then?
  Did you perceive it as an integral element of the Art (and Design) department's teaching or was it located in a different area of the curriculum?
  Was it part of Craft, Design and Technology (CDT) timetabled with Home Economics, or, an aspect of Design Technology (DT)?
  Did it part company with Craft and if so why?
  How is design defined in secondary education now and in what ways is this definition still shifting?

You may identify a lack of consensus and begin to appreciate the problems associated with attempts to classify this elusive and changing subject. In order to understand how this situation has arisen it is important to look briefly at the recent history of design education. Even today teachers working in different departments in secondary schools use the term 'design' to mean different things; sometimes as a verb, sometimes as a noun. This confusion has no doubt arisen from the way design has been shifted from one subject area to another. The field, too, is vast, including architecture, fashion, ceramics, silver, furniture, interiors, town planning, graphics, environments, systems.

In the 1970s design was, in the main, perceived as part of CDT where the work undertaken by CDT teachers was seen as a separate discipline from art. Here, an emphasis was placed on identifying particular aspects of the design activities and this was evidenced by the teaching of specified skills, materials and outcomes; a process model. Pupils were taught the purposes of products and systems and how this knowledge might be applied to their own designs. Through this, pupils improved their *manual* dexterity and developed an ability to solve specific design problems. This rigid model was not wholeheartedly endorsed by art and design teachers.

The CDT GCSE Guide for Teachers (1986) emphasises particular aspects of the design activity through its assessment objectives which focus on the following sequential development:

- identify and analyse problems
- generate ideas and potential solutions
- plan the production
- produce the selected solution
- compare and evaluate against its specification.

(p. 1)

Integral to this is a requirement to foster economic and industrial understanding. With this in mind pupils are taught to identify needs, both in terms of the uses and purposes of products and systems, and how this knowledge can be applied to their own designs. Additionally, stress is put on discovering the efficacy of different materials, tools and techniques. Although CDT encompasses a wide variety of activities and teaching styles, it subscribes to the principle that a pupil's level of understanding can only be manifested through designing and making in response to a perceived need and with a prescribed range of materials and tools.

The late 1980s saw the growth of large faculties in some secondary schools which incorporated art, craft, design technology and home economics, and in some cases the performing arts. These faculties often operated a carousel system, which some schools have retained. They function by dividing the curriculum into discrete, six to nine week units. Sometimes these are brought together with an overarching theme or common topic, entitled, for example, 'local architecture' or 'old and new'. This arrangement gives all pupils a taster course working with different methods and materials, but it tends to remain superficial. For it to succeed, it requires rigorous co-ordination and time for teachers to plan, review and evaluate. Such arrangements are not without their problems. Teachers may have different degree specialisms, philosophies and ways of working. An uneasy relationship can be formed between subject areas which may have little in common. However, most educators now regard carousel arrangements to be no more than an administrative convenience, guaranteed to ensure that there is little depth, coherence or progression. As Thistlewood (1990) remarks: 'Design is not easily taught; it is far simpler to teach a form of pseudo designing in which pupils draw and make pastiches of existing consumer products' (p. 6).

Design and the National Curriculum

> It [Design and Technology] is a messy story of political and ideological meddling combined with a good deal of confusion and frustration in schools.
>
> (Kimbell 1996: 29)

With the development of the NC Subject Orders following the Education Reform Act (ERA) (1988), the carousel approach was largely abandoned in favour of discrete areas. The ERA proposed that art and design, and design and technology (D&T) become independent. However, the legacy remains in some schools, for example specialist rooms and teachers. Following many discussion documents and views fielded by the Design Council, the Engineering Council, the Crafts Council and the NSEAD about the ways in which art and design could contribute to the proposed NC Design and Technology Order, design became decoupled from art and design and embedded within a separate and often discrete new NC subject, design and technology. It is interesting to note that D&T was not originally intended as a subject in its own right, but more as a vehicle for cross-curricular work. This is in line with the Design Council's view that design is a whole school issue.

The NC, D&T Order (DES 1995) has driven a specific approach to design through knowledge of engineering and technical instruction linked to industry via vocational pathways. It promotes a particular way of working, that of playing the role of 'designer as producer'. Less emphasis is placed on a critical understanding of the manufactured world.

This is not to devalue what can be taught successfully. Through group work, pupils can actively investigate a design brief and by working with materials and processes, create a 'quality' outcome which draws on their existing knowledge and experience. Such an approach develops an understanding of research as well as fostering inter-personal skills. This mirrors the way a real life design team works, where each designer co-operates and collaborates by 'chipping in' something from their own experience and imagination to add to the final outcome. Pupils can be empowered if they are given control over the process of designing through the teacher fostering an environment for creative questioning, intelligent making and pertinent evaluation.

All too often design in schools leans towards the proliferation of limited and pre-scriptive design work reflecting the need for quantitative and comparative assessment outcomes rather than process-led open briefs. If a simulated 'neo' design for public view is to be made by pupils it needs to be costed, trialled, evaluated and made to work, activities which have social and educational benefits.

The problem-solving approach

> 'Problem solving' has, over the years, had a very full treatment from psychologists, educationalists and philosophers, seeking to describe and explain the processes of thought. Thompson (1959) outlines the characteristics of problem solving behaviour; Kubie (1962) relates it to a cybernetics theory of learning; Vygotsky (1962) links it to the development of language and particularly to symbolic representation; Rogoff (1968) relates it to schools and education, 'a difficulty is an indispensable stimulus to thinking'.
>
> (Kimbell *et al.* 1996: 30)

Since the 1980s a particular approach to design teaching and learning has become a recognised model of good practice in secondary schools. Green (1982), Baynes (1982; 1990), Kimbell *et al.* (1996) and Eggleston (1996; 1998) have all promoted variations of the problem-solving approach. It has been successfully adopted by both art and design, and technology departments.

Problem-solving can provide a framework for experimentation and learning, where pupils are active participants. It is the teacher's role to provide a design brief, which replicates, within the classroom, the idea of the client who requires an identi-fied need to be resolved. This may be, for example, a logo design for a brand of sports shoe, a local council requiring its bus shelters to be redesigned, or a wall hanging for a local hospital. Pupils at this stage begin by brainstorming and concept mapping using both verbal and visual means. This can be done individually, or in pairs, and can be discussed later with the whole group in the classroom. Pupils need to map their ideas, keeping in mind the possibilities and limitations of suitable processes and materials along with historical precedents, ergonomics and social, economic and environmental concerns. After considerable discussion, interviews, research, reading and drawing, pupils can reflect upon their findings in order to propose a number of tentative solutions to the given problem. They may find it useful to present these to small groups within a whole class situation. Reflecting upon their findings and the views of others can lead them to select materials and processes and produce a prototype.

This model of design learning combines the active with the reflective. As the problem-solving task becomes clearer the constraints on the solution are subject to critical analysis. Kimbell *et al.* (1996) point out that there is an astonishing similarity between the design process and more general thought processes. They suggest that the introduction of a problem disrupts routine and forces the learner to '. . . stop drifting and think about what they are going to do . . . this creates the cognitive conflict that is essential for a subsequent reformulation of a new, more comprehensive schema' (p. 30).

This process of reflection on action is described by Schon (1987) (see Chapter 2). Often the problem-solving process is perceived as a linear model which starts with the identification of the problem and ends when the solution has been evaluated. Figure 8.2.1 illustrates this diagrammatically:

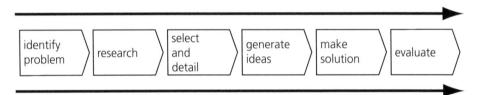

Figure 8.2.1 Problem-solving process (after Schools Council Design and Craft Educational Project, 1972)

The strength of this way of working is that thought- and decision-making processes are made concrete. Kimbell (1996) explains how, by looking through pupils' design folders, their decision-making processes are revealed. He suggests this facilitates the following:

- You can go back over it with the pupil to examine where critical decisions were made.
- You can look to see the basis of evidence for that decision.
- You can examine points at which alternatives would have been possible.
- You can use these as jumping off points into new lines of development.

(p. 31)

We consider that all the above points should be recognised as a two-way process or dialogue between pupils and teachers, extended through group evaluation and paired discussion between pupils.

However, this activity has its limits. Pupils are often given a design brief without their having knowledge of the socio-economic and cultural values through which real designers operate. You need to ensure that problem-solving encourages pupils to stand outside themselves and consider the needs of others, recognising differing tastes and values.

Art and design and/or design and technology: their positions in schools

Unless you advocate the role of design in art and design most headteachers are unlikely to recognise your practice as a key contribution to technology in the curriculum. If art and design is to be adequately resourced to maintain its present wide range of activities and experiences it is necessary to argue this case by providing evidence of good design practice within art and design departments. Teachers of art and design are keen to identify the contribution that they can make to D&T, in particular, teaching pupils to formulate design proposals, to apply aesthetic judgements, to evaluate products and to work with materials and ICT. D&T can be seen as an important component in the art and design curriculum and teachers have sought to establish the relationship between these subject areas and ways that one might complement the other. However, others have suggested it is not which 'bridge to build but whether there should be a bridge at all' (Buchanan 1993). Buchanan suggests that the D&T curriculum is, 'too dull, too mechanistic, too linear and not enough about design' (*ibid.*).

However the D&T Working Group Interim Report (DES 1988) identifies the role art could play within this foundation subject:

a) . . . a greater emphasis on creativity, curiosity, imagination and the aesthetic dimension.

b) a recognition . . . that the way of thinking which governs the route leading from ideas to tangible designs often involves a considerable amount of intuition. Many of the most innovative ideas are initially intuitive and subsequently are rationalised to meet a variety of constraints.

c) an ability to analyse the relationship between design and environmental and social concerns.

(ibid.)

Six years later Steers (1994) points out that in D&T the creative and aesthetic dimensions are still neglected. He claims: 'It is doubtful whether the [technology] proposals, constrained as they are by specified skills, materials and outcomes, would inspire adequately pupils to experiment, challenge assumptions and take conceptual risks' (pp. 3–4).

Art and design Newly Qualified Teachers (NQTs) often find themselves working in art, craft, design, technology faculties as specialists in either creative textiles, 3-D or graphic design. Therefore it is essential that you are familiar with the D&T Orders and examination syllabuses.

Task 8.2.2 Design in schools

In your placement schools look for evidence of design practice:

- in the art and design curriculum;
- within school documents;
- on public display;
- in the technology department.

Task 8.2.3 Questioning the status of design

Consider the following statements and use them as a starting point to help you devise a questionnaire for both the Head of Art and design and the Head of Technology.

- Design is future-orientated.
- To design is always to develop some form, structure, pattern or arrangement for a proposed thing, system or event.
- Design is problem-solving.
- Design education must address values. It should be informed by environmental, social and moral issues: not which bridge to build but whether there should be a bridge at all.
- Design is a real-life activity.
- Design is as much about aesthetics as function.
- Design is designer labels.
- Design requires an understanding of skills and techniques.
- Design makes the boring beautiful.
- Design is central to all our lives.

This questionnaire should enable you to identify the position on design that has been established in your PTE schools.

How is design education perceived (high or low status, from differing perceptions and philosophies)?

Where is it located in the curriculum (as a discrete subject or cross-curricular)?

Share your findings with other members of your tutor group. Is there any consensus of opinion?

What are your views of design and how it should be taught? Find an opportunity to discuss these issues with your SCT or other colleagues.

Baynes (1990) claims: 'art and design are not synonymous. They are not identical. Art and design is only one pair. Technology and design is another of equal importance' (p. 51).

Rigid boundaries between art and design do not exist, nor would we want to draw strong demarcations between them in this chapter. The 1980s was the time when design filtered into every aspect of urban culture and an obsession with labels continues to pervade the classroom. It is important that you ensure that you do not become 'prisoners' of the present, unable to foresee alternatives or recognise possibilities of choice. You need to ensure that pupils are introduced to design histories including their wider social and cultural implications.

Task 8.2.4 Craft and design links

In discussion with other student teachers reflect on what you have read about craft earlier in this unit to consider the links between craft and design.

How might you contrast and compare these terms which are often used indiscriminately in schools? Both terms can be used as nouns or verbs.

How have you used these terms in your planning and teaching?

As a noun, verb, interchangeably, independently, separately? Why?

Drawing and designing

Perhaps one easily identifiable difference between craft and design is that designers are more likely to start with drawing. Drawing, it is generally agreed, is a fundamental element in the design process, and as Kimbell *et al.* (1996) so aptly remind you: 'it is no coincidence that designers frequently talk of themselves as "thinking with a pencil" ' (p. 30). This is in direct contrast to the craftsperson who often starts by working directly with materials. Design drawing takes the form of exploration and communication, it enables the designer to both organise thoughts and express ideas to others. Garner (1990) points out that the drawing strategies employed by the designer aim to: 'explore problems, manipulate information and visualise responses', and that these have no clear divisions between them; he states:

> Whilst a designer may be exploiting drawing creatively and personally, these same drawings may communicate form, detail, scale, or other information quite readily. Similarly much sketching activity is used to simultaneously clarify conceptual development, facilitate evaluation and provoke the further generation of ideas.
>
> (p. 51)

Although new software packages (CAD) enable designers to draw and visualise electronically, the designers' sketchbook is still one of the most important tools used for generating ideas.

Task 8.2.5 Drawing and design

Design a SoW in which a need is resolved through a process of sequential drawing.

Socially responsible design

This is an age of mass production in which everything is planned and designed. Design has become very powerful in that it demands a high social, moral and ethical responsibility from the designer. Design must become more responsible, more creative and more responsive to people's real needs, and not merely respond to the demand for designer artefacts, labels or more throwaway objects. Now that many things have become possible through technology, you as a teacher need to alert your pupils to the ethical and moral issues surrounding design and its marketing power.

The most public champion of green design is without doubt Papanek. His seminal text *Design for the Real World* (1971) calls for a responsible design movement. Papanek is concerned about the mismatch between the power and influence of design and the lack of social and moral responsibility shown by many professional designers. He bemoans the fact that design is dominated by consumerism and profit. He proposes an agenda for design which comprises the following five priorities:

- design for the Third World;
- design of teaching and training devices for disabled people;
- design for medicine, surgery, dentistry and hospital equipment;
- systems design for sustaining human life under marginal conditions, e.g. polar icecaps, underwater, deserts;
- design for breakthrough concepts, a radical rethinking of approaches to design, rather than the more conservative approach of continual refinement of existing products.

Papanek has been described as unnecessarily reactionary and sensationalist by some design critics who dismissed him as 'a cult figure' while ecology was fashionable during the early 1970s. Others have heeded Papanek's call for social responsibility. These include: the Unit for the Development of Alternative Products (UDAP) based in Coventry; Sheffield Centre for Product Development and Technological Resources (SCEPTRE); London Innovation Limited (LIT) and the Centre for Alternative Technology, Wales.

Task 8.2.6 Designing for social responsibility

Working in small groups, develop a visual resource pack for use in school which promotes socially responsible design. For example, you might include products or systems designed for the physically disabled, promotional literature for a specific local environmental concern, products which rely on natural power to function.

Using this resource as starting point devise a SoW for Year 10 students.

Critical, historical and contextual design studies

Consider Attfield and Kirkham's (1995) distinction between art and design:

> Designed objects surround us and form part of our culture in a much more intimate way than art — we interact differently with artefacts like clothes, cars, buildings, streets and food packaging directly in our everyday lives. But

because we encounter these in a much more mundane way – we wear shoes, walk on carpets and drink coffee from mugs – design tends to become invisible and absorbed into the less thinking part of our daily existence. Art, on the other hand, has to be sought in special places such as galleries where it is set apart for contemplation and appreciation. While the fashion and style aspect of design is so self-conscious and visible, depending as it does for its very existence on ephemerality, the fascinating depths of meaning it embodies are disguised and hidden beneath its apparent triviality.

(p. 1)

Design understanding and decisions are often taken by pupils in school in a vacuum. The need for a design education that links the study of formal and functional elements with training in visual language, communication and contextual studies has been recognised for some time. Baynes (1982) advocated this when he said: 'I want to look towards social and cultural aspects of design education and away from the debate about the content of the curriculum. I am disenchanted with it, it seems to me that it has rapidly become sterile' (p. 106).

Buchanan (1990) reiterates the importance of an extended notion of design education when, nearly a decade later, he bemoans the fact that, 'Little has been done in most schools to introduce pupils to the work of designers . . . to explore the range of influences on their work or to judge or value directly, and in a critical way the outcomes and effects of their work' (p. 9). Design history is an extensive field but it is only since the late 1980s that it has become a subject in its own right at degree level. It is not surprising therefore that teachers have not felt comfortable introducing and integrating this into their SoW. Pupils are often asked to design without the benefit of understanding either the history of design or the socio-economic and political context in which designers operate. It is widely believed that as art and design educators it is our responsibility to encourage pupils to view their internal (domestic) and external environment and receive messages critically rather than passively. The concept of design as described in this unit and promoted by the NC Art and design Order is not universal. You should be aware that the attempt to demarcate definitions between art, craft and design is a particularly western preoccupation. It would be wrong to assume that western definitions have been accepted globally. Design needs always to be contextualised in time and place.

Design history is concerned with analysis of the forces that influence visual and material culture. Design has for the most part escaped critical analysis especially within the art and design school curriculum. Designed artefacts surround us in our everyday lives but it is far rarer that teachers and pupils choose to investigate these in the classroom in a critical way. Although such objects are ubiquitous in art and design departments as part of still-life groups, few teachers make use of them as resources for pupils to learn about their values and significance. As Conran (1993) states: 'The history of the world can be documented by the design of objects and the study of those objects gives a clear message about the changes that were taking place in society' (foreword). If Conran's word 'by' is substituted with the word 'through', then there is ample room for critical enquiry. It should be noted that contexts do affect the perception of artefacts. An artefact takes on different connotations depending on its location; imagine the same object displayed in a museum, shop or classroom. Designed objects of whatever type: chairs, shoes, computers, cars, telephones, may be examined for their significance and values. It may be more relevant for teachers to provide such objects for handling and discussion, or ask pupils to bring in objects which are central to their lives. Alternatively, artefacts collected by the teacher or borrowed from museums alert pupils to different and unfamiliar types of design. These act as triggers for the investigation of function. Pupils' mobile phones, bags and drink containers might be objects to start the ball rolling. Appropriate questions should be asked so that pupils can analyse and make 'use' of objects and specialist language can guide pupils towards asking effective questions about their artefacts. Formulating the 'right' question is as important as its answer.

Schofield (1995) has revealed that by using a carefully constructed framework, pupils' perception and understanding of an object will change over time. She suggests that five domains are appropriate, within which questions and ideas about objects can be located. These are: history and time, function, form and style, materials and senses.

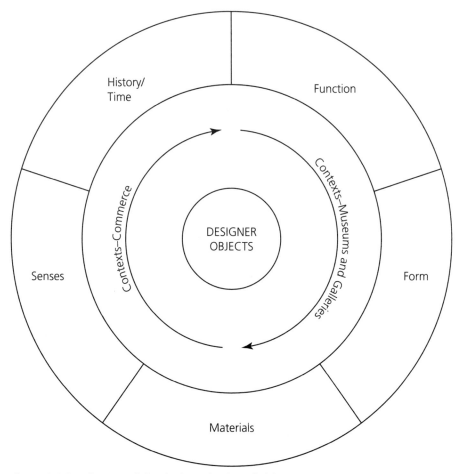

Figure 8.2.2 A framework for the investigation of objects

Schofield (1995) has shown that these domains are not exclusive and that no one domain should be seen in isolation when discussing or investigating a designed artefact. Emotive, aesthetic and sensory responses to designed artefacts can be manifested in both verbal and visual ways. Pupils may record and document their reactions and perceptions by communicating their feelings and knowledge of the object in question.

Task 8.2.7 Reading design

Collect a selection of designed artefacts chosen by yourself or your pupils.
Produce a SoW which makes use of the domains suggested by the framework. What other domains would you like to include?
How would you link critical/contextual work with practical making in this SoW?

The position of the Design Council

It is no coincidence that the Design Council and the Design Museum have concentrated their attention towards NC Design and Technology rather than NC Art and design where the traditional western fine art canon continues to dominate.

In an attempt to clarify the place and role of design in mainstream education the Design Council (1994) convened a consultation forum. It invited representatives from a wide range of interest groups such as HE and FE. The resulting paper identifies four key issues which the Design Council believed form the core of the current range of design opportunities and problems in education. They are:

- the status of design in schools;
- the relationship of design to other subject areas;
- the values which should inform learning and teaching in design;
- working towards a total design policy for schools.

However, rather than locate design firmly in one area of the school curriculum the Design Council calls for an expanded definition:

> We recognise that 'design' as a word is problematic within schools. But where we use it, 'design' should be read as covering both subjects of the National Curriculum (Design and Technology and Art and design) and *indeed a much wider remit* [our italics].
>
> (Design Council 1994: 4)

They go on to suggest that what is missing in design education is an element of risk taking. It called for design-related activities that deal, 'with subjective and interpretative issues without the safety net of a fixed knowledge base. Such activities offer pupils a challenge and an opportunity, but teachers must allow pupils to take supported risks in their own learning' (*ibid.*). This raises the pertinent question whether these approaches can be encouraged in initial teacher education and in in-service training. The lack of a fixed, static definition for design can be recognised as a strength as well as a cause for confusion.

The position of the Design Museum (London)

The mission of the Design Museum Education department is to encourage and enhance the study of design using its collection and temporary exhibition programme as a primary teaching resource. It aims to broaden public perceptions of design, including both the products that are used and the environment in which people live. The Museum believes that design can be defined as the purposeful use of knowledge, skills and physical resources to create products that meet a perceived need or opportunity. The Design Museum's programme provides publications, resources, workshops, lectures, seminars and professional development courses for all levels of education. It suggests that pupils can contribute effectively in a technological society by being able to:

- create practical solutions to problem-solving;
- communicate effectively and work in teams;
- develop visual awareness by appreciating the importance and value of design;
- appreciate the relationship between design and environmental issues;
- understand the relevance of design to industry.

Task 8.2.8 Visiting the Design Museum

How would you prepare pupils for a Museum showing contemporary design?
What issues would you ask them to investigate?
Design a worksheet for pupils to complete during their visit.
How would you follow up this visit in the classroom?

Future: the twenty-first century

As you can see a rather two-sided view of design in education has emerged:

- the vocational training of a 'neo'-designer; the creation within the classroom of mini workshops, 'pseudo' factories or design studios where cardboard mock-ups act as substitutes for real prototypes;
- design activities within a broader cultural context that promote a critical understanding of the visual and made world.

Task 8.2.9 Future gazing	

Divide into three groups, each selecting one statement to discuss.

Designers for the most part are forward looking types. Part of their job description is to spot trends and to make them. They tend to care little for what has come and gone except as a source to raid and plunder in scrapbook fashion.

(Design Review 1994: 39)

Intuitive flair will have to be backed up with technical know-how and marketing nous and worldly savoir-faire if we are to make design relevant.

(Design Review 1994: 40)

Designers can no longer take refuge from responsibility for their own actions and continually repackage the same old type of consumer goods at a time when issues about consuming and its relationship to the world's resources and energy needs to be acted upon.

(Whitely 1993: 3)

In a plenary, debate your findings. To what extent does design practice in schools reflect the changing practices and philosophies of the professional community?

Further reading

Many of the books mentioned in Unit 8.1 on craft don't provide a clear line of demarcation between craft and design and should be recognised as relevant for both areas of study. Likewise many of the authors who have written about craft also write about design. Such is the overlap between these two closely related, interrelated spheres.

Attfield, J. and Kirkham, P. (eds) (1995) *A View from the Interior: Women and Design*, London: Women's Press.

Dormer, P. (1991b) *The Meanings of Modern Design*, London: Thames and Hudson. Dormer suggests that contemporary design is more restricted than contemporary art, it is always wary of its audience. If design moves too far ahead of what people understand then it fails them as consumers and they stop consuming.

Papanek, V. (1995) *The Green Imperative*, London: Thames and Hudson.

Sparke, P. (1995) *As Long as its Pink: The Sexual Politics of Taste*, London: Pandora.

Walker, J. (1992) *A History of Design and Design History*, London: Pluto.

9 Attitudes to Making

UNIT 9.1 SKETCHBOOKS AND ARTISTS' BOOKS

James Hall

> A sketchbook is . . . a personal visual memory bank that can be used as a
> resource for collecting and developing ideas . . . a repository of ideas which
> come faster than they can be realised.
>
> (Robinson 1995: 14)

> I don't call them sketchbooks, I don't like the word sketch. They are just 'my
> books'. I number them and I'm up to No.10. They are collections: ideas,
> thoughts, quotes, photos, drawings, images, newspaper cuttings, messy and
> very personal.
>
> (Artist quoted on the Drumcroon Gallery website, 2006)
>
> (see Plates 8 and 15)

Introduction

The sketchbook is regarded as the visual notebook and key research tool of the artist,
craftworker or designer. Sketchbooks can offer insights into the personal vision,
source material, ideas and working processes of the artist. The artist quoted above also
emphasises the highly personal, often autobiographical nature of the sketchbook –
looking inwards to the private self, contrasting with the outward-facing public self, as
represented by a final exhibited piece. Equally, the educational value of sketchbooks
has long been recognised. The National Curriculum (NC) Art and design, GCSE
and GCE 'A' Level syllabuses all emphasise the development of pupils' personal
responses, their recording and observation skills and their abilities to investigate and
analyse ideas. Sketchbooks provide evidence of pupils' realisation of these aims. As

well as supporting pupils' art education, the use of a sketchbook can help pupils to develop self-awareness and skills as independent learners and critical observers.

OFSTED inspections have shown that poor standards in art and design are evident in schools 'where pupils have not been taught how to draw or use sketchbooks to record their visual research' (AAIAD 1994). Taylor and Taylor (1990) found that it was possible for a pupil to reach the sixth form without ever having kept a sketchbook. For pupils to be taught and encouraged to use and keep a sketchbook, it is necessary for teachers to have a clear understanding of the purpose of sketchbooks in art and design education.

Sketchbooks support all aspects of the NC Art and design and pupils should be encouraged to keep sketchbooks so as to function, from an early age, as researchers (Robinson 1995: 95). Pupils can be taught to develop their sketchbooks in increasingly diverse and personalised ways as they mature and their confidence and skills grow. Furthermore, the development of critical studies in art education has increased the amount of written work which pupils produce in their courses and submit for examinations. There are examples of some highly original 'personal study' submissions for 'A' Level examinations, which challenge any division between so-called 'academic' and 'practical' work and cross boundaries between text and image, essay and artefact, into a synthesis of written and visual work. Such personal studies and sketchbooks often come closer to the artform known as 'artists' books', which will be discussed in the concluding section of this unit.

Objectives

By the end of this unit you should have:

- reflected upon and considered the educational value of sketchbooks;
- developed awareness and understanding of the sketchbook as an invaluable research tool for pupils;
- investigated ways in which sketchbooks can be used in a variety of ways to realise aims and learning objectives in art and design;
- planned for, taught and assessed pupils' sketchbooks.

The role of sketchbooks

Why should sketchbooks be used by pupils in secondary-school art and design? One of the key functions of art and design in the secondary-school curriculum is to give pupils the opportunity and the skills to respond to their personal experience of the world in a visual way. There are other areas of the curriculum in which the learning content is prescribed. However, it is in art and design where pupils' personal responses, ideas and experiences are highly valued. In many departments, pupils work increasingly to their own personal agenda within the framework of the programme of study offered by the teacher and department. By the time pupils reach their GCSE or 'A' Level years, they should be encouraged and expected to generate their own ideas and themes as a focus for their studies. Robinson (1993), in her research into pupils' use of sketchbooks, saw them as a means for pupils 'to sustain and celebrate their individuality' (p. 74).

The sketchbook is a creative tool: its use should encourage information-gathering, experimentation and risk taking in the search for a creative solution to a self-generated idea or problem. Six key functions of the sketchbook are to encourage pupils' development of:

1   *Personal responses*: Pupils are encouraged to develop and express their own unique and individual vision and experience;

2   *Investigating and making skills*: Pupils develop their ability to record their responses as keen and critical observers. They build their confidence in handling and controlling a variety of media in a spirit of experimentation. They should also develop their visual vocabulary through using the elements of art and design to communicate their ideas and intentions;

3   *Critical and analytical skills*: Pupils are encouraged to develop a critical and inquiring approach to their own work and to the work of artists, craftspeople and designers. They develop their ability to select and reject information and to evaluate their responses. This leads to informed choices of approach and media which pupils can justify as their intentions become clearer;

4   *Self-awareness as independent learners*: Through these processes, pupils are developing knowledge about themselves and confidence in their own ideas, interests and abilities as a self-directed and independent learner;

5   *An active and creative approach to learning*: Pupils are encouraged to develop an active and creative approach to learning, in which thinking and planning is conducted visually and actively within the investigative process and does not precede the action;

6   *Documentation skills*: Pupils develop the skills and habits of collecting and storing data to provide source material for work in the classroom, to suggest starting-points for ideas. The sketchbook can be a valuable aide-mémoire, a repository of ideas in which to record complex events or phenomena simply and quickly for future reference.

In relation to each of the above functions, pupils should compare their own work with that of artists, craftspeople, designers and other makers.

Investigating the role of the sketchbook

Task 9.1.1 Your own use of a sketchbook

- How do the functions of a sketchbook outlined above relate to your own experience?
- Do you value or use your sketchbook in other ways?
- Discuss and compare your use of sketchbooks with those of other teachers or fellow student teachers.

Task 9.1.2 Your placement school's approach to sketchbooks

Do the functions of a sketchbook described above fit with:

- your observations of how sketchbooks are used by pupils?
- the department's written policy for sketchbooks?
- how teachers see the functions of sketchbooks?
- how pupils see the functions of sketchbooks?

Do pupils of different ages use sketchbooks for different purposes?
Are pupils taught ways in which they might use and develop their sketchbooks?
What is the nature of the tasks given for sketchbook work?
Would you say the tasks are largely pupil directed or teacher directed?
What might you expect to see in a 'good' sketchbook?

Task 9.1.3 A broader picture of the role of sketchbooks

Read reports from OFSTED inspections for their references to the use of sketchbooks. You could read the report for your placement school or a summary of national findings (OFSTED 1995b; AAIAD 1996).

Read Clement's (1993) chapter on drawing, Robinson (1993), AAIAD (1994) or the Drumcroon Gallery website (2006) and compare their ideas with your findings.

The form of sketchbooks

The term 'sketchbook' may suggest it is only used to 'make sketches' e.g. 'a preliminary, rough, slight, merely outlined or unfinished drawing or painting, often as an experiment to assist in the making of a regular picture' (*Concise Oxford Dictionary* 1985). This definition implies that a sketch is a preliminary drawing done in advance of and in preparation for 'the real thing' or 'finished piece'. Is this the only way in which a sketchbook can or should be used? It implies that the creative process is a linear development of ideas in which the artist moves sequentially from preparation, to execution and conclusion. The ways in which artists research and create artworks are highly personal, diverse, complex and idiosyncratic. Indeed, artists may not be conscious of the processes they engage in. Some artists make drawings, in books or otherwise, in their own right and during or after the creation of the 'final' or 'finished' piece. Does it not demean drawing for it to be seen merely as a preparatory process? The Campaign for Drawing (2006) (http://www.drawingpower.org.uk/menu2.htm) and the annual Big Draw (http://www.thebigdraw.org.uk/) event have helped to counter any perception of drawing as a subordinate activity, affirming and publicising the value of drawing.

As the artist quoted at the beginning of this introduction suggests, the term 'sketchbook' is something of a misnomer, belying rich and diverse practices. The Drumcroon Gallery website (2006) suggests alternative terms for a sketchbook: journal, visual diary, drawings file, work diary or workmate; definitions that highlight the personal, autobiographical approaches and the concept of a collection. Where a sketchbook had historically been regarded as a set of preliminary sketches, a rough drafting of ideas, it is now regarded as a creative artefact in its own right. An artist may work solely through sketchbooks. Paul Ryan's exhibition at the Imperial War Museum 'Drawing for Survival' records his experiences as a war artist though a series of sketchbooks as final outcomes (Drumcroon Gallery website, 2006).

A sketchbook therefore may include drawings annotated with notes, text, cuttings, clippings, collaged papers, collections, objects glued or stapled into the book. It can be argued that the shift from visual record to personal journal challenges the role of visual observation and recording as a primary aim in art education, especially when a student is invited to keep a diary or daily record of thoughts, feelings and experiences rather than simply a visual record. Furthermore, there are no rules or given methods to the sketchbook, anything being possible, though pupils need to be taught new methods and encouraged and supported in their experimentation. However, whilst there is scope for openness and risk-taking with the sketchbook, you may also come across examples of narrow and conservative practice, where predictable and uninspiring subjects are prescribed as homeworks and may represent the only use of the sketchbook. Equally, an uncritical, 'scrapbook' approach to collecting for example, leaflets or postcards from an exhibition, is inadequate: there has to be evidence of a personal response, the beginnings of an analysis of what has been experienced.

It is well known that GCSE pupils often produce their preparatory studies after they have completed their final piece. This practice is supported by the notion of the creative process as a cyclical rather than a linear process. Ideas are informed by the process of making and handling clay, paint, print, plaster or whatever the chosen medium. Sketches can be made in the chosen medium, not simply as preparation for it. Pupils could be encouraged to use their sketchbooks not only to prepare for, but to document the evolution of their ideas.

Task 9.1.4 What's in a sketchbook?

Look at some pupils' sketchbooks to discover whether they contain 'sketches', done in preparation for a 'finished' piece, or whether they are used for other, more varied purposes. For example, is the sketchbook used:

- as a visual diary – a daily record of observations, thoughts or feelings?
- as a travel journal – a record of a journey?
- as a personal museum – a scrapbook, an album of memorabilia?
- as a visual notebook – a record of visual data, noted and stored?

Task 9.1.5 Where do pupils use their sketchbooks?

Survey the sketchbooks of a whole class to discover where they have used them, e.g. in the classroom, at home, in their locality, on holiday, in a museum or gallery.

Feelings of self-consciousness or preciousness can inhibit the use of sketchbooks in an open and experimental way. Most of you, no doubt, will have experienced similar emotions leaving you bashful about drawing in a public place or worried about what teachers or other pupils might think about your sketchbook. One can feel the compulsion to 'edit' out works which do not come up to a predetermined standard. These are natural inhibitions, so how can teachers help pupils to overcome them?

Task 9.1.6 Overcoming inhibitions

Think of tasks or strategies which might help pupils to get over these inhibitions or timidity and adopt a truly experimental and creative approach.

Experimenting with media

If it is a function of sketchbooks for pupils to explore and experiment with a variety of media, how will you encourage and teach them how to do this? If pupils are working on the theme of the built environment, they could be asked to investigate an aspect of their own home or locality and to record their observations. As the teacher, you could help the pupils to select from the following list:

- drawings from direct observation, memory or imagination in a variety of media – pencil, pen, charcoal, crayon, pastel, felt-tip, ballpoint, or freehand drawing using computer software;
- collecting images using photography or a digital camera, to be manipulated using computer software;
- collage – using scrap or found materials, photographs, cuttings, perhaps worked into with drawing or other media;
- printmaking – crayon rubbings, monoprinting, block-printing and stamp-printing techniques can all be employed directly in sketchbooks;
- writing – drawings can be annotated with notes, poems, diary entries, diagrams, etc.

Pupils are increasingly using electronic sketchbooks; the teachernet website distinguishes between 'actual' and 'virtual' sketchbooks. Actual sketchbooks might include, along with the traditional media listed above, printouts of scanned or digital images, and word processed text and scanned articles. The virtual sketchbook is 'a compendium of original and scanned images, words and references sorted into several categories' (Teachernet website 2006). Virtual sketchbooks can be saved as JPEG files with text copied onto word-processing files and, eventually, saved on discs for departmental archives.

Using a sketchbook to develop research skills

Gilbert (1998) promotes student teachers' use of sketchbooks as research tools that seek to synthesise artists' processes with teachers' pedagogies. One of the main purposes of the sketchbook is to encourage pupils to develop research skills. Robinson (1993: 104) suggests the four steps in the scientific and mathematical process identified by Poincaré can also be applied to the artistic process:

- Preparation: investigating the problem, gathering relevant data. Sketchbooks develop skills of data-gathering and action planning, and provide an arena for the concrete working out of ideas.
- Incubation: consciously getting away from the problem and waiting. Sketchbooks develop skills of assimilation.
- Illumination: the sudden insight or breakthrough when the solution comes. Sketchbooks offer opportunities for reflection and moments of illumination. Drawing in sketchbooks is a means of discovery.
- Verification: evaluating and testing the solution before applying it. Sketchbooks foster skills of self-criticism and self-evaluation.

> **Task 9.1.7 Research skills**
>
> Do you recognise these stages of the artistic process in your own work or in pupils' work? Could you use this four-stage model to help plan work for pupils?

Sketchbooks and assessment

Sketchbooks can be key documents for assessment: they inform the dialogue between teachers and pupils which should be at the heart of the assessment process in art education. Sketchbooks can provide evidence of pupils' progress within a project or course, revealing their developing ideas, skills and ability to work independently. Contributing to a process of continuous assessment, sketchbooks can help teachers to teach more effectively through understanding pupils' ideas, intentions and abilities and to plan for individual needs accordingly.

How pupils' sketchbooks should be marked is more of a vexed question and will depend upon the department's aims, policies and practices. Indeed, teachers' views differ as to whether sketchbooks should be marked at all. On the one hand it is argued that if teachers want to encourage pupils to value their own thinking then sketchbooks should not be marked and no adverse comments should be written in them by teachers. On the other hand, there is the view that pupils need and enjoy receiving marks or grades as an indication of the standard they are reaching and as an incentive to improve and develop. Many schools have developed systems to track or monitor pupils' progress and achievement and this is often done using quantifiable data such as grades or marks. Target setting involves teachers setting or negotiating targets for pupils' improvement, such as higher grades or marks. Thus, art and design departments may be required to follow whole-school policies on assessment. In fact, both views serve to highlight the nature of assessment in art education: that negotiated and formative assessment are at the heart of the dialogue or conversations between teachers and their pupils. Assessment in art and design is a continuous process, both a flow of views and evaluations between the teacher and pupil, and an internal conversation of the pupil with her or himself. In this process, the sketchbook can serve as a valuable self-evaluative or self-critical tool.

This raises the question of when pupils' sketchbooks should be assessed or marked. If sketchbooks are used continuously for work undertaken both in school and elsewhere, they are likely to be looked at by teachers every time they teach a pupil. Teachers may look at and mark sketchbooks during the lesson as they move around the class talking to pupils. Alternatively, work could be collected for teachers to assess outside of the lesson, which has the obvious disadvantage of depriving pupils of their sketchbooks.

Task 9.1.8 Assessment

How frequently do teachers look at pupils' sketchbooks in your department?
    What form does the assessment of sketchbooks take? For example, does the teacher discuss them with pupils individually, in groups, or as a whole class?
    How are pupils' sketchbooks assessed and/or marked in your department?
    If marks and/or grades are awarded, how frequently is this done and are criteria for the marks or grades published and known to pupils?
    Are teachers' comments written in sketchbooks or only given orally?
    Do pupils evaluate their own work in their sketchbook?

Monitoring pupil progress through sketchbooks

Task 9.1.9 Monitoring progress

Select a number of pupils in a class which you are observing or teaching. Monitor their progress through their use of sketchbooks, asking yourself the following questions:

- How are the pupils responding to the sketchbook tasks set by the class teacher or by you? For example, how well do they understand and interpret the task?
- Are they realising the objectives of the project or task?
- Do they use work done outside the lesson to support their classwork?
- How could the pupils be helped to use their sketchbook to make more progress in the project?

Sketchbooks can reveal progress or development over longer periods of time than a project or scheme of work. Sketchbooks produced in Year 7 or even in primary school can be compared with those produced during a GCSE or GCE 'A' Level course. Michael Rothenstein looked back over many decades to the sketchbooks he produced when aged 4 to 9 years old and found that the imagery reflected his concerns and obsessions at the end of his professional career in his seventies and eighties (Rothenstein 1986).

Artists' use of sketchbooks

In simple, physical terms, a sketchbook will usually comprise a collection of sheets or pages of drawing paper bound between covers. However, the ways in which the sketchbook can be used are infinite and there is no single or orthodox method for using them. The ways in which an artist or art and design student uses a sketchbook can be as diverse, creative, imaginative and original as the ways in which artworks themselves are realised. One major difference is that while the finished art and design work represents the artist's ideas realised and made public, the content of the sketchbook often represents the artist's thinking, musings, seeds of ideas and observations – the raw material and data which may be used by the artist to work towards more complete statements.

Turner's sketchbooks provided him with a means of painting watercolour sketches wherever he travelled. Constable kept sketchbooks to record his beloved Stour Valley:

> During the years 1813 and 1814, he carried two minute sketchbooks around with him which contain the seeds of most of his mature paintings, sometimes with the whole idea there to be barely altered in the paintings of ten and twenty years later.
>
> (Taylor and Taylor 1990: 59)

For Picasso, the sketchbook appeared to have been a means of capturing fleeting ideas on paper. Picasso filled innumerable notebooks, often composed of lined writing paper, with drawings, notes, collages, shopping lists (Glimcher 1986).

Smith, writing about the drawings of Joseph Beuys, offers insights into the value of sketchbooks:

> It is in his still active drawings which may be found the most poignant and accurate insights. These works on and of paper were torn from squared jotters, used notebooks and diaries, or were simply scraps. They were often combined, sometimes later, to produce new images . . . in these works we can experience the faint scratching-around the thought-made sculpture . . . the drawings are in pencil, watercolour and unorthodox media such as blood and stains, and were used to search out, to depict and to document. They are often annotated.
>
> (Smith 1995: 181–182)

Whilst there may be art and design departments in which the 'sketchbook habit' is not fully developed (Robinson 1995: 26), there are others in which teachers have encouraged pupils to develop highly personal and idiosyncratic approaches. Many schools and pupils use hard-backed varieties of sketchbook which can be 'doctored' and personalised. For example, the cover can be collaged or faced with materials such as papier-mâché, leather, fabric, found materials, even wood or metal. The pages of hard-backed sketchbooks are often made of medium- to heavy-weight cartridge paper which can cope with the range of media and the transformation of surface to which the pages are often subjected. These sketchbooks, when displayed in GCSE or GCE 'A' Level exhibitions, are often sculptural in quality, tactile, object-like and highly individual and personal in nature.

Within the typical 'A' level syllabus pupils are required to make a personal study of an artist or art movement. This provides students with an opportunity to link critical and contextual study with their practical coursework. Students learn about an artist's work through researching the literature, in some cases interviewing the artist, critically responding to original artworks, and through practical and theoretical investigation. These investigations can result in ingenious ways of integrating different modes of investigation, combining text with image or object, and theory with practice. However, the personal study can also be narrowly interpreted as a 3,000–5,000 word illustrated essay.

Artists' books

So diverse, individual and creative have some pupils' responses become to the notion of the 'personal study' and 'sketchbook' that they relate to that growing field of the visual arts known as artists' books. It is difficult to provide a reliable definition of the artist's book as the contemporary genre traverses traditional boundaries between image and text, book and art and design object. However, artists' books refer not to literature about artists nor sculptures constructed from books but to works by visual artists that assume book form. Courtney (1995) has described artists' books as 'meeting spaces', a particular form of collaboration between a story or idea and a succession of images unavailable in more traditional artforms such as individual paintings, prints or sculptures. Some interesting material on the British Library website (2006) draws inspiration from the Lindisfarne Gospels and encourages us to 'think about books as sacred objects; physical things you can handle and cherish, storehouses of provocative, powerful ideas'. Some of the following characteristics of artists' books may have implications for developing the concept of the sketchbook in art education beyond the obvious. If pupils acquire the sketchbook habit early in their school education and are taught ways in which they can develop increasingly individual and diverse approaches, their natural curiosity, enthusiasm and need for personal realisation and expression could find a locus in the sketchbook.

Task 9.1.10 Artists' books

To what extent do the following elements of artists' books relate to the characteristics of pupils' sketchbooks:

- The viewing or reading of an artist's book requires the interaction of the viewer or reader.
- The experience of the work is kinetic and temporal in that the work or cover must be opened and the pages turned.
- The work offers a tactile and spatial experience: the book may fold or open out in unexpected ways; the paper, binding and materials have their own smell and sound as the pages are turned.
- Only one person can experience the work at a time: it is an intimate and private experience; the scale of the work is small, hand-held.
- The element of surprise is an important characteristic: our expectations may be confounded or confirmed; the notion of chance which was so important to the Dadaists and the collaging and recycling of materials which informed the work of Schwitters are significant concepts and processes for book art.

Task 9.1.11 Alternative forms

Think of ways of extending the given definitions and uses of sketchbooks and artists' books taking into account other cultural forms, e.g. scrolls, new technologies, time-based media for recording a journal.

Further reading

On art education

Association of Inspectors and Advisers in Art and Design (AAIAD), Midland Group (1994) *Sketchbooks*, Bromsgrove: AAIAD. This is one of a series of pamphlets on Art, Craft and Design in the National Curriculum produced by the Art and design Advisers' Association, offering a useful summary of the principles of good sketchbook practice in primary and secondary schools.

Association of Inspectors and Advisers in Art and Design (AAIAD) (1996) *Art and design: A Review of Inspection Findings 1995–96*, AAIAD.

(The) Campaign For Drawing (2006) online. Available HTTP: <http://www.drawingpower.org.uk/menu2.htm> (accessed 11 April 2006).

Clement, R. (1993) *The Art and Design Teacher's Handbook*, (second edition) Cheltenham: Stanley Thornes. The chapter on drawing offers useful implications for the role of sketchbooks.

(The) Drumcroon Gallery Website (2006) online. Available HTTP: <http://www.drumcroon.org.uk/Sketchbooks/sketchbooks.html> (accessed 11 April 2006).

Gilbert, J. (1998) 'Legitimising Sketchbooks as a Research Tool in an Academic Setting', *Journal of Art and Design Education*, 17, 3: 255–266.

OFSTED (1995b) *Art and design: A Review of Inspection Findings 1993–94*, London: HMSO.

Robinson, G. (1995) *Sketchbooks: Explore and Store*, London: Hodder and Stoughton. This book focuses on the primary age range but is also relevant for secondary teachers of art and design, particularly at Key Stage 3 of the National Curriculum. It sets examples of sound practice within a philosophical framework informed by Robinson's research into pupils' use of sketchbooks, which is also described in: Robinson, G. (1993) 'Tuition or intuition? Making and using sketchbooks with a group of ten-year-old children', in *Journal of Art and Design Education* 12 (1): 73–84.

Taylor, R. and Taylor, D. (1990) *Approaching Art and Design: A Guide for Students*, London: Longman. This is a practical guide and resource written for students of art and design particularly at 'A' Level. Section 4 is entitled 'Personalised use of the sketchbook'.

Teachernet website (2006) Online. Available HTTP: <http://www.teachernet.gov.uk> (accessed 11 April 2006).

On artists' use of sketchbooks

Berger, J. (2005) *Berger on Drawing*, Aghabullogue, Cork: Occasional Press.

Cézanne, P. (1985) *A Cézanne Sketchbook*, London: Dover Publications.

Drumcroon (2006) Online. Available HTTP: <http://www.drumcroon.org.uk/Sketchbooks/sketchbooks.html> (accessed 11 April 2006).

Glimcher, A. and M. (eds) (1986) *Je Suis le Cahier: The Sketchbooks of Picasso*, New York: The Pace Gallery.

Moore, H. (1972) *Henry Moore's Sheep Sketchbook*, London: Thames and Hudson.

Rothenstein, J. (ed.) (1986) *Michael Rothenstein: Drawings and Paintings Aged 4–9, 1912–1917*, London: Redstone Press.

Smith, M. (1995) 'Joseph Beuys: life as drawing', in D. Thistlewood (ed.) *Joseph Beuys: Diverging Critiques*, Liverpool: Liverpool University Press and Tate Gallery Liverpool.

Turner, J. M. W. (1987) *The 'Ideas of Folkestone' Sketchbook, 1845*, London: The Tate Gallery.

Van Der Volk, J. (1987) *The Seven Sketchbooks of Vincent Van Gogh*, London: Thames and Hudson.

Victoria and Albert Museum, London (1985) *John Constable: Sketchbooks 1813–14*, London: Victoria and Albert Museum.

On artists' books

The Artists' Book Fair is held in London every November.

British Library Website (2006) online. Available HTTP: <http://www.bllearning.co.uk/live/sacredbook/object/SacredBook> (accessed 27 October 2006).

Courtney, C. (1995) 'Private views and other containers: artists' books', reviewed by Cathy Courtney for *Art and Design Monthly 1983–1995*, London: Estamp Editions.

The Hardware Gallery, 162, Archway Road, Highgate, London N6 5BB, specialises in artists' books.

Lyons, J. (ed.) (1985) *Artists' Books: A Critical Anthology and Sourcebook*, New York: Visual Studies Workshop Press.

Phillips, T. (1980) *A Humument*, London: Thames and Hudson.

Turner, S., Tyson, I. and Courtney, C. (eds) (1993) *Facing the Page: British Artists' Books, A Survey, 1983–1993*, London: Estamp Editions.

UNIT 9.2 WALKING THE LINE – DRAWING THE LINE

Claire Robins

Each academic year, as a PGCE tutor, I receive a request from many art departments asking to be sent 'a student teacher who is good at drawing'. The prospect of interpreting this request always fills me with a sinking feeling. It's not that I don't want to be obliging, nor that I feel particularly irked by the predictability of the demand. It is more a reflection, that with every year that passes I become increasingly unsure whether it is possible to define 'good drawing' without acquiescing to an outmoded, modernist paradigm of assessable technical ability. Much like the activity of 'sucking pebbles' in Beckett's *Molloy* I nevertheless find myself turning the words 'drawing,' 'good' and 'skills' about in a futile attempt to achieve a 'satisfactory [twenty-first-century] solution', a definition, or a perfect match of student teacher and departmental stipulation. Just as Molloy systematically moves his sixteen stones from pocket to mouth to pocket, I too engage in a circular game of hermeneutics, searching for some elusive 'common' interpretation. I might be tempted to reach the conclusion that all art and design PGCE students are 'good at drawing' and, in a sense, this is probably true.

As postgraduates, all of you, whether your background is film and video, history of art, textiles, sculpture or any other categorisation or combination of art practice/ theory, will have developed and consolidated capabilities for generating visual images using line (I will return to this definition later). This aspect of your professional practice may be realised as an end in itself, as a process of recording, mapping, visually analysing, communicating, experimenting, planning, narrativising and/or note taking. You might use a computer, a sketchbook, a wall, rolls of *Fabriano* or the proverbial 'back of a fag packet', all the above can facilitate drawing.

There are many reasons to celebrate drawing, not least because it is arguably more accessible, widely practised and easily disseminated than any other art form. Requiring no more than a mark-making implement and a surface, its democratising potential is undoubtedly a key factor in its popularity and omnipresence both historically and geographically.

It is little wonder that drawing has seen no fall-from-grace since its inception in the fund-conscious terrain of the secondary school. But accessibility and economics are not the only reasons for its enduring nature. Once set in place, drawing has remained the backbone of the art and design curriculum from Year 7 though to Year 13. Moreover, art and design teachers in schools and colleges will generally agree on the importance of learning to draw. It is only when questions about how such learning should be shaped or shaded, and what exactly should be learnt, are posed, that the consensus diffuses. Aesthetic conventions allied to philosophical affiliations, social, cultural and political aspirations, all effect the theory and practice of teaching drawing; no more so in the twenty-first century than they have for centuries beforehand.

Drawing in an expanded field

During the Renaissance the Italian term *disegno* came to mean the bodying forth of the creative idea using line, and this is still perhaps the closest and most succinct definition of drawing; line has remained its essential element. But a brief etymology or scan of dictionary definitions seems too restrictive when considering drawing in contemporary art and design practice. Here, there has been a substantive extension to the scope of what constitutes drawing. Its 'field' of activity has followed a trajectory not dissimilar to that mapped out in Rosalind Krauss' seminal essay 'Sculpture in the Expanded Field' (1978), which proposes that sculpture, unrestricted by media, opened out from the 1960s onwards to encompass an expansive range of art practices such as land art, performance, etc. Similarly, drawing has been pushed and stretched, through the machinations of minimalism and conceptualism, to emerge at the limits of a useful definition. Tate Modern curator Emma Dexter, for example, suggests that 'Bruce Nauman's entire oeuvre can be seen as a form of drawing whether the results appear in video, film, sculpture or print or actual drawings themselves, his art works are all drawings in the sense that they are the results (and documentation) of his various experiments with material space and language' (Dexter 2005: 7).

Drawing in a restricted field

On beginning to teach art and design in the secondary school some of you may observe that the broad definitional stance taken by Dexter and the diversity of practices to be sourced in galleries, events and web pages are not necessarily echoed in the classroom. Compare the contents of Dexter's classificatory net with what you will find in most art and design departments, and two spectral ends of drawing's definitional continuum are revealed. Film, video and sculpture are certainly not considered drawing in most secondary schools where the terms 'experiment with materials, space and language' may not be comfortably aligned with the objectives of a drawing lesson either.

You may well have already observed that drawing in schools is characterised by a perceptualist model, where pupils are taught to develop their hand and eye co-ordination in order to achieve an increasingly convincing two-dimensional representation of chosen facets of the three-dimensional world. Still life or observational drawing is a common art lesson through which formal values such as line, tone and texture become inculcated alongside the representational schemata of linear perspective. Even before pupils start secondary school, perceptualist approaches may have become synonymous with drawing. By then many pupils' views on who can and can't draw have already been formed, with verisimilitude most commonly used as the yardstick.

Alongside and often aligned to perceptualist approaches, is the (unconvincing) appropriation of historical styles, known as pastiche. This too is a popular model for teaching drawing, whereby pupils draw from reproductions of the work of a narrow range of artists (less frequently designers) imitating stylistic codes and conventions. In both pastiche and perceptualist models, there is a perceived need for pupils to acquire a 'vocabulary' of drawing techniques before they are able to 'progress' to making meaning through drawing. As the acquisition of such skills can take a very long time, and pupils have a very short period of specialist study in art and design, drawing *qua* skills acquisition may overshadow drawing as a meaningful activity in the experience of many young people.

The rationale for pupils to learn about and value certain approaches and applications for drawing needs, perhaps, to be seen in the light of the resistance to change in schools. The characteristics of 'school art', as a distinctively anachronistic genre, endangering art's relevance, is something that many art and design educators (Adams (2002), Addison and Burgess (2003), Atkinson (2006), Hughes (1998), Robins (2003), Swift and Steers (1999), plus many others) have pointed to. A gap between school practice and professional practice is to be expected, as is a gap between practices in higher education and schools. Dexter's all embracing characterisation of drawing has emerged from the legacy of minimalism and conceptualism's destabilising and widening effect on art's classificatory systems which has been embraced in Higher Education. School's definition of drawing, follows a trajectory too, but one which since its inception in the nineteenth century seems to have been a little shaken but not substantively stirred.

Task 9.2.1 Drawing practices

Drawing has expanded to incorporate a range of practices and media. How would you situate your own drawing practice?
How might this practice translate to the secondary school?

Serving the Commonwealth

When drawing became part of the mainstream elementary school curriculum in the latter part of the nineteenth century, there was a very programmatic motivation for its inclusion; drawing would serve the needs of the *Commonwealth* by ensuring high-quality artisanship. As minister for education in the mid-nineteenth century, it was Henry Cole who was largely responsible for championing the importance of learning to draw as an entitlement for all young people. His aspirations for introducing *Drawing for Children* were published in an article of the same name in the *Journal of Design and Manufacturers* 1849.

> The time is not far distant when drawing will become part of elementary education in schools of all grades for the working classes, where writing is taught. We think every carpenter, mason, joiner, blacksmith, and every skilled artisan, would be a better workman if he had been taught to see and observe forms correctly by means of drawing.
>
> (Cole in Bermingham 2000: 233)

The specific objective of producing skilled manual workers determined the particular aspects of observation and understanding that Cole intended should be made manifest as lines drawn onto crisp sheets of paper by all English school children. It was a curriculum that advocated the drawing of geometric shapes, straight lines and simple perspective. Uncomplicated still-life drawing, focusing on individual objects and learning from the exemplars of others would be ideally suited to pupils' progressive skills acquisition and would be easy to correct and assess. The exercises, advocated by the manual 'Teaching Elementary Drawing' seem regularised, dull and mechanical to most contemporary sensibilities; and this tended to be the opinion that many pupils acquired of their art lessons. Object drawing characterised pupils' experiences of art well into the twentieth century, and inaccuracy in representing objects, mistakes in perspective and untidiness generally, were almost the only, and certainly the main grounds for assessment of pupils' work.

I would suggest that several principles that underpinned Cole's nineteenth-century edict for the teaching of drawing in schools, have continued to exert an influence to the present day. Like many good founding ideas, the practices they spurn can continue long after everyone has forgotten the original reasons for their inception.

Learning to 'observe correctly' by training the mind through hand/eye co-ordination

Cole's principle that looking and drawing are inexorably linked and that the 'harder' the pupil 'looks', the more 'truthful' the drawing s/he produces will be, is an enduring value in the teaching of drawing in schools, but one that needs to be considered carefully.

An exhibition of drawings curated by Derrida in 1990 suggests that drawing is linked more intensely to memory than to perception. Derrida made his point with his first exhibit in *Memories of the Blind* at the Louvre: Joseph Benoît Suvée's 'Butades', or 'The Origin of Drawing' (1791). In the work, Butanes, a young Corinthian woman, is making a drawing of the silhouetted shadow of her lover, directly onto the wall behind him, in an attempt to retain his image during their forthcoming separation. Derrida utilises this piece to highlight the easily missed realisation that it is almost impossible to look at an object and look at what you are drawing simultaneously.

Still reflecting on looking, Atkinson recounts a still-life drawing lesson experience from the start of his teaching career in the secondary school in which he questions this monolithic wisdom:

During the lesson I would try to *correct* those drawings that appeared to deviate too far from the family of projection systems known in the West as perspectival projection, assuming, wrongly, that this should be the form of representation which students should aim for and try to achieve. If I came across a drawing which appeared strange or incomprehensible one of my responses I remember vividly was to insist that the student 'looked more carefully', thereby assuming that this was not happening.

(Atkinson 2002: 47)

Atkinson acknowledges that his approach to his students' drawings 'was predetermined by a particular representational expectation, that of linear perspective which determined or influenced [his] judgement' (*ibid.*). He is not suggesting that linear perspective should never be learnt or that pupils should not draw from observation, but questioning the theories and practices of learning, and the history and context of training in which such values have been produced and repeated until they have assumed a 'natural' or given status. The 'truth' or 'accuracy' of the drawing, perceived by the teacher, as Atkinson illustrates, is almost always related to a particular set of expectations. Davies (2002) also notes how often these expectations are shaped by perceptions of external regulation and scrutiny. He refers to the limitation of drawing practice in schools as inherently linked to an education system that places a disproportionate value on continually assessing and evaluating pupils. This, as he suggests, tends to militate against opportunities for experimentation and less predetermined outcomes. 'Drawing, in most of our partnership schools retains its position as the cornerstone and widely considered evidence indicator most suited to external verification' (Davies 2002: 290).

Task 9.2.2 Assessing drawing – the criteria employed

Select three examples of pupils' drawings – one that you consider to be good, one satisfactory and one poor. Compare these with others selected by members of your tutor group. Consider the criteria you are using to make value judgements. How do they correspond to those that others have employed?

Now consider how the intentions of pupils may relate to your assessment criteria.

Drawing comes first?

By the time that pupils reach GCSE many will have gained an impression that there is a linear order that must be followed when making art which should start with drawing and progress rationally and systematically to a 'finished piece'; rarely a drawing. Drawing is encouraged as a support medium, whether this is for three-dimensional work or more often before progressing to painting. Cole's pragmatic nineteenth-century rationale for teaching drawing was that it should come first, and therefore must be learnt first, to facilitate other (more economically viable) art and design activities. Drawing was positioned by Cole at the service of object manufacture and taught to assist a future generation of artisans. The careers of the 'working class children' that Cole referred to were preordained and subsequently many aspects of art and design knowledge and practice were perceived as 'unnecessary' for this social group. In the twenty-first century such narrow, class distinctions and hierarchical views of pupils'/'nation's' needs seem anachronistic. Teaching drawing for the sole purpose of object manufacture in the context of Britain's twenty-first-century economy seems faintly absurd. But has the art and design curriculum responded to the future needs of young people in a pluralistic global economy?

Drawing and communicating

For both young people and adults, drawings often have a strong and direct sense of purpose, more so than other aspects of art making. In everyday life people will turn to drawings when words or photographs aren't adequate for their purposes, an obvious example might be drawing a map or diagram for someone, or making a drawing because they are trying to make sense of remembering something inaccessible or intangible. In this sense drawing seems set apart from other forms of image making in its immediacy, relationship to memory/idea/hypothesis and ready employment as a signification system.

The often varied and combinational drawing conventions used to these ends have something in common with the drawings of young children. Children's 'eclectic' uses of combinatory drawing conventions within cognitive and communicative develop-ment, are recognised by Atkinson (2002) and Matthews (2003) as imperative to their needs for relating a story or imagined landscape. Likewise, artists and designers employ a range and mix of drawing conventions in the service of a great many different personal and professional needs.

Drawing the line

The twentieth century emergence of an approach to drawing a line no longer tied to predetermined purposes, reflected a belief that freedom to experiment was needed to serve the needs of both of artists and designers. 'An active line on a walk moving freely, without a goal. A walk for a walk's sake. The mobility agent is a point, shifting its position forward' (Klee 1968: 16). This quote, from Paul Klee's pedagogic sketch-book, written in 1925, is indicative of his proposals for teaching drawing at the Bauhaus in Germany, proposals which were adopted by Foundation courses in the UK, but by few secondary schools. Throughout the twentieth century the experimental aspect of a walking line provided inspiration, and took drawing right off the page to physically and conceptually embrace context. A line made by walking by Richard Long (1967) is just one of a range of expanded drawing practices that posited possibilities for drawing unconnected to traditional media. In the blurred distinctions between process and product, drawing has continued to feature strongly. But since the 1990s it has experienced a mini-renaissance. Publications on drawing currently abound. A possible backlash to the spectacle and portentousness of many of the (then)

Young British Artists' production and display methods, drawing has emerged as the 'quiet contender' for a next generation of artists. An intimate practice, often revelling in a bed-sit aesthetic, it is now strongly represented in contemporary exhibitions where its diaristic subjectivity and celebration of fantasy can often align it to the world of the adolescent. In this same period of time undergraduate and postgraduate courses dedicated to the specialist study of drawing have proliferated. It appears as if drawing is becoming both more general (wider definitions, wider participation) and more specialist (HE courses, research projects, etc.) simultaneously.

Task 9.2.3 Approaches to drawing

Discuss the diversity of twentieth- and twenty-first century-approaches to drawing. For example, consider the following list and add to it.

Edwina Ashton, Fiona Banner, Louise Bourgeois, Frank Ghery, Rebecca Horn, Paul Noble, Robyn O'Neil, Roy Oxlade, Dan Perjovschi, Grayson Perry, Raymond Pettibon, Elizabeth Peyton, David Shrigley, Silke Schatz, Julie Verhooven, Daniel Zeller.

As you start to plan your SoW and lessons, consider whether you too are automatically adopting some of the principles for teaching drawing that were mapped out in the nineteenth century, but, most importantly, whether the models that you are using are appropriate to meet the needs of young people in a pluralistic, technological twenty–first century.

Task 9.2.4 Approaches to learning through drawing

Make a list of the approaches to learning through drawing that you have observed in schools.

What do students learn through drawing?

What possibilities can you see for introducing alternative drawing strategies to promote relevant learning?

Just as writing in the secondary school is not limited to English lessons, neither is drawing confined to the art room. If drawing is fairly widely practiced across the curriculum, are there special characteristics of drawing that should be fostered in art and design?

Emergent signs of meaning

Looking back to earlier attempts to integrate drawing into education, arguments in the seventeenth century proposed drawing as a form of scientific enquiry and an aspect of writing. Drawing as a means to record, evaluate, hypothesise (outside, as well as inside, the perimeters of art and design practice) continues, as does drawing's ability to extend the process of communication beyond writing's fixed sign system.

Walter Benjamin considers the relationship between drawing and writing in a short piece written in 1917. He guides his readers to consider both the 'making' and the 'reading' of drawings in relation to the wider domain of signs. Benjamin identifies the graphic line's meaning-making potential within signifying systems. 'The sign is always linked to its background', in Benjamin's terms it is 'on something' unlike painting, which to Benjamin smears and occludes its surface making a total composition: 'The graphic line is defined by its contrast with area. This contrast has a metaphysical dimension, as well as a visual one; the background is co-joined with the line. A drawing that completely covered its background would cease to be a drawing' (Benjamin 2004: 83). He makes the distinction between the way that understanding and making sense of the world might be aligned with distinctive models of image-making emphasising the symbolic characteristics of drawing.

> We might say that there are two sections through the substance of the world: the longitudinal section of painting and the cross-section of certain pieces of graphic art. The longitudinal section seems representational; it somehow contains the objects. The cross-section seems symbolic; it contains signs.
>
> (2004: 82)

Although these statements seem over-determined, in the light of more recent developments in both painting and drawing, there is perhaps an aspect of drawing's interconnectedness with the process of thought and meaning-making that is still relevant for understanding the process of learning through drawing in, and outside, art and design practice.

It is significant that the term 'making marks that have meaning' was selected by the 'Campaign for Drawing' for its educational research project 'Power Drawing' (Adams 2002: 222). Through three strands of activity the campaign aims 'to show how drawing can be valued more highly in both education and everyday life' (*ibid.*). Some of you may already be familiar with the first, the annual Big Draw event that has been organised since 2000. These activities are often sited in museums and galleries and the Big Draw connects hundreds of venues for a week of drawing activities for people of all ages. Second, the campaign has promoted the significant role of drawing in schools. In a climate where more and more time is dedicated to developing children's ability to communicate using words, often at the expense of developing visual literacy, the campaign's Power Drawing programme seeks to redress this balance. Its third strand of activity is a research network where those in higher education who are involved with teaching and researching drawing can share initiatives and information.

Making drawing meaningful

Within the plethora of options for making images in the twenty-first century many, such as Bermingham (2000), argue a case for the increasing position that photography and video have gained as ways of seeing and recording the world around us. The corollary of which means that drawing may be in danger of becoming an increasingly specialised activity, lodged precariously within education. In UK households, photos, particularly since the widespread use of mobile phone camera technology, define many young people's relationship to image making much more than drawing. But contemporary understanding of the photograph has also moved on. Images captured by a camera are no longer regarded as pure or fixed entities. Manipulation through ICT programs has, for some time, enabled a more fluid, less linear relationship for image making, where drawing and photography co-exist within the field of the digital.

In order to define a vital and relevant role for drawing in a twenty-first-century art and design curriculum it is necessary to acknowledge the assimilation of new technologies into everyday lives and act on the changes and developments mediated through such technologies. This does not mean that traditional definitions of drawing must be thrown out; it may well mean that essential and enduring characteristics of learning through drawing can be re-asserted. But it does mean that they need to be questioned, extended and reconfigured in order to attain a relevant contemporary rationale for drawing's place within the secondary school curriculum. Recognising and valuing drawing as a vehicle through which pupils can come to make sense of the world by literally making meaning through drawing seems an essential aspect of these aims. And if ideas of how this might be possible demand engagement with the 'extended field' of drawing practices used by contemporary artists and designers, then this can only be an improvement on relying on the educational value of drawing inherited from a bygone age.

Further reading

Adams, E. (2002) 'Power Drawing', *International Journal of Art and Design Education*, 21, 3: 220–233.

Dexter, E. (2005) *Vitamin D: New Perspectives in Drawing*, London, New York: Phaidon.

Kovats, T. (ed.) (2005) *The Drawing Book. A Survey of Drawing: the Primary Means of Expression*, London: Black Dog.

UNIT 9.3 SCULPTURE IN SECONDARY SCHOOLS: A NEGLECTED DISCIPLINE

Andy Ash

Introduction

> If you know exactly what you're going to do, what's the point in doing it? Since you know, there's no interest in it . . . It's better to do something else.
>
> (Picasso 1974: 71)

An interesting insight into one artist's way of *making*. Yet, Picasso's humble materials and modest proportions revolutionised the formal assumptions that had dominated western sculpture for a millennium, even when deviating from classical norms.

To a degree this is what I set out to do in this chapter and reconsider the formal assumptions which dominate secondary-school education. Sculpture has been neglected due to assumptions held by art and design teachers about what it is, what it entails and how to teach it: too much bother, too much to learn, too much time taken! In this chapter I form a picture of the daily reality of sculpture in schools and the problems it faces. I realise that not everybody is sure of the value of sculpture and there are those who need convincing of its place in the art and design curriculum. I devote some time to highlighting the educational benefits of teaching sculpture and use examples from a recently completed action research project in collaboration with Tate Britain.

Objectives

By the end of this unit you should be able to:

- consider the possibilities for sculpture in the curriculum;
- examine why sculpture is neglected;
- identify ways forward for sculpture in schools.

Sculpture

Sculpture is a developing, growing thing. Any definition needs to be organic. I use an artist's perspective, for artists speak intimately on the subject. Sculpture can be found in many forms and made from a diverse and often unexpected range of materials, particularly with modern and contemporary sculpture. For instance, at the start of the twentieth century, traditional media and techniques were used such as bronze casting and stone carving (Rodin and Brancusi are typical examples). However, from about 1911, and especially in contemporary practice, artists have used just about anything to make sculpture, including everyday found objects, manufacturing processes, light and sound. Due to this openness, sculpture is a notoriously difficult area to define.

> Over the last twenty years, sculpture has increasingly become art's dustbin . . . anything and everything that moves away from or no longer fits into the concept of its parent *language*, has a *sculpture* label slapped on it . . . Sculpture's nest is heavy with cuckoos.
>
> (Williams 1994: 2)

In this chapter, sculpture was embraced in its broadest sense, in the spirit of Krauss (1986) as an 'expanded field' and an all-encompassing terminology meaning: 3-D works which can be seen 'in the round', created using materials shaped or positioned primarily by the artist. It includes installations, new technologies and staging events and performances.

If you think of sculpture as a living and growing entity rather than fixed and static, you can appreciate that it changes with the addition of each new 'thing': this is why definitions are confounded by experience. 'The key I had discovered was this: a sculpture, like any work of art, is a thing, a made thing, and more of a thing than any other thing' (Tucker 1974: 22). Tucker believes that attempting a definition is not futile. By saying what sculpture is or is not, in setting out its limits, you are inviting others to break the law. It is important that pupils understand that orthodoxies can and should be challenged. Therefore it is necessary to include developments in sculpture, areas such as construction, assemblage, found objects, installation, happenings, performance, time-based and environmental sculpture.

> With the rapid changes that sculpture has passed through since 1945, there is a particular interest in what other arts, or disciplines outside the arts, sculpture butts up against, what attitude it takes to subjects as diverse as history, memory, landscape, theatre, architecture, the museum, the art market, the manufactured object . . . Sculpture in this period borrows its terms of reference from many other areas; its resonances and inflections come from countless sources. It is a peculiarly open discipline.
>
> (Causey 1998: 7)

Why do we need sculpture in schools?

I am a multi-disciplined practising artist who trained in Fine Art Sculpture at Trent Polytechnic. This was a very important and influential time for me. I was extremely fortunate to be learning in a department which valued and nurtured the individual's right to expression and exploration. Art was seen as a process, a way in which an individual organised experiences, not just simply making an artefact.

> A way of broadening personal consciousness . . . a sense of situation and awareness of self and the world around, whether that purpose is conceived of as a large abstract (broadening personal consciousness), a personal search (everything came from what I wanted to do), a particular need (getting particular energies out of your system).
>
> (Gentle 1990: 269)

I slowly began to realise that the word 'art' should be interpreted as meaning 'to do'. I chose to specialise in sculpture at a very early stage, after experiencing next to nothing of it in school. Why? After all, appreciation of sculpture depends upon the difficult task of responding to form in three-dimensions:

> That is, perhaps, why sculpture has been described as the most difficult of all arts, certainly it is more difficult than the arts which involve appreciation of flat forms, shape in only two dimensions. Many more people are 'form-blind' than colour-blind. The child learning to see first distinguishes only two-dimensional shape; it cannot judge distances, depths. Later, for its personal safety and practical needs, it has to develop (partly by means of touch) the ability to judge roughly three-dimensional distances.
>
> (Moore 1952: 7)

I think I was seduced by sculpture, and perhaps I still am. I remember the releasing feeling of being able to look at, and appreciate art which existed in the same space as myself. I could walk around it, look at it from 360 degrees, view it close up, or from great distances, touch it and smell it. It was an opportunity to interact with it, to have a relationship. Sculpture demands active participation on the part of the viewer, rather than passive enjoyment. I can stand next to it and measure myself against it, literally and metaphorically.

Art educators establish values and foster opportunities for learners. I want pupils, in sculpture, to actively challenge beliefs, partly through knowledge and appreciation, and partly by developing personal responses through making. I hope the result is that the learners make some sense of their physical world. 'It is this process of discovery and realisation which is at the base of art in education' (Barrett 1982: 14).

Golomb (1989) realises that sculpture has a special and important role to play in the educational process:

> The tactile contact with the material fosters a more intimate involvement
> in the making process and a special appreciation for volumes and surfaces.
> Modelling (and sculpture generally) also requires more persistence and a
> re-thinking about the object to be represented and it may well lead to a
> deeper understanding of the object and of the self.
>
> (Golomb 1989: 117)

There is a need for a broad and balanced art and design curriculum acknowledging both two and three dimensions. Pupils come into contact with volume, space, plane, form, movement, materials and tools, in all aspects of their daily lives. Sculpture is vital because it helps decipher complex sensory and intellectual experiences in a tangible and physical way. Individual experience depends upon the kinds of qualities the sensory system picks up. The kinds of meanings an individual produces are determined by experiential factors: the character and distribution of qualities in an environment and the particular focus an individual brings to that environment. When two-dimensional representation is emphasised, only one particular sensory system is utilised. The kinds of meaning this type of representation expresses are limited. If the skills necessary for engaging with three-dimensional form are not taught, the individual is denied access to the full range of meaning. As an educator providing opportunities for these kinds of experiences, it is important to remember that the relationship between the pupil and the environment is an interactive one:

> It is simply not the case that the qualities themselves determine what will be
> selected; nor is it the case that the individual will entirely project his [sic]
> internal conditions on the environment. There is a give-and-take process.
> Each factor makes its own contribution, and out of the interaction experience
> is born. This point is particularly germane for designers of curriculum.
>
> (Eisner 1982: 55)

By preparing pupils with alternative ways to understand experience, you equip them with tools to decipher the concepts of the world and their behaviour in it (by tools

I do not mean literally an implement for working upon something, but a conceptual framework).

From a very early age children develop and learn naturally through sensory exploration. They desire to experience the three-dimensional world around them. A child's sensory exploration automatically leads them to touch, grasp and even place in the mouth objects which they investigate. As children develop into young adults they are guided away from direct, spontaneous and egocentric responses. They are constantly told, 'Do not touch' and 'Keep off the grass', yet there remains a primitive need within, which, if starved, hinders the development of the whole person, both cognitively and artistically.

> Although as adults we grow away from the need to experience directly the sensation of things that are part of our environment, it is apparent that our need for physical, sensory experience is very strong, even though we may not equate it directly with our capacity to learn and earn a living.
>
> (Gentle 1988: 3)

Sculpting affords the 'pupil/artist' the opportunity to consider the totality of objects and to represent their major aspects. Both the pupil and the viewer are free to vary their position *vis-à-vis* a sculpture which, unlike a drawing, can be viewed from diverse angles (see Plate 26). A sculpture can offer a more comprehensive statement about the nature of the object, and such working through of the possibilities of this unique vehicle for representation has significant implications for pupils' understandings of objects and of the symbolic activity of creating equivalences.

> Such 'working through' may affect not only the child's cognitive and artistic development but, perhaps, also his [sic] personal sense of self. To see oneself as a true 'maker' in the basic sense of creating shapes, involves the mind, the heart and the body. Because of the revisability of the medium, it offers the child new opportunities.
>
> (Golomb 1989: 117)

In a culture and a society that provides much visual information in the passive mode of watching television and playing computer games, sculpting encourages an active and constructive approach and a different experience of the dimension of time. However, more activity, if not directed towards some end, may result in developing muscular strength, but it can have little effect on mental development. Pupils need activities that reproduce the conditions of real life. This is true whether they are studying things that happened hundreds of years ago, solving problems in mathematics, or learning the properties of clay.

> When a pupil learns by doing he [sic] is reliving both mentally and physically some experience which has proved important to the human race; he goes through the same mental processes as those who originally did these things, because he has done them he knows the value of the result, that is the fact . . . They [practical activities] are necessary if the pupil is to understand the

facts which the teacher wishes him to learn; if his knowledge is to be real, not verbal; if his education is to furnish standards of judgement and comparison.

(Read 1943: 244)

Building sculpture relies on tools and machines, and an understanding of the physical properties of structural materials such as wood, clay, plastics, metal, paper. However, this should not be interpreted as the need to teach basic manual skills, rather like the traditional teacher of woodwork who insists on teaching pupils all the ways of working in wood. It is about allowing pupils to apply visual skills immediately, relating skills to a *real task*.

> The abstract nature of much basic design teaching, and the assumption that 11/12 year olds can make that intellectual jump from theory to practice, can all too often only lead to rather low level pattern making in materials.
>
> (Clement 1986: 201)

The quality of the material used for expression should be closely related to the subject matter. The subject matter expressed and the materials used to portray this expression should have unity. This becomes apparent if you look at the function and design of one object. Lowenfeld uses the example of a single spoon:

> Such questions as 'why isn't a spoon made out of cloth, or paper, or glass?' can begin to stimulate thinking about the relationship between material and expression. 'Would the addition of holes in the handle of the spoon, or bumpy roses embossed upon the spoon bowl, help or hinder its function?' The development of aesthetic growth should not be minimised, and an increased awareness of tangible examples of both good and bad design should be explored.
>
> (Lowenfeld and Brittain 1966: 249)

Making sculpture in school

Task 9.3.1 Sculpture in schools
Examine your placement school's PoS and identify the kind of sculpture produced. List the materials, processes, related concepts and type of learning that takes place. Establish how much time is spent on sculpture. Does this change at any time: if so why?

'Well over half the schools do no 3-D work apart from ceramics' (DES 1991b). They highlight the absence of sufficient provision. OFSTED's continued examination of schools across the whole country for over a decade identifies a picture of neglect. The inspections of schools in 93/94 revealed: '3-D work tends to be weaker than other aspects of art in all Key Stages. Schools need to review the balance of work to make sure work in 3-D is adequately covered' (OFSTED 1995b). This warning did not appear to have any impact. In the following round of inspections OFSTED noted 'Where expenditure is low, three-dimensional work, and ceramics in particular, is most affected. In too many schools facilities . . . are lying idle for want of teacher expertise or funds' (OFSTED 1997). By 2000 they report 'there are too many art departments where resources are underused or not used at all' (OFSTED 2000: 4) and in 2001 there is a particularly demanding picture where 'a narrowly-defined curriculum in Key Stage 3 remains a significant concern, with one school in ten providing little more than drawing and painting, sometimes augmented with print-making and work in mixed media. Even in more ambitious departments, work in three dimensions is often too little represented' (OFSTED 2001: 372). A narrowly defined curriculum for pupils in art and design remains a significant issue to-date with opportunities to make sculpture a 'rare experience[s] for pupils' (OFSTED 2004b: 31).

My research findings paint a similarly patchy picture. I was able to gain insight into the current state of sculpture teaching. Imbalances exist in many schools, raising issues for consideration especially: How much sculpture is being taught and by whom?

1   Most art and design teachers identify with painting and drawing. Attracting sculptors into schools appears to be a factor, although not all teachers who teach sculpture need to be specialists.
2   Time constraints are a deterrent.
3   Materials are usually limited to card and paper, scrap and junk, clay (roughly in that order); more traditional materials: stone, wood and metal, are rarely used. This is because of:

  a   Cost: card and paper are cheap, general-purpose materials which can be used in a variety of projects;
  b   Familiarity: knowing a material's properties and limitations makes individuals feel 'safe';
  c   Equipment: basic, or no tools required;
  d   Preparation or technical assistance: very little needed.

Task 9.3.2 Benefits and deficits

What benefits and difficulties have you experienced in teaching sculpture? Discuss in your tutor group the various issues and share solutions.

Teachers' knowledge base is a considerable issue. What they know and who they know produces some shocking results. Some teachers when interviewed had difficulty in naming contemporary sculptors. Examples offered to pupils in classrooms are Dead White European Males (DWEMs). By perpetuating the sole use of DWEMs and refusing to consider contemporary, female, and non-Western sculpture, teachers restrict possibilities. All of these factors generate an image of a restricted and unbalanced diet of sculpture education, a particular and peculiar aspect of 'School Art'.

School Art: school sculpture

The term 'School Art' can be traced to an article by Efland (1976), which differentiates between pupil production and the art made by artists. Taylor (1986) quotes Rose: 'What does it [school art] relate to? Only itself. I hear the argument "But you have to teach the basics". Basics to what? School Art bears little relation to the Art of the past and none to the Art of the present' (p. 91). School pupils' sculpture, 'school sculpture' is no different now from these observations made two decades ago, it is obsessed in a superficial way with 'basics'.

Ways forward

Contemporary sculpture bears little reference to dictionary definitions and school sculpture. Hughes (1999) finds this problematic and a concern for the future of art and design education: 'Much of the work taught and examined as art is too often depressingly trivial and bears scant resemblance to the vital, radical and often anarchic activity to which this name is attached by the rest of society' (p. 129). Looking to the world of contemporary practice is not a new suggestion and Hughes points out that Efland (1970), Rosenberg (1967) and Bruner (1960) all advocated similar ways forward.

If you take the work of Cathy de Monchaux for instance, you can see her interest in exploring elements of the human condition. Her sculpture draws upon a range of stimuli: religious paraphernalia, fantasy and reality, sexuality, human anxieties, the beautiful and the grotesque; all these elements are explored through her treatment of both concepts and materials. Her work draws on a wide range of cultural references which she personalises:

> In a way I see all my work as a form of cultural plundering in an attempt to evoke a culture of my own. I suppose it's a kind of world – not like a world view or a world truth, but just a world that has its own internal logic. Each piece . . . has a language which runs through it that's very much my language.
>
> (de Monchaux 1997: 5)

While de Monchaux's sculptures create a personal world, she invites the viewer to extract their own imaginative interpretations (Plate 21). It is here that pupils can engage and ask questions, fundamental questions perhaps, questions about the very nature of sculpture, and to wonder how we can define the unique experience of this art form.

The plethora of centrally driven government publications and websites provide disappointing examples of sculpture which perpetuate safe orthodoxies rather than challenge; they are predictable, of questionable quality, and yet are set up as a kind of national bench mark. Consider the table below as a starting point in your planning.

Table 9.3.1 Twelve elements of sculpture

• form	– space
• material	– process
• colour	– surface
• weight (balance)	– volume
• rhythm (composition)	– gravity
• scale	– environment

These twelve elements are by no means definitive. They serve as a starting point and could serve as stimulus for inquiry and discussion.

Task 9.3.3 Elements of sculpture

Consider Table 9.3.1: what elements are missing?
How might you relate contemporary practice to these elements?

In learning about these elements it is not necessary to produce finished outcomes. Some might be investigated more meaningfully through problem-solving or play. For instance: stacking blocks, rearranging and re-siting objects, balancing weights, pushing materials to their limits, placing things under tension and so on. Importantly the will to discover is retained; the will to invent, try out and experiment avoids the stifling effect pure instruction has upon learning. By moving away from the mass production of skills-oriented artefacts to speculating and taking risks, a closer relationship to contemporary practice and learning in sculpture can take place.

'Forms of Experience: Engaging with 20th Century Sculpture'[1] (Ash 2002) was a three-year sculpture education initiative which provided an opportunity to explore new critical approaches and engage with a rich diversity of twentieth-century sculpture. This action research project was a partnership between Tate Britain and its collection, the Institute of Education, contemporary artists and ten London secondary schools. The project set out to identify issues, to provide strategies for teachers to challenge their knowledge, understanding and practice in teaching sculpture, and to develop guidelines for good practice and teaching resources. In this next section I will describe some of the outcomes, experiences and processes the participants engaged in which might serve as an aid to discussions in the classroom. Four facets served as a starting point for the dialogue:

---

[1] Forms of Experience: Engaging with 20th Century Sculpture was the title I gave to the action research project. The project ran from 1998 to 2001.

- *Handling objects*: By using a range of touchable objects and surfaces you can develop tools for developing looking and language skills.
- *Language*: Through looking, touching and describing these objects discussion can evolve around the kinds of language sculptural objects elicit. This serves to highlight differences in specialist language, its limitations, and how sensory experience can enable a different way of learning about sculpture.
- *Materials and meaning*: Describing objects was an important part of the process. For example, whether they were hard, rough, smooth, angular, heavy, organic or made, guided discussion about how the inherent qualities of materials affects the ways people relate to objects. This can lead to further discussion about artists' choices of materials and how they inform meaning.
- *Material qualities and associative connections*: The handling objects can later be used to support understanding of the media and techniques used by sculptors. The handling objects can also be used to encourage associative connections, such as personal interpretations, tangential links and familiar points of reference (see Table 9.3.2).

Table 9.3.2 Sculpture vocabulary

MEDIA	PROCESSES	TOOLS AND EQUIPMENT	CORE ELEMENTS	KEY WORDS/ AREAS OF EXPLORATION/ THEMES/ISSUES/ CONCEPTS
found objects	assemblage	clamps	colour (light, tone)	anthropomorphic
newspaper	piercing	goggles	environment (context, site, time)	bust (portrait)
wax	weaving	protective gloves	gravity	kinetic

Task 9.3.4 Finding the under-represented

Examine the sculpture wheel. Try to research and discover as many sculptors as possible to fit into the outer circle whose works may relate or are concerned with the particular issues in that segment.

Once you have completed this, revisit and establish whether you have represented the following: contemporary, non-Western, female sculptors.

The real task is to see sculpture as a process of problem–solving. Here, work is determined by problems set or defined. These can range from experimenting with materials, exploring general principles, to working with clearcut problems which can be solved through designing, experimenting, thinking and making an artefact. The curriculum should provide opportunities for pupils to become conscious of the

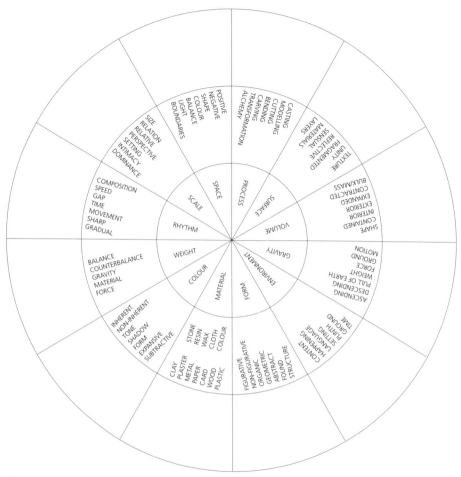

Figure 9.3.1 Sculpture wheel

world and themselves through exercises which simultaneously develop the intellectual and emotional spheres.

Task 9.3.5 Talking sculpture

Examine the incomplete sculpture vocabulary table on page 236. Add terms you feel are appropriate to your schools' context. Add words that relate to historical, multicultural and contemporary practice. Would you want to make separate lists for the different stages, e.g. KS3 in comparison to 'A' Level? (See Plate 13.)

Look through the NC for Art and design and underline all words relating to sculpture.

Repeat the exercise and focus on how the Order can be delivered through sculpture using a less literal interpretation. For instance, set yourself the task of considering different aspects one at a time, and debate how they might be addressed through sculpture.

How can colour be taught through sculpture? (See Plate 23.)

Conclusion

There is a need for local and national forums to debate the significance of sculpture in the art and design curriculum. It is important to stress that you are the agents of change; you can develop and push forward sculpture in schools.

> It could be argued that art teachers need to behave more like real artists and less like bureaucrats. School art, at its worst is the art of the bureaucrat: neat, safe, predictable, orthodox and amenable to MOT type testing. School art adds up: the real thing rarely does.
>
> (Ross 1993: 161–162)

Politely, but relentlessly, confront and advocate the place of sculpture in schools. Its current situation will continue and, I suspect, proliferate, as long as you tolerate it. Address the imbalance and neglect. Challenge the orthodoxy, and produce meaningful making and discourse for pupils. In the words of Joseph Beuys: 'Thinking is sculpting'.

Further reading

Golomb, C. (1989) 'Sculpture: the development of representational concepts in a three-dimensional medium', in D. J. Hargreaves (ed.) *Children and the Arts*, Milton Keynes: Open University Press.

Tucker, W. (1974) *The Language of Sculpture*, London: Thames and Hudson.

UNIT 9.4 MISFITS: TECHNOLOGY IN ART AND DESIGN

Pam Meecham

> We must surely choose to adapt technology that will ensure that the classroom will fit the child, and buck the growing trend for technology to be used to make the 21st-century child fit the classroom.
>
> (Greenfield 2006)

The above quote from Baroness Susan Greenfield's speech delivered in the House of Lords, encapsulates her concern about the lack of a conceptual framework for negotiating the often fragmentary and rapid communication that takes place with technology. Greenfield argues that without knowledge of other more traditional frameworks and narrative structures, young people's relationship to 'screen culture' could lead to:

> . . . a premium on the most obvious feature, the immediate sensory content – we could call it the 'yuk' or 'wow' factor. You would be having an experience rather than learning. Here sounds and sights of a fast-paced, fast-moving, multimedia presentation would displace any time for reflection or

any idiosyncratic or imaginative connections that we might make as we turn the pages and then stare at the wall to reflect.

(Greenfield 2006)

Greenfield's speech was a call for a co-ordinated approach, through initiatives such as ESRC's seminar series 'Collaborative Frameworks in Neuroscience and Education', to understand the relationship between learning and technology. A collaboration between neuroscience and education, Greenfield contends, can help us consider 'learning and create an evidence base on which 21st century education can be built . . . We need to consider how 21st century technology can help us deliver a 21st century education system' (Greenfield 2006).

Although the fruitful research collaborations that Greenfield proposes, typically through science and education, could lead to greater 'evidence-based' understanding of the perceived problem and the development of frameworks for understanding our relationship to technology, it is unsurprising that a scientist should seek enlightenment through recourse to the rationality and empiricism of science. While I wouldn't countenance the false dichotomies Baroness Greenfield puts in motion, there is a germ in her rejection of immediacy and sensory pleasure rather than presumably *hard-won learning* that made me rethink the debate about what art and design can contribute to *real* learning rather than the presumed, mindless, sensory gratification of screen culture. In part this rethinking arises precisely because so much *looking* at art can be immediate sensory gratification, the awe that precedes resonance (Greenblatt 1991) and none the worse for that. Indeed the wow factor in art may be the first step towards learning. Leaving aside for a moment my conviction that young people are discriminating, and unlikely to remain in thrall to the wow of screen technology if the technology does not deliver something more, I wondered just how art can and has responded to the challenge of new media.

Greenfield's speech also reminded me of the limitations often imposed on an art curriculum that could potentially contribute to a framework that critically examines the fragmentary and the rapid, as artists have always incorporated new technology into making or made a point of *not* doing so. Paradoxically, the possibility of developing an understanding of the effects of screen culture arises as a result of the crisis in the traditional discipline of fine art brought about by the introduction of new media and technologies into the most medieval curricula that still values the slow-paced rhythm of drawing with a 4B pencil and crumbling charcoal, and allows for sensory learning through such antediluvian practices as ceramics and textile-weaving, as well, of course, as the gratification of finger-painting. Moreover, to *look* at art also takes place in an often-antiquated space involving a ritual of silence and contemplation. In terms of creating a framework it is also worth remembering that effective learning in the twenty-first century is made of the past; what is unique to the twenty-first century is only on the margins.

I am not going to make a Heideggarian plea for the authentic embodied experience over the ersatz experience of the screen or for traditional drawing and painting to be used as a foil or even a repost to technology, although a case could be made for both. However, I am going to suggest that the art and design room is well placed through its engagement with materials for you to engage critically in a discussion that could uniquely contribute to issues raised by Greenfield and many others. Singularly, art clings tenaciously to Renaissance techniques and media, some of which have changed little since the fifteenth century. However, rather than be seen as an irrelevance or anachronism (which the curriculum steers very close to at times) art can be viewed as a site for exploring material culture in a range of guises. From an understanding of handmade-printing technologies to identity in cyberspace the art curriculum can help develop the self-reflexivity that young people need to discriminate successfully in a complex visual world, precisely because a structured curriculum can offer immediate sensory satisfaction as well as critique.

A brief detour into art history might offer some elaboration of the issues at stake. In 2003, Hans Belting reconsidered his philosophic ramblings detailed in *The End of the History of Art* (1995) and revised the project to *Art History after Modernism* (2003). In his elegiac epilogues for Art and for Art History, he urges a reconsideration in part because: '[W]e have recently begun to admit the changes that affect even the canon of art history, which now reappears as a local and Western concern, despite its universal pretensions. This does not mean that the traditional discussion of art history is on the verge of collapse, but it invites us to reopen that discussion to communicate with others from non-Western traditions' (Belting 2003: vii 'Preface to the English Edition'). However, merely opening up the dialogue to 'others' is not a transparent, unproblematic issue, especially if technology is evoked. Robert Mugerauer, for instance, reminds us that technology is not a neutral tool and that there are tensions between modernity and traditional environments and peoples (Mugerauer in Alsayyad 2001: 90–110). It is worth making a parallel through to the subject in question, the traditional art curriculum and technology.

What is clear is that the world has changed and many of the certainties surrounding art's position were predicated on unsustainable myths. It is paradoxical that in a period that has seen a marked decrease in art's claim to a higher moral truth, and a collapse of the belief in art as a universal form of communication, visiting art galleries has never been more popular. A glance at the rapidly multiplying global biennials, triennials and so forth confirms the return of the repressed. That is to say the European–North American axis has been recolonised by once marginal cultures, transforming not just our sense of what constitutes artistic endeavour but a sense of identity that placed the West at the centre of the cultural world. We now accept that the art production of 'others' is as legitimate as our own. What has been termed the New Museology has gone some way to giving access to a wider range of 'other' art often facilitated in the museum, gallery and classroom by technology.

However, the introduction of technology into museums and galleries has not been a seamless transition. The most obvious contribution that art can make to issues of quality and sensory experience as opposed to the presumed ersatz screen experience is that of experiencing the authentic, real and hard-won in an embodied encounter with art objects in their naturalised habitat: the gallery or museum. There has, how-ever, been some resistance to technology (which usually means a screen) in the museum excepting the handheld mobile which is often used as a crowd control or navigation strategy. Most notably, museums such as the British Museum and Tate Modern keep screens away from 'the real thing' in a technological apartheid, arguably, as their presence signifies as *in-authentic*. The screens also signify as education that in many art galleries is still separated out as a significantly different process from the unmediated experience of looking at paintings and sculpture.

If galleries are still chary of placing technology into the gallery itself there has been an explosion of online support for schools and independent learners. The digital offering for the remote visitor is often seen as an adjunct to, rather than a substitute for, the encounter with 'the real thing'. Gere (2001) suggests the web:

> . . . can offer students access to work other than that in more mainstream forms of representation, and to other historical narratives. It also offers them a way of making their own kinds of representations, representations that can reach others perhaps more effectively than traditional means.
>
> (p. 60)

But a closer look at some galleries' online materials suggests a far from radical rethink but rather useful and interactively compelling access to *information*.

The information download can be seen most easily through an example from Tate Modern's permanent collection: Cornelia Parker's (1991) *Cold Dark Matter: An Exploded View*. Parker's mixed-media installation lends itself to what technology can do best in terms of a multi-dimensional exploration of art works. The Digital Programmes team at Tate Modern have put together a range of materials that can be accessed online. They offer an auditory, as well as written accounts from an archaeologist's perspective to one by the Army Major responsible for the demolition through controlled explosion, of Parker's subject: a garden shed. The artist's subjective view of her own creation is also well represented. Moreover, the online site proffers classroom activities: a long way from a formally held belief that the art gallery was no place for school-based education as contemporary art and child art were not compatible.

Recent innovative exhibitions have explored further what technology can 'legitimately' offer to an understanding of art so that rather than being dispatched to a dark corner as a competing distraction, or to online activity, technological displays are an integral part of learning in an exhibition. This can be seen in three, insistently pre-technology/pre-modern (and metropolitan) 2006 exhibitions: *Constable: The Great Landscapes* at Tate Britain, *Michelangelo Drawings: Closer to the Master*, British Museum, and *Leonardo da Vinci: Experience, Experiment and Design* at the Victoria and Albert Museum. All three exhibitions use technology to give access to diverse ways of thinking and experiencing the artwork. Through working processes, by giving access via x-ray to Constable's repainting of his 'six footer' canvases: to simulate (albeit on a smaller scale) a context for the Michelangelo drawings so that the placement of the final works in the Sistine Chapel ceiling can be 'experienced' remotely: to animated da Vinci sketchbook drawings that give movement to that which is only suggested by static line, these exhibitions coherently use technology to enlighten audiences. Of course much can be done with the catalogue: the weighty intellectual support for most exhibitions, but usually written for scholarly activity. The recent exhibitions used technology's ability to convey scale, x-ray and animation unavailable in standard print. What is evident from the examples above is designers working across disciplines with historians and educationalists to use technology in a convincing, accessible rather than gratuitous way to illuminate our understanding of process and materials from the past. Cries of instant gratification and theme park can be heard from the House of Lords, perhaps, but observation at these exhibitions would suggest otherwise. All three exhibitions are 'traditional' in subject matter and the canonical reputations of the artists make them prone to hagiography (evident in the use of 'Master' and 'Great' in the exhibition titles). The technology goes someway, however, to make the artists' work accessible beyond the mere absorption of canonical art history. Still carriers of the flame, these galleries' use of technology does not offer a radical critique of artist and work but in a quiet revolution lays bare *process* in comprehensible non-word based ways. In doing so they could be seen to uncover some of mythologies that support the continuation of reputations. The National Portrait Gallery's online activity is useful here. At www.global-leap.com it is possible to book a videoconferencing session between pupils and gallery educators. The site is quick to point out that the sessions are not straight presentations but allow for reciprocal discussion. Sessions develop curriculum links via portraiture with Citizenship, Science and History. Key Stage 3 Art sessions that explore 'The Modern Portrait' and 'Self-Portraiture' are linked to 'Images of Power' and the possible future development of parliamentary democracy. And of course, it is the democratic potential of technologies used by galleries (see websites for the Tate Galleries at http://www.tate.org.uk/learning/ and the Whitechapel Art Gallery at www.whitechapel.org) to offer a range of ways for you to help pupils in developing understanding that is important, rather than exclusively promoting aesthetic responses to artworks.

The use of technology in the ways described above, however, does little to alter canonical art history which has been found wanting in a post-colonial culture. It could even be said to sanction and validate through some kind of digital authority. Raymond Williams is still a relevant voice in urging caution when faced with technological innovations, not in Greenfield's puritanical terms but because technology, too often, replicates the existing order and rarely offers a critique of existing or developing practices as it moves hot-foot in pursuit of the new. That technology has not significantly contributed to shifts in society has been a commonplace criticism of technological 'progress' since Williams' critique of technological determinism and television back in the late 1970s (Williams 1979). As we discuss new media perhaps a timely reminder from Hall and Fifer (1990) on television is useful. They maintain that it:

> . . . was not the communications medium it claimed to be but, rather, a one-way channel, broadcasting programs that sanctioned limited innovation and whose very means were invisible to the home consumer. Television through its management by corporate monopolies or state run systems had become a seamless hegemonic institution.
>
> (p. 71)

To move to a more recent example, in web development, the cyber libertarian's ideology of *get-wired* as a way of casting off the past and getting into a new world has been tempered by familiarity and the encroaching corporatisation of all forms of technology. Geert Lovink is less than sanguine about the claims for emancipation through the Internet projected by some optimistic forecasters. In *Dark Fibre: Tracking Critical Internet Culture* (2002) Lovink examines the inability of cyber-libertarians to forestall market forces in their efforts to create a decentralised, accessible communication system. In seeking to understand the development of online life, he concentrates on the activities of hacker groups, Internet activists and artists. Crucially for this unit a determined understanding of the ways that technology can operate to wrest the Internet from corporate and state control is a key plank of his work. In this quest Lovink allocates a central role to arts and culture which he suggests have a central role to play in the growth and development of new media, but believes users will have to move beyond merely,

> . . . installing a few terminals in technologically deprived spaces. It is not about bringing computers inside the museum or digitising cultural heritage. What is at stake here is the acceptance that we are living in a [sic] 'technological culture'. Culture and arts should not be instrumentalised as spiritual compensations for the technological brutality of the everyday.
>
> (Lovink 2002: 10)

He calls for a more radical cultural policy on new media than that witnessed for instance in the British context: the familiar culture online and a computer in every classroom of recent government policy. What you can glean from Lovink's approach is that it is not just the *content* of new media that requires critique but an *understanding* of the way the medium can be shaped.

Techno-cultural pessimism is not appropriate for those working with young people whose relationship with technology is conditioned, rightly, by enthusiasm for the possibilities that new media open up. Because of their historical role it can be seen that artists are well-placed to help pupils explore the potential and pitfalls of digital technologies as they have always involved themselves in testing the possibilities of new media. Often involved in witty, engaging, eccentric, if hopelessly impractical projects, artists operate at two levels: one that describes or records technological innovation; the other offering open-ended critiques that are sceptical but not necessarily cynical. Artists and pupils in the art and design room can use imagination and playful ways of visualising that the evidence-based science and technology curriculum cannot. Just a few examples demonstrate the range of responses artists have engaged in. Depicting technological progress, from Joseph Wright of Derby's (1768) *An Experiment on a Bird in an Air Pump* to Else Driggs, the American Precisionist's enthusiasm for industrial forms best seen in her 1927 work *Pittsburgh*, which was seen as a marker of modernism through its engagement with new industrial forms. The collage and montage experiments of early modernists such as Hannah Hoch and John Heartfield can be seen as critique as well as experimentation. More recently, Turner prize winner Jeremy Deller's (2006) work *The Steam Powered Internet Machine* comes to mind. Working with collaborator Alan Kane in a field in Kent in the shadow of redundant technology, a disused power station, Deller and Kane suggest the work 'connects the industrial revolution and the digital revolution . . . worlds apart but there is also a proximity' (Higgins 2006: 11). The work consists of an Apple Mac powered by a model steam engine created by model maker Alan Gibbs.

To return briefly to Lovink and the call for a better understanding of the workings of the web there is a further point I want to draw from his work. If people are to retain a free and open Internet, any sense of independence from corporate culture or government, they will need to think beyond the merely technical. In this way, as Lovink suggests, they will need to bring together 'aesthetic and ethical concerns and issues of navigation and usability while keeping in mind the cultural and economic agendas of those running the networks, on the level of hardware, software, content, design and delivery' (Lovink 2002: 15).

A tall order perhaps, and in terms of the curriculum easier to hand over to media studies and hope that the storming of the Winter Palace takes place quietly in some-body else's curriculum. But I want to return you to a point left hanging earlier: the issue of 'otherness' and technology. As well as using technology for information gathering, important as that is, working with net-artists can broaden your understanding of technology and its relationship to traditional materials, cultures and practices. For instance, an artist such as American-born, Keith Townsend Obadike highlights the ways that digital technologies, rather than offering emancipation from commercial activity (an early net aspiration), reflect the ambitions of privatised, corporate culture, 'branding' all aspects of life. Obadike auctioned his blackness on eBay with an opening bid of $10.00 as Item 117601036 under Collectibles/Culture/Black American. The 'heirloom' was put up for sale with the caveat not to use his blackness when, for example, 'seeking employment, making or selling "serious" art, shopping or writing a personal check, making intellectual claims, or while voting in the United States or Florida'. The site was prematurely closed by eBay. Obadike's work is calling into question some of the more utopian claims made for technology in general and cyberspace in particular: that it is a space free from the construction of gender, class, age, sexuality and race.

Another example of the exploration of values that are constructed around new technology can literally be seen through the prism of traditional dress in the Korean-born conceptual artist Nam June Paik's recent and final work (2006) 'Om-Mah' *mother*, a collaged video of three children playing that is glimpsed through a sus-pended, almost ephemeral, traditional Korean garment. This elegiac work has the possibility of opening up discussion around the ways that powerful technological cultures developed in the West often subordinate the spiritual and sensory values of others. Moreover, in this artwork aesthetic considerations are played out with ethical concerns.

If a curriculum imperative is needed surely a critical citizenry can be called upon across a new subject alliance to develop an understanding of the machinery of the net beyond the safety-first chat-room obsession of the media. The art room and the gallery with their blend of aesthetics, materials and intellectual questioning is a unique site for you and pupils to debate and produce works that call into question, in a positive and crucially visual way, your relationship with technological progress.

To state the obvious, in art gallery and classroom it is still possible to reflect on a range of visual objects made from such diverse elements as elephant dung, pixels, oil-paint, 120 firebricks or stone. Developing critical skills in visual culture is not just about critiquing power-structures and so on but can also be about understanding the different meanings that accrue to different materials: the ambition of the impermanence of the performance artist's ephemeral offering against the aspirations of say brass, bronze or marble. Too often perhaps the *stuff* of art, the paint and the photograph, both with complex histories, are left unspoken to the detriment of pupils' learning.

In order to advance understanding of technology and art, a useful separation for you to make might be to distinguish between art made with so-called traditional tools and that made through new technologies.

An illustration can be found in a recent (2005) photographic project at Dartford Grammar School. Drawing on Greenfield's 'traditional frameworks and narrative structures' in this case the traditional photographic portrait, sixth form pupils worked with medium format cameras under the watchful eye of the German photographer Bettina Von Zwell. Given control of the release cable they produced self and double-portraits. Although by no means the only outcome of the project, the pupils who used digital photography downloaded onto computers (for speed and cheapness) compared the quality of digital to medium format images.

The quality of images was further explored through the examination of their own and other photographers' work. Traditionally displayed, with pupils' involvement on the walls of the school's Mick Jagger Centre as well as incorporated through multiple-reproduction into a range of art practices, students engaged in issues around aesthetics, control of production, ethics of display, cost of production as well as what constitutes a self-portrait within their own interpretive community.

Consciously addressing materials, (technological or otherwise) and the way that meaning is informed by an understanding of who controls the medium and what historic and contemporary values are placed on them will go someway to ensuring that young people enjoy both sensory and visual gratification, the wow factor, and the ability to discern.

The enthusiasm with which young people embrace new media in all its digital varieties is part of the cultural revolution that has seen an overhaul of traditional practices to less skills-based, more conceptual work. With style and periodisation as the controlling structures of art history now redundant, a cheerful zeitgeist of mix and match prevails with all products and discourses of equal if incompatible merit. Leaving aside the issue of a marking system locked into an antiquated meritocracy that renders issues of quality essential but indefinable, the new media do present both utopian and dystopian possibilities that you need to address within a critical framework.

Task 9.4.1 The participatory potential of new technologies

In small groups discuss the participatory potential of new technologies as decribed for instance in the talk-back sessions with museums such as the National Portrait Gallery.

How can pupils build on debate about the democratising of technology to work collaboratively across different countries/counties/schools to create visual and audio repre-sentations that talk-back about how they relate to others via new technologies?

10 Critical Studies

Nicholas Addison

What do you understand by the term critical studies in art and design? Does it refer to a type of studio-based critical review, a supplementary form of art history, or does it imply a more interdisciplinary approach to the study of visual and material culture?

How can you use critical studies to extend and broaden your own and pupils' subject knowledge?

What methods of analysis and investigation can you employ to develop your own and pupils' understanding of visual and material culture?

How can you ensure that the relationship between the reception and the production of art, craft and design in your SoW is balanced and productive?

OBJECTIVES

The units in this chapter should help you to:

10.1   understand the purposes of critical studies in art and design;
10.2   explore different methods for investigating art, craft and design to develop visual and aesthetic literacy.

UNIT 10.1 THE PURPOSES OF CRITICAL STUDIES

By the end of this unit you should be able to:

- understand the purposes of critical studies in art and design;
- examine the range and suitability of critical approaches and methods;
- evaluate the orthodoxy of transcription and pastiche;
- explore ways to integrate critical approaches within making practices.

The purposes of critical study

No curriculum subject is an entirely discrete or autonomous phenomenon despite the fact that some secondary school teachers, especially in art and design, like to suppose that their subject is entirely unique, separate and incommensurable. You may well have noted that such singular claims produce a territorial culture in which pupils' experience is fragmented by abrupt changes in site, teaching styles and expectations. Indeed, the way most pupils adapt to these discontinuities is remarkable. Nevertheless you can help pupils make sense of these fragmented experiences by linking art and design to other curriculum subjects as well as to their own experiences. Critical studies can be the vehicle through which a web of ideas is constructed, enabling pupils to connect the seemingly disparate approaches and dimensions of their lessons: technical, aesthetic, interpretative, social and personal.

The term 'critical' can be off-putting to some pupils as it suggests a negative approach to the objects of study (Trowell in Addison 2003). But here it is used to signal the notion of enquiry, a continual questioning of assumptions and the challenge to discover through investigative and interpretative practices why visual and material culture has been, and remains, so significant for so many people. Within art and design 'critical' is often appended with the words 'contextual' and 'historical' which has had the effect of encouraging some teachers to locate study within a particular type of popular art history. Here, an accessible modernist, Eurocentric canon has become the privileged source for transcription and pastiche (Downing and Watson 2004). When attention does stray from this territory, perhaps in response to the National Curriculum's advice to engage with the art of today and of world cultures, such study is unevenly directed at contemporary practices and often focuses on stereotypical and ahistorical exemplars from non-western sources (see the following chapter for further discussion of this point). But the 'critical' in critical studies suggests more than a cursory and superficial knowledge of the art, craft and design of others, a sort of 'pic "n" mix' art history of preference or contingency: this is what I like, this is what is available. What the term indicates is a reflexive process in which production and reception, that is the cycle of making and reflection, discussion and interpretation, work together to help pupils understand that art activities are not solely exercises for developing practical skills but, significantly, are ways to make meaningful and useful things.

Developing a discursive environment

As the making of art, craft and design is an expressive, communicative and purposeful act it is not surprising that people have wanted to understand the social and cultural significance of such practice both in terms of the objects/events in themselves but also in relation to the uses to which they have been or might be put. The convivial and critical act of discussion is the usual vehicle through which this takes place, a social process supplemented and extended through written and filmed discourses in which participants (in the contemporary field: artists, audiences, collectors, communities, critics, curators, historians) propose and defend, champion and critique practices that reach into every corner of people's lives (from the built environment to the digital screen). It is only through participating in discussion that pupils can begin to consider the questions that arise from these discourses in order to understand how their own practice relates to wider concerns and to enable them to contribute to contemporary debates in the field, whether they be on matters of aesthetics, economics, environmental sustainability, morality, purpose or taste.

Language, particularly in the form of discussion, is therefore central to developing an understanding of this meaning making process. The divisions between word and image that have been reinforced by a making-led curriculum often relegate language to an instructive and regulating function when it could be used more constructively as a tool for communication, the development of ideas and reflection (see Unit 3.6; Stibbs 1998; Franks 2003). In this way language can be used to help you and your pupils make sense of the experiences that form the basis of making but also how it is that they and others come to represent those experiences. This means that you must try to help pupils recognise the way works of art, craft and design mean differently to the way language means, and this difference should be celebrated.

Some works do have a parallel with language in that they represent something that is absent to us; the pictorial, for example, can describe a place, person or 'tell' a story. However, works of art, craft and design often mean by embodying rather than representing things and values and because meaning is embedded in this way, they are often used by people to signal status, aspirations, commonalities and differences, all those things that can be summed up by the term identities. It must be remembered that the phrase 'works of art, craft and design' can conceivably include everything that is made by people, all cultural artefacts including the built environment. Values are therefore also signalled through uses, the way domestic spaces are organised, or the way artefacts are displayed in a museum. This renders the field dangerously large but also incredibly rich and I shall offer advice as to how you might select from this inexhaustible richness.

A critical approach should therefore enable pupils to investigate the purpose and meaning of their own and others' work, to look outside school art, beyond formulaic exercises and the insularity of formalist practice. Such an approach ensures that the study of art seeks meanings additional and complementary to those of immediate perception or personal judgement, meanings that are culturally specific and ideologically conditioned and this demands a willingness to investigate art as a social and cultural practice inextricably bound to its historical contexts. In Chapter 5 Lesley Burgess outlines the multiple resources that can be used to enable pupils to study art and design in an investigative and inclusive way and you may wish to return to her chapter when planning for critical activities.

Contexts

Over a number of years David Thistlewood (1989; 1992; 1996) was the most persuasive advocate for critical studies in art and design and he proposed the heresy that one day making and the history of the maker might serve the needs of critical study rather than the other way round. He realised that the criterion for inclusion into the critical canon of school art 'was (and substantially still is) the notion of what, from the realm of experience, stimulates the creativity of the representative practitioner' (1996: 4). This argument is a reminder that 'the realm of experience', the contexts in which making takes place, is as legitimate a focus for critical inquiry as the work of art itself. However, much 'critical' study begins with the artwork rather than its contexts and therefore investigation tends to be focused on formal and stylistic features. Just by shifting the starting point from the work to the context you can radically shift orthodox practices. For example, a GCSE student wishing to produce work in response to a theme such as 'inside/outside' could begin by exploring the multiple associations such a binarism brings to mind, whether in relation to the body, society, the built environment, and so on. If s/he chooses the latter as of particular interest s/he might consider in which architectural moment, or under what type of weather conditions, the concept of inside/outside is of particular importance. With guidance s/he could be led to investigate architectural practices in which this opposition is broken down, as in some of the early experiments by Rietveld and De Stijl (Overy 1991), or how in classical Japanese architecture, the fluid movement of partitions and views through to landscape dissolve any clear demarcations (Young *et al.* 2004). Then again the opening up of the ceiling to a heavenly vision in Baroque churches might prove captivating and s/he could explore how painting, sculpture, decorative architecture and light combine to produce such effects. In terms of her/his practice s/he could explore moveable units, the folding of one space into another producing a series of architectonic/sculptural models, or s/he might use projections to transform a specific interior creating an installation in which illusion breaks down definite boundaries. Alternatively s/he may wish to shore up such divisions, producing drawings of, or transforming, a specific site by finding ways to demarcate and reinforce hierarchised spaces.

The following table indicates something of the range of contexts that determine and condition cultural production and the types of experience and artefact that can stimulate responses and form the basis for investigation.

Table 10.1.1 Contextual framework (many of the categories overlap)

- **structures determining practice:**
  1) time/space: historical continuities/intercultural trade routes;
  2) environmental: climate-specific architecture, sustainable resources;
  3) social/cultural: gendered practices/forming national identities;
  4) political/spiritual: means of production/religious iconography;
  5) somatic/psychological: facial expression/desires, denials;
- **experiences that stimulate practice:**
  1) cultural/receptive: e.g. watching a film, viewing historical artworks;
  2) cultural/productive: e.g. a commission for a public memorial, responding to the needs of others through a design brief;
  3) social: e.g. an age-related rite of passage, a conversation;
  4) aesthetic: e.g. the play of light through foliage, eating a meal;
  5) somatic: e.g. swimming under water, being confined;
  6) psychological: e.g. memories, dreams, emotions;

- **representational practices, mimetic and symbolic:**
  1) images: e.g. photographs of a city, flags;
  2) texts: e.g. manifestos, religious liturgy, diaries, poetry;
  3) multimodal texts: sketchbooks, music DVD, Internet;
  4) performance: e.g. theatre, dance, ritual;
  5) artefacts: e.g. portrait busts, body adornment;
  6) simulations: these are a special and increasingly pervasive kind of representation mixing the artefactual and the performative, they are either 'fake', e.g. the First World War Trench Experience at the Imperial War Museum in London, and/or decontextualised, e.g. a reconstruction of a still life by Cezanne in a classroom.

Inviting pupils to investigate experiences or forms of cultural production as the starting point for 'critical' study provides an alternative to the orthodoxy of copying or pastiching the work of canonic artists. The experiential categories relate strongly to the thematic approach encouraged at GCSE but the structural categories indicating the contexts in which making takes place are often ignored altogether and, depending on the structural focus, may help you to choose a critical approach or tool suitable to the investigation. The representational categories extend the types of cultural production to which pupils can refer and encourages them to consider bringing in skills and knowledge from other areas of the curriculum. When exploring these categories as a starting point it is advisable to deploy the strategies outlined in Unit 3.6 to help you consider possibilities other than the orthodoxies of school art (Steers 2003). But just what are these orthodoxies, how do you recognise them?

The orthodoxy of transcription: critical study or pastiche?

When following a SoW which has a clear historical dimension a preferred 'contextual' homework is to ask pupils to find out all they can about a favoured artist and write it up in their sketchbook. In this way it is hoped that knowledge and understanding can be addressed and lessons reserved for the 'fundamental purpose' of making. The result is usually a copy of the introduction to that artist's life from a popular art history or the Internet and serves little educational purpose. Alternatively, critical study is delivered by a programme of copying and pastiche (Hughes 1989: 71–81). It is important that you critically examine what is fast becoming an orthodoxy at KS3 and beyond, the 'transcription' and its companion the critical and contextual sketchbook.

For the past few years critical and contextual study in art and design has been managed at 'A' Level in sketchbook form, and as a result is filtering down to GCSE and in some instances KS3 (see Unit 9.1). These books are filled with found images, an eclectic mix of the popular and the canon, a multicultural and historical pot-pouri. In addition pupils copy from reproductions and to a lesser extent objects from galleries and museums. This item is often separate from the 'true' sketchbook in which they draw from their environment, from life. The book also includes art historical 'investigation', often arbitrary extracts from the Internet quoted verbatim, or personal responses which, however heartfelt, lack even the most rudimentary analytical method. They often contain impressive evidence of hours of patient copying and annotation and some pupils take the opportunity to explore the potential of materials; but as evidence of critical investigation they are thin. What they provide is a 'look', a collection of images that act as the sign for investigation which can be replicated from pupil to pupil and from year to year.

The transcription is a singular and extended piece of work which mirrors this dependency on the 'look'. At KS3 it is often a copy from a postcard, enlarged through squaring up. It allows pupils to mix and match colour, consider the relationship between the formal elements in composition and possibly imitate an artist's handling of paint (reproductions are nearly always of paintings). Pupils who choose their own reproductions are likely to have a very different experience of these skills if, for example, one chooses a Mondrian and another chooses a Kahlo.

Why is such practice consuming so much of pupils' time? Copying is something that most young people do; indeed the mimetic process is innate, a fundamental stimulus for learning requiring time and patience. The choices young people make are evidence of their developing tastes, those aspects of visual culture with which they identify and through which they begin to construct aspects of their identity. However, in art and design teachers often limit such choice to legitimised reproductions of canonic artworks. What teachers seem to anticipate is that the means of representation, the style, the handling of paint and so on, will transfer to the pupils' own work, particularly their observational studies. But the process of transcribing an image from a two-dimensional surface to the same is perceptually at odds with the process of representing the three-dimensional world in two dimensions; this is a process of transformation not transcription. Transcription is imitation of surface, outcome not process. It entirely decontextualises the artist's work so that its original meanings and mode of production are ignored. It also privileges painting as the dominant mode of representation, albeit in the form of photographic reproduction. Transcription in schools attempts to elevate pastiche to a critical method (for an evaluation of parody see Unit 11.3, Popular Culture). It is not without its benefits in developing technical skills and in helping to acculturate pupils within dominant discourses on taste, but it is not a critical practice and in my opinion should be saved for pupils' own time, not taught sessions.

Mapping approaches to critical study

While on your course of initial teacher education (ITE) you may find some students and teachers who believe that theory bears little relationship to what happens in practice. This division may be familiar to you from your own experience of art, craft and design education where, to put it crudely, there can be 'those who do' and 'those who think'. It is however important as a teacher that you develop a sense of the interrelatedness and interdependence of the visual/textual, theory/practice continuums and the mediating role of language. Historically these relationships have been fundamental to the way art, craft and design have been separated and developed as distinct disciplinary categories in western culture. What you choose to focus on, and to a certain extent the way in which you do so, is left very open in the NC and the examination syllabuses and a critical approach is possible within any of the three disciplinary domains. If an understanding of contexts is one approach, how can you develop the critical skills necessary to make this meaningful? Investigation (what is often referred to as research in art and design) is the key to this.

Table 10.1.2 presents the domains of subject knowledge in art and design. They are categorised into separate and value-laden approaches to indicate difference and complementarity. The technical and productive elements may seem marginalised as they appear to be subsumed within two categories. However, there is no reason why all approaches cannot be interpreted from the point of view of making. To a certain extent they repeat the contextual elements of Table 10.1.1 but here the approaches are related to specific traditions within the school subject rather than to the wider field of cultural production. Particular investigative approaches are also signalled, e.g. historical, formalist, semiotic, and some of these will be discussed more fully in the subsequent sections.

Task 10.1.1 Balancing the diversity of methods and approaches

In your tutor group discuss the suitability of the approaches in Table 10.1.2.
See how you might combine them so that within the three years of KS3 you can develop SoWs that introduce pupils to a diversity of approaches.
Think of omissions to this table.
Consider how your background, experience and strengths have conditioned your choices.

Planning and integration

When planning lessons you should avoid the temptation to tack on a critical and contextual dimension to what is essentially a techniques-based exercise. This may at first seem difficult, for you are likely to witness formulaic SoW designed to ensure regular and acceptable outcomes based on the replication of exemplars, a process that can provide excellence in terms of outcomes but that does little to develop pupils' independence or critical skills. Often such practice is supplemented by a sort of retrospective research where pupils produce an 'academic' complement to their making practices. Try instead to work out how a discursive and critical approach helps pupils to develop their practical work and how, by placing their practice in a relevant context, they can come to understand the social significance of their practice, for example: its environmental impact, its relationship to history, its communicative or market potential.

In order to ensure that the critical dimension of SoW is meaningful and productive try to conceive it as an integrated part from its inception. Consider the project in Unit 3.7, 'Audience and Site-Specific Problem-Solving'. Here you see that the context, the site-specific needs of under-fives, is an immediate and necessary starting point for problem-solving as the brief requires pupils to use critical skills to understand how

Table 10.1.2 Critical and contextual studies in art and design

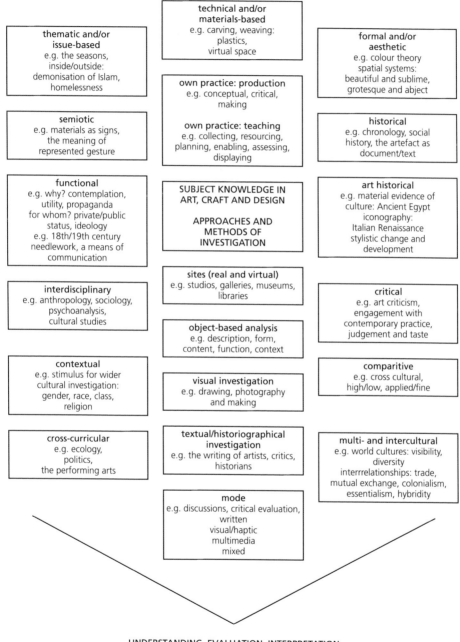

thematic and/or
issue-based
e.g. the seasons,
inside/outside:
demonisation of Islam,
homelessness

technical and/or
materials-based
e.g. carving, weaving:
plastics,
virtual space

formal and/or
aesthetic
e.g. colour theory
spatial systems:
beautiful and sublime,
grotesque and abject

semiotic
e.g. materials as signs,
the meaning of
represented gesture

own practice: production
e.g. conceptual, critical,
making

own practice: teaching
e.g. collecting, resourcing,
planning, enabling, assessing,
displaying

historical
e.g. chronology, social
history, the artefact as
document/text

functional
e.g. why? contemplation,
utility, propaganda
for whom? private/public
status, ideology
e.g. 18th/19th century
needlework, a means of
communication

SUBJECT KNOWLEDGE IN
ART, CRAFT AND DESIGN

APPROACHES AND
METHODS OF
INVESTIGATION

art historical
e.g. material evidence of
culture: Ancient Egypt
iconography:
Italian Renaissance
stylistic change and
development

interdisciplinary
e.g. anthropology, sociology,
psychoanalysis,
cultural studies

sites (real and virtual)
e.g. studios, galleries, museums,
libraries

critical
e.g. art criticism,
engagement with
contemporary practice,
judgement and taste

contextual
e.g. stimulus for wider
cultural investigation:
gender, race, class,
religion

object-based analysis
e.g. description, form,
content, function, context

comparitive
e.g. cross cultural,
high/low, applied/fine

visual investigation
e.g. drawing, photography
and making

cross-curricular
e.g. ecology,
politics,
the performing arts

textual/historiographical
investigation
e.g. the writing of artists, critics,
historians

multi- and intercultural
e.g. world cultures: visibility,
diversity
interrrelationships: trade,
mutual exchange, colonialism,
essentialism, hybridity

mode
e.g. discussions, critical evaluation,
written
visual/haptic
multimedia
mixed

UNDERSTANDING EVALUATION INTERPRETATION

the re-formation of materials can develop safe spaces for play. Had the brief been conceived from a different starting point, for example, an exploration of materials and their potential in constructing open forms, continued by a consideration of the ways in which these forms might be developed to construct climbing frames and then, as an extension, how such constructions might be applied to the needs of five year olds, the emphasis would be different: aesthetic experiment, design construction, application to a specific context. There is nothing intrinsically wrong with this sequence as an exercise, but the later stages can easily be abandoned through lack of time, thereby decontextualising practice. The integrated approach ensures that all dimensions are covered and that the social needs are taken into consideration from the start.

Consider for a moment how you can transform a school orthodoxy, the still life, which remains a favourite genre (Allen 1996). In order to contextualise still life, teachers often relate the displayed objects to well-known exemplars or invite pupils to employ a favoured technique or approach: Cezanne for objects, Picasso for form, Van Gogh for application of paint, Matisse for colour. This encourages pupils to develop formal and imitative processes but does little to suggest the significance of the genre beyond mimetic and pastiche-led practices. The exhibition, *Objects of Desire* (Hayward Gallery 1997) demonstrates how, during the twentieth century, the 'genre' was extended to encompass sculpture, photography, the readymade and installation, approaches that provide alternatives for use in schools. But still life has a long and varied history and can also be investigated iconographically for its symbolic content, or indeed more immediately for its different social and aesthetic uses, that is its embrace of the unavoidable 'conditions of creaturality, of eating and drinking and domestic life' (Bryson 1990: 13). Schneider (1990) presents an extraordinary range of still life from the sixteenth to the eighteenth centuries in Europe. One example, Sanchez-Cotan's well known *Quince, Cabbage, Melon and Cucumber* (c.1602) demonstrates how a symbolic approach can encourage pupils to explore objects beyond their formal and perceptual properties.

> [it] contains only a small number of fruits so that they seem almost sanctified by virtue of their geometrical arrangement. Indeed, the festive presentation of these everyday objects was inspired by mystical notions surrounding St. Teresa of Avilla (1515–1582) or St. John of the Cross (1542–1591), who were close to the people and emphasised the sacred nature of a simple, ascetic lifestyle, radically opposed to the wastefulness of the royal court.
>
> (Schneider 1990: 124)

Bryson (1990: 63–70) suggests that Cotan, by impelling the viewer to witness the intensity of his own looking, reverses the expectation that painters should choose the spectacular, heroic or apocalyptic as subjects worthy of their attention, or indeed, with still life, the sensuous world of sustenance and abundance. Rather, Cotan turns to what is normally overlooked, the contents of a larder, to conceive an abstract, mathematical and distanced space quite different in its quiet stillness to the rhythms and flamboyance of contemporaneous Italian and Dutch paintings with their emphasis on tactility and the familiar space of the table.

In his compelling book, 'Looking at the Overlooked' (1990) Norman Bryson discusses with unfailing insight how still life has been used differently in different social and cultural contexts. In antiquity, he explores its function as a marker of class within decorative schemes, in Baroque Spain as a critique of worldly splendour and virtuoso performance, in Holland as an index of wealth and abundance and how in the seventeenth and eighteenth century the spaces of still life configure a feminised domain at once socially marginal but suffused with the politics of power, specifically around gender. He also touches on its use in the twentieth century as a vehicle for formal and material invention. If you are asked to develop SoW in which still life plays a part, Bryson's text provides innumerable thought-provoking ways in which you can engage critically with the genre, helping you to plan investigative, as well as technically and formally challenging projects and encouraging you look again at possibilities for what is all too often a predictable and dismal exercise in schools. After investigating and discussing historical still life and the ways the genre not only reflects but produces cultural meanings, pupils can construct and record a still life to explore their understandings in relation to contemporary life. For example in relation to the Cotan they might explore, necessity and indulgence, thrift and waste, in relation to Picasso's *Still Life with chair caning* (1911/12), different modes of representation, the popular and the elite. Choice of object is no longer exclusively formal as pupils set up and juxtapose objects in an attempt to construct meanings other than balance or colour contrast. This is not to suggest that formal properties should be ignored, only that they should be used in conjunction with an object's symbolic or social significance. It may be that the constructed still life turns out to be the final outcome and that it is recorded in ways other than drawing and painting: e.g. photographed in different lights, or cast in plaster.

> **Task 10.1.2 Contextualising still life**
>
> Select examples of still life that approach representation in different ways.
> Investigate their meanings and the context of their making.
> Record this information so that a pupil at KS3 can understand it.

Further reading

Bryson, N. (1990) *Looking at the Overlooked: Four Essays on Still Life Painting*, London: Reaktion.

Hickman, R. (ed.) (2005) *Critical Studies in Art & Design Education*, Bristol: Intellect Books.

Hughes, A. (1989) 'The copy, the parody and the pastiche: observations on practical approaches to critical studies', in D. Thistlewood (ed.) *Critical Studies in Art & Design Education*, Burnt Mill, Essex: Longman.

Thistlewood, D. (1996) 'Critical development in critical studies', in L. Dawtrey *et al.* (eds) *Critical Studies & Modern Art*, Milton Keynes: Open University.

UNIT 10.2 EXPLORING METHODS FOR INVESTIGATING ART AND DESIGN: DEVELOPING VISUAL AND AESTHETIC LITERACY

By the end of this unit you should be able to:

- compare and apply different critical methods to works of art;
- consider the possible role of semiotics in developing visual and aesthetic literacy;
- differentiate between the ways that images, objects and words are used to construct meaning.

A critical approach to visual and aesthetic literacy

> **Task 10.2.1 Investigating art**
>
> The following analytical method, 'Ways into the Object' offers a set of questions that you can use to investigate the objects of art, craft and design. You are invited to consider whether these questions can be used to analyze all objects of visual and material culture, be it a contemporary installation, a film, a commercial textile, a chair, a building, a magazine illustration?
>   What additions and subtractions might be necessary to accommodate different types of object?

Table 10.2.1 is a full, if not comprehensive, set of questions which mixes different methods in a somewhat liberal way. However, despite the mixture, it provides a cumulative method of investigation, parts of which can be managed empirically (directly from the object) while others, the contextual elements, may need recourse to further objects and texts. The emphasis in the interpretative section is on a semiotic mode of analysis because this ensures that the interpretative act acknowledges the relation between form and content as productive of meaning rather than as separate elements such as style and subject matter that can be analysed independently. As you and your pupils become more familiar and expert in your observations you may be able to bypass the analytical categories at the beginning of the table and interpret the artwork in an integrated way from the start.

Further reading

Charman, H., Rose, K. and Wilson, G. (eds) (2006) *The Art Gallery Handbook: A Resource for Teachers*, London: Tate Publishing.

Hatt, M. and Klonk, C. (2006) *Art History: A Critical Introduction to its Methods*, Manchester: Manchester University Press.

Harris, J. (2001) *The New Art History: A Critical Introduction*, London: Routledge.

Table 10.2.1 Ways into the object: object-based analysis

0 - Initial responses (optional) (Charman *et al.* (2006) provide a rationale for beginning this way) What is your initial response?

1 Empirical analysis

Record your observations and cite evidence.

a *Description* (recognition and identification)

What can you see?	• things, participants, forms
How do these parts or different representations relate? What is happening?	• the relationships between the participants or the object's constituent parts/forms

b *Form*

Which formal elements are used?	• e.g. line, colour, texture, space, time
How are they used?	• e.g. is the colour local, descriptive, evocative, symbolic, arbitrary?

Is a particular perspectival system used? Is the space defined by solid or empty forms, mass or plane?
What are the relationships between the formal elements and how do they interact?
composition

Try using binary questions.	• e.g. simplicity/complexity, sameness/difference, unity/fragmentation

c *Materials and techniques*

What is it made from?	• e.g. bronze, fabric, light
What processes are used to make it?	• e.g. casting, embroidery, performance

Identify the material evidence for this.
Is it permanent (or intended to be so) or is it transient and ephemeral?

d *Your relationship to the artwork*

How do you gain access to the artwork?	• e.g. as a customer, maker, owner, visitor
Where are you positioned?	• height, angle, distance, etc
What does this position suggest?	• e.g. subservience, dominance, privilege

Do you have to be static or can you move round the object or interact with it?
Do you need to view the object over a period of time (temporality) or can you view it, or its various parts, at any one moment (simultaneity)?

2 Functional analysis

a *Patronage*

For whom was it produced?	
What did the patron determine?	• e.g. its content, form, materials

b *Purpose* Why was it made and what purpose did it serve? Has its function changed: is it used in ways other than intended?	• e.g. propaganda, decoration, ritual, habitat, self-expression • i.e. decontextualized and/or recontextualized, from sacred object to specimen
c *Provenance* Where was it originally? Has it moved? Where is it now?	• If so, trace its journey.

3 Contextual analysis

a *Art historical classifications*

Iconography What is the art object about, what do its images/references mean? Is it: an <u>literal</u> representation?	• e.g. the man standing with an axe through his skull represents a martyr • concerned with appearances, e.g. a portrait bust, a visual record
an <u>index</u>?	• a causal or associative connection, e.g. a palette and chisel suggest a traditional artist
a <u>symbol</u>	• an analogy or <u>metaphor</u>, something other than itself defining its character, e.g. the lotus blossom symbolises union
an <u>allegory</u>?	• a symbolic story, often with moral overtones, e.g. a classical myth
Is it historical, literary, utilitarian, genre, etc.?	
Style Does it belong to an identifiable <u>tradition</u> identified with a culture, nation, period, etc? Is it composite?	• e.g. Islamic, Ming, Baroque, Finnish, Popular, Folk, International, Postmodern • intercultural, <u>hybrid</u>
Mode Does it conform to a specific way of representing/making? Does it conform to a specific aesthetic category?	• e.g. mimetic, abstract, diagrammatic, functional, documentary, narrative • e.g. beautiful, sublime, grotesque, abject
Influences What might its influences be?	• e.g. intercultural, generational

(continued)

b *Contexts* How does it relate to its historical and contemporary contexts? What is its relationship to broader cultural structures, including class, gender, race, sexuality??	• e.g. religious, political, economic, social • It would be useful here to consider the methods of other disciplines, e.g. anthropology, sociology, psychoanalysis (<u>interdisciplinarity</u>)

c *Interpretation: significances*
How do the various elements, (forms, techniques, materials, images) work together as signs to suggest particular meanings? (<u>semiotics</u>)

Are there any differences between its explicit content (denotations) and your personal and cultural associations (connotations)? Are there any hidden messages? If so are they?	• e.g. intentional or subliminal, celebratory, subversive?

How might its changing site/function alter its meaning? (<u>framing</u>)
How has it been understood by others in its own and other times?
What is its significance(s) today?
What do you think it might mean?
What do you bring from your own experience that might condition the way you make sense of it?

On semiotics

It is helpful, if challenging, to introduce semiotics, the study of sign systems, as a means to develop visual and aesthetic literacy. This is because the systems of cultural production in the twenty-first century, in shifting from the verbal to the visual, requires pupils to understand multimodal forms of communication. Semiotics provides the possibility of a unified method capable of addressing the interrelationship between the visual, verbal, aural and kinaesthetic modes provided by multimedia technologies and sites. Pupils are already likely to be using semiotic methods, in practice if not by name, in English, where images are often employed as a stimulus for descriptive, analytical and creative writing. Semiotics is, in addition, a system which promotes an inclusive study of different forms of visual and material culture, from wrestling matches to fashion (Barthes 1957). It is sometimes given a bad press in secondary art circles because it is perceived as having its roots in linguistics, particularly the pioneering work of the structuralist, Saussure (1974), and is therefore deemed antithetical to visual phenomenon: this tradition can be more legitimately termed semiology (Sturrock 1993: 71–73). However, the theories of Peirce (Hoopes 1991), who formulated a system for categorising images rather than words, may seem more accessible to you because they are more immediately applicable to art and design in schools. But before exploring his system it is worth spending time examining the relationship between words and images from a semiotic point of view.

Words are the sense stimuli, the formal means, abstractions or *signifiers* which in combination with a concept, the *signified*, refer the user to a thing or an idea, a *referent*. For example, the letters a-p-p-l-e in combination with the concept apple (all possible apples) refer the reader to an actual apple. A *sign* is produced when there is a repeatable relationship between the signifier/signified and the referent, thus the word 'apple' is the sign that leads us to the actual fruit. Diagrammatically this can be represented as:

$$\textbf{Sign} = \frac{\textbf{signified}\ (\text{concept})}{\textbf{signifier}\ (\text{medium/form})} \left.\right\} \quad \text{relate to a } \textbf{referent}\ (\text{event, thing, idea})$$

to produce a **signification** (meaning(s))

Grammatical phrases and sentences enable you to combine words within a syntactical structure to describe interactions in the world and to form propositions. For example, words in combination identify more clearly the type of apple or context: e.g. 'they like to eat crunchy green apples', or allow you to speculate on why these people like them.

Likewise, in the visual arts, the image or object is a kind of sign or complex of signs produced by a relationship between signifiers (colours, lines, sounds, textures, volumes) and signifieds (associations, representations) which lead the viewer to specific referents (things seen, experiences felt, actual events). For the viewer this recognition suggests the significance of the signs for the maker and invites her or him to make sense of them, that is to give them specific meanings (significations) and these may well be different for different readers. The meanings of visual forms depend on a whole series of interconnected systems including spatial placement, modality, affectivity, materiality, etc. (Kress and Leeuwen 2006). The parallels between the verbal and visual are not absolute (*ibid.*: 16–44, 76–78). One difference relates to the way language tends to construct meanings in a temporal, linear fashion, syntactically over time, whereas an image can provide information simultaneously providing a complex matrix of signs that can be grasped in a moment: thus the popular adage, 'a picture tells a thousand words'. Perhaps a more fundamental difference is that the relationship between a visual sign and its referent often seems more direct; a drawing of an apple looks more like the fruit than the letters a-p-p-l-e which thus makes the latter appear somewhat arbitrary. However, just as with language, the signified and the referent are not one and the same thing and it is the signifier that holds the key to this difference.

Magritte teasingly invites you to consider the paradoxical relationship between signs and their referents in his painting *The Treachery (or Perfidy) of Images* (1928–29) in which a pipe is portrayed appended by the statement 'Ceci n'est pas une Pipe'. The point here is that an image, whether a pipe or a pope, is only a representation not the represented object itself. This is frequently forgotten in art and design teaching where the perceptual model of education slavishly encourages pupils to produce the 'essential copy' (Bryson 1991), an 'unthinking' imitation of appearances suggesting that the relationship between signified and referent is absolute and universal rather than contingent and culturally specific. Gregory (1977: 161–162) examines assumptions about the nature of visual perception elucidating the difficulty of perceiving perspectival systems by peoples who live in environments in which there are no right angles. Although the research cited was undertaken in the 1960s, and today it is likely that few can avoid the right angles of industrial production, it proves that such systems are learned. It follows that all forms of visual description are mediated through systems of representation, and these systems are not necessarily as 'universal' as is sometimes supposed. At the same time visual representation tends to provide more semantic freedom than the verbal to make associations between the perceived image and an endless chain of other referents drawn from memory or fantasy. Words, of course, encourage this as well, particularly in poetry, but language is often more conditioned by intended messages than the image: this may be a cultural rather than an intrinsic phenomenon.

Task 10.2.2 Comparing visual and verbal meaning-making

Collect a series of images or objects and texts which have the same referents, e.g. a Biblical story and a painting of the same, a written recipe instruction and its imaged counterpart, an obituary and a monument to the dead.

In pairs or small groups try to work out by what means and to what effect your examples are communicating.

What are the differences between the ways words and images communicate?

What can words do more efficiently than images and vice versa?

Plan a SoW that incorporates this exercise as a starting point and leads pupils to produce a work in which images and words function together but in complementary ways.

Semiotic approaches are often practised in ways analogous to linguistic analysis, thus a narrative picture or a magazine advertisement is decoded in terms of who is being represented, what they are doing to one another, whether the action is literal or allegorical, what type of audience is being targeted and so on. But art, craft and design are not purely representational and their other modes of communication and being must be considered. Thonet's bent wood chair (1830s to today) represents nothing; it is a chair, ubiquitous throughout the cafes of Europe. But the materials and forms with which it is made are signifiers that produce multiple meanings. Its material, wood, is natural, presented simply and unadorned; its silhouette is clean, unpretentious. These are qualities, signs of utility and functional elegance which are common today, but if compared to a typical mass-produced, nineteenth-century household chair, replete with machine-carved ornament, they seem remarkably spare. After identifying signs of difference the aesthetic and economic conditions for its production can be investigated leading to any number of associated investigations: form=function; ornament is crime; the private/public sphere; the body and domesticity. Equally, you discover very different meanings if you investigate the material basis of an oil painting rather than just its iconography.

Icon, index, symbol

Peirce's tripartite definition of the sign may prove a useful differentiating tool for you to use in the classroom. He separates the visual sign into three types:

1)   the icon refers to something through likeness, a similarity of forms: most representational art is iconic, e.g. a portrait is like a sitter, a circuit diagram is like the components that together make up an electrical system;

2)   the index refers to something through association, perhaps through a relationship of cause and effect, e.g. a footprint in the sand is an index of a person, a moustache is an index of masculinity;

3)   the symbol refers to something through a code or rule that has to be learned, e.g. colours in a particular formation make up a flag and refer to, e.g. a nation; a cross can refer to the Christian faith but in another context addition.

Not all signs are discretely iconic or indexical or symbolic, they may have components of all three; thus the wing/arm of Picasso's *Nude Woman in a Red Arm Chair* (1932: Tate Modern) is iconic in that it is like a wing, indexical in that it indicates softness and sensuality, and symbolic in that it signifies as 'feminine' to the polar opposite of the 'masculine' paw on the other arm. Identifying the occurrence and significance of these three types of sign in a work of art is a much quicker process than the 'Ways In' table outlined in Table 10.2.1. It provides a framework by which pupils can bypass judgements of quality and taste and helps them interpret a work by looking at the way in which meaning is constructed (for an example of this type of interpretation see the article on Marcus Harvey's painting of Myra Hindley, Walker 1998).

Pluralist cultures tend not to fix symbols, they educate people to understand diversity, e.g. of languages, which are symbolic systems. In time, new symbols are constructed. They have the effect of unifying the knowledgeable group but excluding others. As such, they are frequently used to construct outward signs of identity. It is very important that you take the time to investigate the symbolic systems of the work you are showing pupils; this is not something they can understand without a 'way in' to the code. However, you should remember that some of your pupils are liable to have the key to systems you may not know well: invite them to help. Symbolic systems are clearly different and their frequent use should dispel any notion that the visual arts are in any way universal.

Task 10.2.3 Applying Peirce's system

Introduce Peirce's system to a KS3 class and assess to what extent it helps them to investigate the function of art.

How might you use this system beyond a method of categorisation?

Devise a SoW which invites pupils to make these different types of sign: e.g. pupils can design iconic, indexical and symbolic signs intended to orient people around the school; they can then test which is more easily understood.

Denotation, connotation, myth

In combining or juxtaposing signs further relationships are established, producing meanings that require interpretation. Barthes (1965) suggests that meanings are of two

types: the denoted, that is the expected or 'common sense' meaning, and the connoted, associated meanings (whether they are personally or socially specific). After it is
read, the denoted sign, made up from Saussure's signifier and signified, becomes itself
a connative signifier, leading to a connotation (the signified) and thus the formation
of a connoted sign. However, Barthes himself realised that the connotation often takes
precedence over the denotation because it is so personally affecting; in this sense the
sequence of its construction does not necessarily follow the trajectory I have outlined.
Image makers have been able to exploit people's compulsion to seek connotations for
a variety of ends: coercively (as in propaganda or advertising) subversively (as in Dada
or Pop art), or subliminally and unconsciously (as in all things that pretend to be
natural or real but are in fact rooted in ideological and or psychological values or
dispositions). These connoted signs thus work together to establish and reinforce
cultural myths which are shared ways of conceptualising and representing truths. The
three levels of signification outlined here bear some similarity to the system invented
by the art historian Panofsky in 1955 in which the reading of images is divided into
three stages: pre-iconographical description, iconographical analysis (which relates
images to their associated texts) and iconological interpretation (where symbols are
decoded as symptoms of cultural and psychological practices) (Panofsky 1970: 66).
Panofsky was primarily concerned to elucidate Renaissance art, but Barthes' system
is designed to address all cultural production. Barthes' interpretative models and
strategies can thus help you to look beyond the surface to uncover the way images
suggest meanings through shared and seemingly given understandings; these myths
are the matrix from which signs are drawn and which they in turn reinforce. As all
people are tied to specific socio/historical contexts, myths often appear natural and
self-evident rather than as constructs sedimented over the ages through particular
power relations; Barthes' system therefore also enables you to engage with the thorny
issue of who controls the production of images and why.

Task 10.2.4 Denotations, connotations, myth

Select a series of images from a popular genre, e.g. magazine advertising, illustrations in
Reader's Digest, children's cartoons.
   Identify their denoted elements; consider their connotations. What cultural 'myths' do
they conjure?
   Discuss your findings with other students.
   How might you include this process in a SoW?

Binary opposites

The technique of presenting difference in the form of binary opposites is frequently
chosen by teachers because the resulting extremes stimulate lively responses. A binary
opposition is formed by juxtaposing two concepts that are mutually dependent,
e.g. light/dark, wet/dry, and they are often used to order the dynamic experience of

the world into a structure at once divergent but easily understandable. Consider, for example, the way binary oppositions are deployed to structure beliefs and behaviours so as to regulate them: culture/nature, good/evil, male/female, white/black. Nonetheless, binarism has a strong theoretical pedigree as one of the preferred methods of the influential anthropologist Levi-Strauss (1963). He proposed sets of oppositions: profane/sacred, cooked/raw, celibacy/marriage, female/male, central/peripheral through which to question cultural structures. A cultural form, e.g. a myth, is analysed by using one such opposition as an interrogative point of entry. The art historian Wölfflin (Fernie 1995: 135–151) had already produced a similar model for the analysis of Renaissance art. His five pairs of formal characteristics: linear/painterly, plane/recession, closed/open, multiplicity/unity, absolute/relative have proved influential and persistent.

More recently binarism has been a favoured tool of the 'new' art historians, for example Pointon (1990: 113–134) who uses nineteenth-century oppositions: male/female, culture/nature to investigate the meanings of Manet's *Dejeuner sur L'herbe* (1863). Binarism is already adopted as a strategy in thematic or issue-based SoW, usually as a stimulus for practical responses: inside/outside, natural/made, rich/poor; such contrasts are common triggers in art and design. It is an appropriate system for use with pupils because of its neatness and simplicity. But its critics suggest that because it is a system based on antithesis it has the effect of imprisoning investigation into a model dependent on conflict rather than co-operation.

Stuart Hall (1997) suggests that binary oppositions are 'a rather crude reductionist way of establishing meaning' (p. 235) and it is through critical investigation that you can begin to question their validity and normative status. This is especially important in those instances, the majority, where oppositions set up power relations in which one term is dominated by the other: master/slave, man/woman, teacher/pupil, etc. The French philosopher Jacques Derrida (1976) posits the opposition presence/absence as central to the western philosophical tradition and recognises how the privileging of the former plays out in other binaries, e.g. speech/writing, penis/vagina. His method of analysis known as *deconstruction* is one in which the oppositional relationship between the two terms is questioned, and often collapsed. His method has therefore been found threatening by those who wish to uphold the belief systems and values that underpin traditional western culture.

> Deconstruction wants to erase the boundaries (the slash) between oppositions, hence to show that the values and order implied by the opposition are also not rigid. Here's the basic method of deconstruction: find a binary opposition. Show how each term, rather than being the polar opposite of its paired term, is actually part of it. Then the structure or opposition which kept them apart collapses, as we see with the terms nature and culture in Derrida's essay. Ultimately, you can't tell which is which, and the idea of binary opposites loses meaning, or is put into 'play'. This method is called 'Deconstruction' because it is a combination of construction/destruction – the idea is that you don't simply construct a new system of binaries, with the previously subordinated term on top, nor do you destroy the old system – rather, you deconstruct the old system by showing how its

basic units of structuration (binary pairs and the rules for their combination) contradict their own logic.

(Klages 2004)

Through this process, oppositions can be seen to take on a reciprocal or symbiotic relationship, e.g. that between the industrial and handmade (waste, industrial materials are often the source materials for inventive recycling and re-crafting, particularly in the 'developing world').

Task 10.2.5 Using binary oppositions

Working with a partner, develop a series of oppositions and decide how you can introduce them as starting points for critical discussion and investigative making.

On additional methods of investigation: 'Ways In'

Table 10.2.1 suggests an analytical, incremental model that can be used in part or as a cumulative method. Dyson's comparative techniques (see Unit 3.7) are readily applicable to the classroom in that they are easy to resource and the task of investigating into categories of similarity/difference is not difficult for pupils to understand. Using his system as a model you can develop additional categories pertinent to your specific investigation. Targeting signs of difference is essentially a semiotic process, albeit, in this instance, limited by a reliance on decontextualised reproductions.

Taylor's *Form, Content, Technique, Mood* (1989; 1999) is a highly favoured approach. It strongly reflects previous systems (particularly Feldman 1987), but purposely rejects the judgmental character of their later stages. However, Taylor's system does not encourage pupils to consider relationships across categories, as if each term in this particular quartet represented a quality possessing an independent existence. This can reinforce the form/content dichotomy of traditional art history. Taylor's final category, 'mood', is a particularly woolly term and invites pupils to project their own associations onto the work of art in an uncritical and ahistorical way.

Cunliffe's recommendations to avoid the 'formalist labyrinth' (1996) are lucidly argued. He suggests that teachers shift between historical and contextual approaches, 'world to mind', and child-centred approaches, 'mind to world', in which the perceptions of the child are taken as a starting point. He provides a framing system for looking at works of art that he calls the 'Semantic differential Technique' (*ibid.*: 316–318) which invites pupils to identify qualities along a five-point continuum, e.g. between rough and smooth, healthy and sick. Allen's (1996) promotion of montage and simultaneity requires further practical consideration before it can be considered a teaching method, at present it stands as a welcome and salutary call to arms. Burgess and Schofield (Figure 8.2.2) consider a method explicitly devised to analyse the objects of design.

> **Task 10.2.6 Contextualising practice**
>
> Select a SoW that is based on a formalist approach.
>   Revise it so that making is related to a wider context:
>
> 1   using a world to mind approach;
> 2   using a mind to world approach.
>
>   Consider ways in which collaboration with another subject area enables you to explore art, craft and design in wider contexts.

The contexts in which you might place investigation and making are then, endless. It is important to strike a balance between pupil- and adult-centred contexts, between formalist and issue-based explorations, between art or craft or design practices and between art specific and cross-curricular investigation. Above all, investigation can be used to develop pupils' interest in art, craft and design as social and cultural practices which affect the ways they live in a very profound sense. Try to remember how as a child you were forever exploring your environment and relationships with people through the investigation of the material world and by asking questions of others. In schools these innate processes are often stifled by over-regulated behaviours and through the valorisation of legitimised and correct knowledge, pedagogical processes that may inhibit the will to find out. Critical studies in art and design takes this will to know as its basis and provides a range of tools and methods to enable pupils to enquire systematically and in depth into visual and material culture so that their interests, expanded by you as an experienced adult, can underpin making and saying, displaying and communicating.

Further reading

Barthes, R. (1957) *Mythologies*, trans. A. Lavers and C. Smith (1990), London: Jonathan Cape and New York: Hill and Wang.

Chandler, D. (2001) *Semiotics: The Basics*, London: Routledge.

Dawtrey, L., Jackson, T., Masterson, M., Meecham, P., and Wood, P. (eds) (1996) *Critical Studies and Modern Art*, Milton Keynes: Open University.

Hall, S. (1997) *Representation: Cultural Representations and Signifying Practices*, London: Sage in association with the Open University.

Hickman, R. (ed.) (2005) *Critical Studies in Art & Design Education*, Bristol: Intellect Books.

Kress, G. and Leeuwen, T. (2006) (2nd edition) *Reading Images: The Grammar of Visual Design*, London: Routledge.

11 Towards a Plural Curriculum

UNIT 11.1 CONTEMPORARY PRACTICES AND THE MULTICULTURAL CLASSROOM

Paul Dash

> Art education . . . must give children the opportunity to make work using themselves as their starting point and their lives as their inspiration.
>
> (Mitchell 2005: 2)

Planning for diversity

A central concern for beginning teachers wanting to run meaningful projects in inner-city schools is the challenge of making their teaching relevant to the wide spread of ethnicities and cultural heritages in the classroom (Grigsby 1977). Just how do you construct schemes of work that take account of, say, Jamaican, Cypriot, Bosnian, white English, Bangladeshi and Korean cultural backgrounds? Beginning teachers in agonising over this issue, sense the crucial importance of making teaching and learning relevant to the complex demographic and experiential spread of the modern-day classroom. In this professional resolve, they share the ambition and insight of progressive educators since the 1960s who have experimented with and engaged practices described as multicultural or intercultural teaching.

Several experts in the field have since the 1960s listed categories of multicultural teaching in art and design, among them Grigsby (1977), Mason (1995; 1996), Troyna (1992), Efland *et al.* (1996), McFee (1998), Emery (2002), Dash (2005). Chalmers (1996) referencing Zimmerman and Stuhr posits five models:

- The first approach is simply to add lessons and units with some ethnic content.

- The second approach focuses on cross-cultural celebrations, such as holiday art, and is intended to foster classroom goodwill and harmony.
- The third approach emphasises the art of particular groups – for example, African American art or women's art – for reasons of equity and social justice.
- The fourth approach tries to reflect socio-cultural diversity in a curriculum designed to be both multiethnic and multicultural.
- The fifth approach, decision-making and social action, requires teachers and students to move beyond acknowledgement of diversity and to question and challenge the dominant culture's art world canons and structures. In this approach, art education becomes an agent for social reconstruction, and students get involved in studying and using art to expose and challenge all types of oppression. Although this last approach may not be multicultural per se, students will probably be dealing with issues that cross many cultural boundaries.

(p. 45)

Teachers of art and design often draw on the first two models, running projects that build on approaches from other traditions, without requiring the pupils to question values or interrogate notions about themselves or others. This is often seen in schemes where cultural materials from the backgrounds of pupils in the classroom are used to foster goodwill (Efland *et al.* p.79), a policy described by Troyna (1992) as 'the 3Ss interpretation of multicultural education (Saris, Samosas, and Steel bands)' (p. 74). The third approach can be much more in-depth and will often engage pupils in study that draws quite heavily on the work of another tradition. Such approaches would begin to question relative views and assumptions through meaningful exposure to the work of others. Benin bronzes, Egyptian art, Indian Rickshaws, artefacts generated by the Mexican *Day of the Dead* and Aboriginal paintings are resources frequently used in such multicultural teaching, a pedagogy described by Efland *et al.* as 'a haphazard inclusion of various cultures' (p. 43). The fourth model which reflects 'socio-cultural diversity' prioritises the traditions and cultural achievements of other groups. Efland *et al.* (1996) in presenting a list of parallel categories state that:

> When employing this approach in the area of art education, art teachers present a lesson and relate it to members of many different groups through a selection of various social and/or cultural exemplars and perspectives. The unique contributions of individuals within these diverse social and cultural groups are stressed.

(p. 82)

The fifth approach, 'decision-making and social action', provides opportunities for learners to critique ways of making sense of the world. It therefore prepares, '. . . students to challenge social structural inequality and to promote the goal of social and cultural diversity' (*ibid.*). Efland *et al.* quote Sleeter and Grant (1987) who assert that such teaching should: 'educate students "to become analytical and critical thinkers capable of examining their life circumstances and the social stratifications that keep them and their group from fully enjoying the social and financial rewards of the country"' (1996: 83).

This chapter recommends the adoption of methodologies for teaching art and design that promote a better appreciation of 'socio-cultural diversity' and pedagogies appropriate to planning for 'decision-making and social action.' In doing so it proposes contemporary art as a tool for ensuring the democratising of practice by bringing to the centre of learning, pupils from all backgrounds and levels of ability. But in order to explore this further you need first to question what is meant by contemporary or post-modern art.

What is contemporary art?

Burgess and Addison (2004) assert that contemporary art often refuses classification, (see also Dawe Lane, 1996) while Adams (2005) contends that contemporary practices, 'inevitably elide the boundaries between author, spectator, producer, and participant, and call into question individual agency itself' (p. 24). Downing and Watson (2004) circumvent a clear description of such approaches, asserting instead that, 'The definition of what constituted contemporary art practice, or of a school that demonstrated a commitment to this, was left entirely to those identifying the schools' (p. 5). Emery (2002), however, while acknowledging the difficulties inherent to providing a firm definition of contemporary practice is more forthcoming:

> Critics look at the diversity of postmodern art and see a huge range of incoherent and conflicting practices. What is more, postmodern art often seems 'anti-aesthetic'. It is sometimes shockingly confronting but at other times sweetly reminiscent of past practices or even crassly derivative of popular art styles. It is hard to know what postmodern art is about, if indeed it is important to know.
>
> (p. 8)

Efland et al. in providing a framework for postmodern art education from '1990–Present' state that such practices:

> Recycle contents and methods from modern and premodern forms of instruction.
> Feature the mini-narratives of various persons or groups not represented by the canons of master artists.
> Explain the effects of power in validating art knowledge.
> Use arguments grounded in deconstruction to show that no point of view is privileged.
> Recognise that works of art are multiply coded within several symbol systems.
>
> (p. 72)

In keeping with this, Downing and Watson (2004) postulate that contemporary art provides an opportunity 'for the exploration of social, moral and political issues and recognition of art as a visual communication tool' (p. 5). Such practices lend themselves to thematic enquiry to enable personal response through themes such as 'Taboo' 'Identity' or 'Metamorphosis' (Kennedy 1995a). Central to these are notions of fissuring, fragmentation, contestation, controversy, difference and even cultural conflict (Efland *et al.* 1996: 56). Contemporary art provides opportunities for teachers to draw on the work of present-day Black and Asian artists concerned with issues of identity, representation and social justice. As stated by a teacher participant in the Tate/Goldsmiths contemporary art in schools research scheme:

> We should be looking for black and Asian artists . . . Chris Ofili features quite highly, partly because of his politics, which are quite pertinent really – particularly in my school which has a lot of race problems. It's good to look at his work and see what statements he's trying to make.
>
> (Charman *et al.* 2006: 9)

Contemporary art and aesthetic valuation

Contemporary art and the pedagogies that espouse its conceptual principles, prioritise the value of individual and collective subjectivities (Efland *et al.* 1996). Attaching to it modernist values such as an ability to draw could therefore destroy its intentions. Problematising the valuation of non-standard drawing practices, Atkinson (2002: 50) demonstrates that in children's drawing, teachers should 'attempt to enquire into the student's experiential relation with the subject of a drawing and consider how this relation is articulated by the student through the semiotics of the drawing'. In other words it is the responsibility of the teacher to enter into the life-word framed by the pupil's drawing and given articulation in his/her semiotics. By this means the manifold worlds of the classroom are valued and fore-grounded. An alert teacher in critiquing or assessing them would be receptive to the diverse voices inherent to the works.

The notion of an aesthetic defining itself through practice is applicable to the work of progressive artists of any period. Teachers should provide frameworks that allow this uniqueness and diversity of response to emerge in the classroom. This brief overview of contemporary practice demonstrates the futility, as Burgess and Addison (2004) and Downing and Watson (2004) assert, in attaching firm making principles to it. In a similar vein, pedagogies that draw on such art test the conventions of creative engagement in classrooms and the traditional means of assessing good practice. Many may have an overt or implicit political dimension that delegate the power of voice to each individual participant or group of collaborators. It is this democratisation of practice that promotes questioning, one that problematises previous givens regarding social and aesthetic values, alongside notions of self and other, truth and untruth that I regard as central to postmodern approaches to teaching and learning.

Student teacher engagement in inclusive contemporary practice: 'Art Now in the Classroom'

In 2001 Goldsmiths College and Tate Modern Gallery initiated a scheme that took Goldsmiths PGCE students into six primary schools for a week. *Art Now in the Classroom* was designed to explore possibilities for using contemporary art at KS2. To facilitate a better appreciation of principles associated with contemporary practice all involved, the pupils, their teachers and the PGCE students, participated in training sessions at the gallery. These took the form of practical workshop activities led by a team of skilled artist educators. What follows are outlines of two case studies from this scheme.

Pacific – Kilmorie school project

Kilmorie is a primary school in south London, like the others in the scheme, it was selected for its fine record in art and design. Six students were placed there to work in pairs with three classes of 30 children and their teachers. On visiting Tate Modern beforehand the PGCE students were intrigued by Yukinori Yanagi's *Pacific* (1992). Recognising its potential for project development, they opted to use it as the main resource for a tapestry that would involve pupils making appliqué pieces inspired by their life experience and personal identity.

Pacific is a mixed-media piece based on the national flags of 49 countries from around the Pacific Rim. Each 'flag' is made from sand and coloured dyes that are trapped in narrow plastic boxes, the combination of natural and made possibly symbolising the artificiality of human separation on the basis of national identity. Yanagi linked the flags by a network of plastic tubes, through which at some point after completion he let loose a colony of ants. They proceeded to 'colonise' the work, burrowing and tunnelling pathways through it to gain access via the tubular pathways to neighbouring 'territories'. But as the insects moved through the flags, they unwittingly collected on their bodies particles of sand that were released elsewhere – a fine dust sediment in the tubes attests to this occurrence. The notion of territorial or national essence is by this means powerfully challenged in this extraordinary work.

The PGCE students with the classroom teachers involved designed schemes of work (SoW) that required the pupils to function collaboratively while celebrating their interests and personal identities. Each pupil was invited to make a handprint for inclusion in their design. Some attached a favourite flag associated in many cases with parental origins, which seen alongside the handprints acted as a linking device in the tapestry. Many added photographs of parents, siblings, grandparents or best friends. Some inserted football logos, photographs of pop musicians, pets, nibbles, sweets and keepsakes. Pieces of texts written by them were also attached to the designs. As with Pacific, the completed appliques were organised in a vertical and horizontal matrix, but whereas Yanagi linked his 49 flags by lengths of plastic tubes the pupils' designs were joined by strips of dressmaker's tape.

Sandhurst school project – landscape/matter/environment

Working to the same *Art Now* scheme Jacqueline Dear and Gemma Kennedy designed an applique project for a group of pupils at Sandhurst Primary School in 2001. PGCE students Gemma and Jacqueline, on being placed there, devised a SoW with the classroom teacher, which came about 'in response to the Inner Worlds room at Tate Modern that is located in the Landscape/Matter/Environment galleries' (Dear 2001: 275). Jacqueline describes the thinking behind the scheme:

> The worksheets they were using asked them to imagine what this world on display would feel like, and the children were more than able to play this game of the 'magical as if'. Through later worksheets, back in the classroom, they were encouraged to map out elements of their own equivalent inner worlds and to find a corresponding sense of the world they inhabit personally. Hopes for the future, memories of the past and political concerns all played a part.
>
> (pp. 277–278)

In keeping with all *Art Now* participating schools, the pupils visited Tate Modern where they were given a talk by the Gallery staff on how to read and work with artefacts in the space around them. They were also required to make drawings. Back in school the teaching team contextualised the SoW but within the shared framework the children were free to contribute material of their own choosing. As Dear states:

> . . . back in the classroom, they were encouraged to map out the elements of their own equivalent inner worlds and to find a corresponding sense of the world they inhabit personally. Hopes for the future, memories of the past and political concerns all played a part. The designing and making processes for this project dramatised the role of art as a carrier of meanings, a symbolic vehicle for matters of significance, possibly cryptic and private (inner) matters, made tangible for an audience.
>
> (2001: 278)

Poems written by the pupils that dealt with private thoughts were attached to the designs, a few secreting their verse in sealed pockets. Some also stitched into their pieces photographs and memorabilia that celebrated their different families and backgrounds. The children under the supervision of their classroom teacher and the students also had the opportunity to work in new ways, participating in machine stitching and generally combining materials and media in surprising new juxtapositions.

Two project ideas

(a) Sonya Clark Beaded Prayer Project
Sonya Clark, the American textile artist, runs the Beaded Prayer Project. As with the Dear/Kennedy project at Sandhurst, participants in her plan stitch private

thoughts and wishes into textile artefacts. Treatment of the pieces encompasses embroidery with some appliqué elements. The finished artefacts are a diverse range of shapes and forms, many pod-like, geometric or organic, which contain special and 'powerful' treasures culled from the life of each participant. According to Clark (2003):

> The Beaded Prayer Project is influenced by packets containing power objects made by people in Africa and throughout the world. . . Each contributor to the project creates a packet that contains a personal wish, hope, dream, blessing or prayer encased in a covering with at least one bead on the surface.
>
> (Project flyer)

While based on traditions from different cultures, the similarities of scale, thought-ful deployment of beads and communal display of the pieces add to the strength of each artefact. Contributors are of any age and the quality of each piece entirely personal and determined by the skill-level of the maker. The inclusion of at least one bead on each piece, like the handprints in the Pacific project, brings a unified feel to the display, one that conveys a message of beauty in cross-fertilisation and sharing.

(b) Camilla Spare project concept

Camilla Spare, in her final year of the Goldsmiths BEd course (2006), developed a project idea of rich potential for the classroom. Looking at diversity and cultural identity, she fused photographs taken in India with digital images of east London to create a visually harmonious yet conceptually disorienting portfolio of prints that mirror the breadth of experience many children share. In her project rationale she states that:

> I think my exhibition reflects the interchangeable multicultural society today. The images also suggest the subtlety of these cultures, often overlooked, as people . . . nonchalantly pursue their everyday lives. On deeper inspection much more is evident and shows how over the century the two cultures have interwoven. I wanted the images to stimulate many personal questions and reflections for the observer.
>
> (Spare 2006)

Ms Spare's integration of disparate environments demonstrates the collisions that take place for all of us in real life (Efland *et al.* 1996). The concept could be adapted to problematise more parochial contexts. I am thinking, say, of life in the inner-city juxtaposed with scenes from country living, home life and school life, etc. Even the discordant perceptual encounters that take place when objects from different cultural, geographic or utilitarian worlds are placed in close proximity to each other could elicit poignant expressive possibilities.

What these four approaches offer are frameworks that are driven by contemporary artistic concerns and activities. Such schemes implicate a wide range of making possibilities from filmic media to photography and computerised imagery. They can be language-based and would often involve children in group work (Efland *et al.* 1996) that promote practices that privilege risk-taking and the development of new media.

Conclusion

Beginning teachers working with young people from different cultural backgrounds can be overwhelmed by the perceived complexity of the learning context they find themselves in. They want to do well by those children but feel ill-prepared for the breadth of the task before them, that of making teaching and learning relevant to their pupils. But this is an issue for teachers and children in any context, including those in largely mono-cultural settings. As I have demonstrated, teachers of art can overcome the perceived challenges inherent to the diverse classroom by building their schemes of work around a simple starting point, allowing each pupil space in that framework to celebrate their separate experiences and preferences in the making process. That way teaching and learning becomes a partnership between teacher and learner that privileges difference.

Activity

In 1816 a ship captained by the inexperienced Hugues Duroys de Chaumareys went aground off the coast of Africa. As Boime (1990) states:

> . . . attempts to get the ship afloat failed, the captain decided to use the lifeboats and make for shore. Due to his general negligence, however, it was discovered that there were too few lifeboats for the passengers, and it was decided to construct a raft for the remaining 152 passengers and tow it with the lifeboats.
>
> (pp. 52–53)

But conditions were difficult and eventually captain Chaumaray gave the 'order to cut the guy wires connecting it with the boats' (*ibid.*). The rafters were left to fend for themselves and, sadly, most of them eventually died. Gericault's great work produced in response to these events, *The Raft of the Medusa* 1818–19, is therefore a piece that offers a range of possibilities for making. There is for example the issue of betrayal, the subdued colouring of the piece and indeed its intricate structure. Most importantly on this raft there is no ethnic, racial or gender hierarchy, men and women, Africans and Europeans are physically interwoven conjuring notions of shared humanity. The coding devices of the period often placed black subjects in subservient positions relative to whites in works of art, and are here turned upside down.

> **Task 11.1.1** *The Raft of the Medusa –
> a Scheme of Work*
>
> With reference to *The Raft of the Medusa*, develop a Scheme of Work based one of the following:
>
> - teamwork – possibly in a composition that features people from different backgrounds working together to overcome overwhelming odds;
> - flags as symbols of different peoples and nations – these could be twisted, elongated creased or partially folded, etc., to create a rich and diverse visual pattern;
> - a composition that focuses on the dense interweaving of bodies and materials in a congested space in which hands, faces, feet, hair and other areas of the body are interspersed with bits of clothing, netting, scraps of timber, rope and other materials from a notional raft.

New teachers of art and design should be at the cutting edge of these discourses. They bring to the classroom an uncluttered freshness rich in new possibilities and creative potential. It is for such teachers to show the determination to engage new pedagogies relevant to the lives of children in today's learning environments (Robinson 1998; Barrow *et al.* 2005). Without their willingness to effect change many learners would miss an important opportunity to interrogate values and precepts that continue to influence life opportunities. Importantly too, those whose art-making practices fall outside the narrow strictures of academic good practice, may continue to be disenfranchised and denied opportunities to grow through art practice. You should make it your business to employ strategies for teaching which build on the profoundly beautiful spirit of inquiry inherent to much contemporary practice, by adopting a more integrated approach to teaching and learning.

Summary and key points

Teaching that prioritises 'socio-cultural diversity' and 'decision-making and social action' enables you to foster a greater spirit of enquiry through investigation and reflection. Such enquiry must involve the interrogation of 'attitudes' and the building of confidence to intervene in the way people construct each other. To do otherwise is to do all pupils a disservice and remain out of harmony with the fundamental need for change in society. Young people derive enormous satisfaction from making discoveries about themselves and others through art practice. I argue here that contemporary practices are the ideal vehicle for engaging innovative and relevant pedagogies in today's world. You should make it your business to employ strategies for teaching that promote meaningful engagement in these discourses by adopting a more integrated and risk-taking approach to teaching and learning.

Further reading

Doy, G. (2000) *Black Visual Culture: Modernity and Postmodernity*, London and New York: I.B. Tauris.

Dyer, R. (1997) *White*, London and New York: Routledge.

Shohat, E. and Stam, R. (1994) *Unthinking Eurocentrism: Multiculturalism and the Media*, London and New York: Routledge.

UNIT 11.2 HISTORIES AND CANONS AS FORMS OF IDENTITY

Nicholas Addison

By the end of this unit you should be able to:

- identify and evaluate the canons which underpin the curriculum in art and design;
- question the function of current orthodoxies;
- consider the resource implications for teaching a pluralist curriculum.

On canons

When you make reference to historical work to provide a stimulus for learning in art and design what do you select? Should you privilege the 'western European' canon presenting it as a progression of historically specific periods making only passing reference to the 'wider world' (DfEE 1999: 20), or are you more inclusive? (see Addison and Burgess forthcoming). Do you privilege the 'fine' arts at the expense of the 'applied' arts, craft and design and thereby fall into the danger of perpetuating the hierarchical, gender and class divisions implicit in that neglect? (Parker and Pollock 1981). Should you, as Gretton (2003) argues, locate your investigations within the canon so as to provide entry into those forms of art that provide cultural capital for pupils who are often marginalised from its discourses?

The gender and class bias of many reference collections in art and design departments is often compounded by a racial and historical bias in the form of a defence of two 'European Traditions', classicism and modernism. The former is often presented as the great 'western' tradition having its roots in Antiquity, located geographically in Greece and Rome, and now perceived as the seed-bed of modern Europe (Unit 12.3). It would be less misleading to point out that historically these locations constituted the northernmost outposts of a Mediterranean oriented world centred around Egypt, encompassing north Africa, Arabia and southern Europe (Bernal 1987). With the collapse of the Roman Empire, partly as a result of the invasion of 'barbaric' hordes from the north east (the modern ex-Soviet Union), the classical tradition is supposed to have dissipated. The ensuing period, the 'Dark Ages' just happens to coincide with the supremacy of the Eastern Church in Byzantium and the rise of Islam (AD 622). Classicism is supposedly revived in the Humanist, Italian Renaissance, finding its vestigial reappearance in Romanesque and Gothic, the styles of northern Christendom; Islam, as the repository and perpetuator of the classical tradition, is usually ignored. In this trajectory Britain is seen as its partial heir, adopting classicism from the seventeenth century as a unifying tradition visually manifest in the

Neoclassical architecture of Empire, although at home classicism's pagan and continental roots were questioned in the patriotic and Christian Neo Gothic during the nineteenth century.

Task 11.2.1 Mapping from a centre

Devise a SoW in which pupils collectively map a culture using its 'capital' as the centre, e.g. the late classical world seen from Rome.

How might the Roman perception of Britain and other territories be visualised? Provide texts and representations to assist pupils' investigation of these perceptions.

Task 11.2.2 Architectural style: purity and hybridity

Devise an investigative SoW based on a survey of architectural styles in the local vicinity, e.g. the local High Street or shopping mall, where pupils could differentiate and record classical, gothic, orientalist, modernist and other stylistic features in the architecture.

How might you develop this SoW?

The other dominant tradition, modernism, is often perceived as an attack on classicism, and, depending on your position, blamed or praised for the *individualism, essentialism, autonomy and anarchy* of western art in the twentieth century.

Individualism: Teachers often exemplify practice with reference to the work of a relatively small number of individual artists from amongst the pioneers of modernism (QCA 1998a; Downing and Watson 2004). This tends to encourage biographical readings which reinforce the myth of the isolated male genius, the story of a series of misunderstood outsiders producing new and radical forms only to be superseded by the following generation: it makes for good copy with its archetypal anti-hero leavened by scatological anecdote. Walker (1983: 46–48) counters the myth of Van Gogh as a mad genius (the most referenced artist at KS 1, 2 and 3) by focusing on his collaborative and social ambitions. The emphasis in schools on individualism can be countered by investigating collective practice, the relationships between the arts, intercultural dimensions, and by issue or theme-based approaches.

Essentialism: An essentialist approach suggests two paradoxical paths, either a belief in discrete traditions particular to a people or class, e.g. Expressionism as an indivisible part of the German tradition, or, the possibility of a universal language which can be understood uncritically because it stems from a basic biological core, e.g. 'all people see in the same way', therefore 'naturalism is universal'. By ensuring that you investigate modernism as an international and collective endeavour (Blazwick 2001) myths of national purity can be challenged, and by exploring difference across time and culture biological determinism can be questioned.

Autonomy: The notion of modernism as an autonomous entity results in art's dis-association from all contexts and criteria other than its formal properties. This invites teachers to train pupils in the 'basic elements': line, tone, colour, pictorial space, etc., in preparation for an eventual 'mastery' which can be applied to other contexts in the real world at a later stage. The implicit limitations of this approach can be countered by exploring the metaphoric and communicative functions of art and by shifting attention from the fine to the applied arts where it is difficult to ignore function and audience.

Anarchy: At its margins modernism is seen as anarchic and irresponsible: Taylor (1989) suggests that its ideology is strictly anti-bourgeois and that its styles are strategies for undermining academic conventions (p. 103). By extension, you might wonder whether modernism is appropriate for fledgling consumers of a bourgeois democracy? Modernism has indeed been allied to revolutionary politics, from both the left and the right, and has thus been associated with utopian schemes to better the lot of the masses. Constructivist, Futurist, De Stijl, Bauhaus and International modernist architecture and design have radically altered the conditions in which many people live in industrial cities throughout the world. Because modernism has at its centre the notion of perpetual change and the possibility of progress, both individual and collective, it is attractive to teachers who have frequently been educated from within its precepts. The anarchy of modernism is clearly not the whole story and needs to be related to the continuous self-criticism that modernism has put itself through, a reflexive process that is its strength. Foster (1996) calls this the 'counter-project' and believes that this continuing critique is a constructive and reflexive check against the totalitarianism which dominated so much of the twentieth century.

The canons which are the result of classical and modernist genealogies are reflected in available resources and sustained in the reference materials constructed specifically for the classroom, e.g. packs of reproductions (Chapter 5). If you present these canons as givens there is the danger that they will be acritically reinforced. Nonetheless, as well as expanding on the canon to demonstrate the diversity of contemporary practices, the canon can be a productive point of departure, after all you can question canons by investigating how they have been constructed and what and whose values they represent. Additionally, Gretton (2003) argues that it is disempowering to deny pupils access to what is deemed most valuable in society, the canon may provide the cultural capital that a pupil's home situation may preclude. It is vital, therefore, that you transform what may appear limited or pedestrian resources through your investigative tools (galleries, Internet, etc.) and approaches (content analysis, semiotics, see Rose 2001) and invite pupils to add to the available store by collecting and contributing examples from their home lives and personal interests. This way they can begin to relate the content and forms of the school resources to their immediate experiences.

In schools it is usually the originators of modernism who are held up as exemplars, in particular those oppositional artists based in Paris, from the Impressionists to Cubism. However, despite their more overtly transgressive credentials, Dada and Surrealism are not exempt from this attention and many of their stylistic if not conceptual traits have been adopted by the commercial world and are familiar in the classroom. This to-ing and fro-ing between 'high' and 'popular' forms is supposedly typical of postmodernism and pupils often enjoy finding that they can recognise and site sources. But as suggested, such boundary crossing can be seen as a central strategy of modernist artists from the Impressionists and their concern with street life and café culture through to such cultural movements as the Mexican muralisits, the Harlem Renaissance and Pop art and on to the appropriative practices of contemporary artists such as Jeff Koons, Yasumasa Morimura and Chris Ofili or the collaborative work of Dorothy Dedeaux, Felix Gonzalez Torres and Sam Taylor Wood. The opposition that is sometimes set up between modernism and postmodernism can be unhelpful and lead to dismissive attitudes one way or the other; I recommend that you look for continuities as well as ruptures (Foster 1996; Harris 1996) so as to help yourself and pupils navigate the complexities of modernity and avoid simple binary oppositions.

Task 11.2.3 Resourcing an intercultural modernism

Collect resources to enable you to introduce pupils to a wider and intercultural definition of modernism, extending your examples to contemporary practice.

Quality or interest?

Having produced a reference development plan (see Task 5.2.5) one of the difficult choices you encounter is the vexed question of quality. This is considered in relation to pupils' own work in Chapter 7. However, when you investigate the work of others, referencing available canons would appear to pre-empt the question. You have already experienced difficulties in selecting exemplars for discussion and investigation from limited resources with the added pressures of what may appear the burden of political correctness. As recent revisionist histories have demonstrated the western canon can be, and now is, more inclusive. Said, reflecting on this collective endeavour writes:

> It was never a matter of replacing one set of authorities and dogmas with another, nor of substituting one centre for another. It was always a matter of opening and participating in a central strand of intellectual and cultural effort and of showing what had always been, though discernibly, a part of it, like the work of women, or of blacks . . . but which had either been denied or derogated.
>
> (in Hughes 1993: 113)

Although quality will tend to have its day, recognition can be indefinitely postponed. It would seem that the canon is not synonymous with tyranny, although there have been instances of conspiracy. Art and politics may not be one and the same thing, but they are mutually bound in a social process of advocation, reinforcement, subversion, suppression and denial.

Revisionist histories have demonstrated that what constitutes art and whether or not it is considered 'great' are not necessarily the only values that are significant for education. Quality is a value that can be measured only within its own terms of reference; as Goodman (1976) argues the question 'what is art?' can be countered with 'when is art?' implying that both what art is, and whether or not it is good, is dependent on historically and culturally specific interpretative communities, not on some universal 'significant form' as suggested by Fry (1909) and Bell (1914) (both in Frascina and Harrison 1982). Therefore, to develop visual and aesthetic literacy you need to encounter a more comprehensive representation of the diversity of visual and material culture. Add to this the absurdity of applying aesthetic criteria to art objects which hold no direct relationship to those criteria and you have a recipe for reinforcing difference and inferiority: for an examination of the mis-attribution of European influence on the Benin bronzes see Coombes (1994). Criteria can also be determined by generation and gender: Rubin (1969), in defining Louise Bourgeois as a 'tenacious artist of uneven but highly personal accomplishment' (p. 17) assumes that consistency and universality are fundamental criteria (modernist and patriarchal?) by which any self-respecting artist should assess their work. For others, diversity and interest may take the place of consistency and quality: posing new questions may be as illuminating as answering old ones. The lesson here is that you are liable to project your own values onto all objects from whatever culture; to overcome this you must attempt through dialogue, to meet difference on its own terms: 'The cultural object can never be an empty vessel waiting to be filled with meaning, but rather is a repository replete with meanings that are never immanent but always contingent' (Coombes in Preziosi 1998: 489).

Task 11.2.4 Relative criteria

Investigate the criteria by which two different cultures value works of art.
  Evaluate an object from one using the criteria of the other and vice versa.
  Now evaluate each object using the criteria of its own culture.
  How do the results differ?
  Discuss in tutor groups what implications this has for the way you present objects from different cultures to pupils.

Further reading

Gretton, T. (2003) 'Loaded canons', in N. Addison and L. Burgess (eds) *Issues in Art and Design Teaching*, London: Routledge/Falmer.

Parker, R. and Pollock, G. (1981) *Old Mistresses*, London: Pandora.

Pollock, G. (ed.) (1996) *Generations and Geographies in the Visual Arts: Feminist Readings*, London: Routledge.

UNIT 11.3 ENGAGED APPROACHES TO CRITICAL AND CONTEXTUAL STUDIES

Nicholas Addison

By the end of this unit you should be able to:

- differentiate and evaluate the approaches of multicultural, anti-racist and intercultural educational strategies;
- question the gender and class values embodied in the objects of visual and material culture;
- examine your own place within, or relationship to, particular traditions;
- examine art as a form of communication and meaning making;
- consider the social uses of art, incorporating the issues that arise from that use into the curriculum.

Investigation and inclusion

Multiculturalism

If you examine the term 'multiculturalism' you will discover that the concept of race is central to its principles. When using the term throughout this book authors do not intend to collude with essentialist notions in which race is perceived as a biological given. Rather, for us, race denotes a cultural construct which frames the way in which people identify themselves or are identified by others. As such race is not a fixed or absolute entity but a constantly shifting system of identification including notions of hybridity or indeed boundlessness (Hall 1990 and 1997; Ladson-Billings and Gillborn 2004).

Within Britain the origins of multicultural education are inextricably linked to its colonial past and there is much visual material in its cities and ports to serve as evidence for an investigation into such historical issues as trade, empire or slavery (Piper 1997). More immediately, post-war demographic change within the Commonwealth and the European Community has secured a multicultural and plural British society in which the influence of different cultures is no longer peripheral or mediated through translation, but central. Although the cultural effects of eastern European migration have yet to be understood, since the 1950s the impact of diasporic communities from Africa, Asia and the Caribbean has dramatically transformed whole areas of British culture from its cuisine and music to the fashion industry. These influences are most apparent visually in commercial and popular forms for it has been particularly difficult for Africans and Asians to gain recognition in the fine and applied arts (Araeen 1989). In asking pupils to address their immediate environments, the multicultural fabric of Britain cannot be denied (see Plate 24). However, you have already made an audit of the resources available to you and may have traduced pockets of invisibility which you are attempting to fill.

The notion of *visibility* is the key to a multicultural approach and it tends to be pluralist and celebratory, reinforcing difference but in the name of tolerance. Holt cautions against what he terms 'the presentation of a cultural kaleidoscope'

(1996: 131) because spectacular, indiscriminate exposure can distort cultures by mis-representing them through stereotype and by allowing historical artefacts to stand in for today's; as if the cultures of the native peoples of Africa, America, Asia and Australia were unchanging. Alternatively, difference in cultures can be sidelined in the integrationist attempt to find an underlying human unity. This universalist approach to the arts has found many adherents within modernism, indeed one of its most notable features, that underneath the conventions all the arts, and thus all people, are essentially the same, has been held responsible for the increasing hegemony of western culture:

> If evolutionism subordinated the primitive to western history, affinity-ism recoups it under the sign of Western Universality . . . in the celebration of human creativity the dissolution of specific cultures is carried out.
>
> (Foster 1992: 24)

Gombrich's ubiquitous bestseller *The Story of Art* (1950) is amongst the texts most used in schools and it can reinforce the notion that significant art is the preserve of European history. Both Kemp's (2000) and Honour and Fleming's (1982) primarily western art histories, although presented in encyclopaedic rather than narrative form, are much less partial in their surveys. There is a vast literature on art from around the world, but it is often allied to the tourist industry, providing a spectacle of exoticism or domesticated difference rather than historical, theoretical or critical comment. Publications recounting aesthetic theories outside Europe do exist: take advantage of your time as a student to study these. You may hear the often repeated falsehood that 'Art' is a European concept and therefore only western objects can be examined from the point of view of 'Aesthetics'. A reading of Coomaraswamy's (1934) comparative analysis of medieval theories of art in India, China and Europe soon evaporates such misinformed claims. Cheng (1994) provides a semiotic analysis of the tradition of Chinese painting focusing on its pictorial structures rather than its history. In relation to more recent cultural history, Brett (1986), Caruana (1993), Baddeley (1994), Powell (1997), Poupeye (1998), Oguibe and Enwezor (1999), acknowledge participation by artists and designers in the development of modernism outside Europe and the USA. The journal *Third Text* is a critical forum for contemporary, international artists and critics and in the UK organisations such as inIVA and the Diversity Forum promote international artists and have useful websites with extensive links.

Task 11.3.1 Multicultural revision

Review the SoW that both you and colleagues have devised at your placement schools. Analyse them from a multicultural perspective.

If any of them are uncritically Eurocentric how can you revise them so that a multi-cultural dimension becomes integral?

Further reading

Boughton, D. and Mason, R. (1999) *Beyond Multicultural Art Education: International Perspectives*, New York, Berlin: Waxman Munster.

Mason, R. (1995) (2nd Edition) *Art Education and Multiculturalism*, Corsham: NSEAD.

Swann (1985) *The Swann Report*, London: HMSO. (Particularly Annex D.)

Anti-racism

Concomitant to this dual and contradictory attempt to embrace or integrate difference is the continuing legacy of racism. An ideology of difference was essential to the European colonial experiment (Coombes 1991) and was particularly denigrating in relation to Africans and Asians as it relied on the notion of the 'Primitive' (Hiller 1991). The relationship between the visual arts and this concept is somewhat problematic, for the primitive has been embraced as an ideal by many European thinkers from Rousseau to Deleuze, and the process of appropriation evident in Primitivism (Rhodes 1994) has been central to the strategies of western artists from Gauguin to Hesse (Rubin 1984). This is further complicated by the fact that the art of children, the psychotic, even the amateur, is sometimes included within this definition (Goldwater 1986). Nonetheless, an anti-racist approach acknowledges that the colonial tendency to 'primitivise' others has left a legacy that is still apparent in many areas of British and European life and the ability to deconstruct a propaganda of difference is central to its strategies. It is therefore important to engage critically with instances of western orientalism and primitivism especially when they appear from within the canon and the circulation of images within popular culture. For example, before World War I a number of artists working in France, Germany and England appropriated the direct carving and 'open form' practised by west African wood carvers (amongst other sources) and such technical and formal procedures offered new formal possibilities for reviving a sculptural practice that had been largely tied to moribund academic models. But it was equally, say in the instance of Picasso, a strategy for self-aggrandisement, where a racist sign of 'savagery' was deployed to mark out a vanguard position in absolute opposition to bourgeois taste and morality. If you acknowledge how visual culture, including the participation of artists and designers, has contributed to colonial representation and how it continues to produce the visual landscape of globalisation, pupils can begin to understand how formal decisions, so often presented as stylistic choices in the western quest for originality, are in fact signs of difference that have ideological implications. Such choices are not neutral but may well serve the needs of colonial oppression or globalisation by perpetuating stereotypes or, indeed, they may counter them and set up forms of resistance. Said's *Orientalism* (1980) provides a detailed examination of the way European colonial texts are projections masquerading as scientific scrutiny and it is an invaluable historical resource for any investigation of the continuing demonisation of Islam. Oguibe and Enwezor's *Reading the Contemporary: African Art from Theory to the Marketplace* (1999) and Pinder's *Race-ing Art History* (2002) brings these historical debates up to date. In addition the exhibition at the Hayward Gallery '*Universal Experience: Art, Life and the Tourist's Eye*' (Bonami 2005) curated around the theories of Urry (2002) provides insights into the ways stereotypes can be used to exoticise geographical others by fashioning them as both desirable and obtainable and in the process denying them their subjectivities and lived realities.

It may be that you wish to bring younger pupils' attention to these issues by using more immediate and readily available resources: Pietersie (1992) has produced a telling analysis of the image of 'Africa and Blacks' within advertising, and his methods of interpretation could be used in the classroom to interrogate contemporary adverts. Holiday brochures can provide you with an endless image supply and it is an instructive task for pupils to collect images and texts about, for example, tropical countries, categorising them into types and then comparing them to alternative information gathered from geographical and anthropological texts. Pupils would then be able to define those tropes by which the 'other' or the 'exotic' become desirable to a post-industrial/post-colonial society and perhaps present their findings in the form of an annotated photomontage.

As has been noted, a particularly pernicious legacy of colonial racism is the perpetuation of difference through stereotype. Hall argues that stereotyping is a process through which people define who they are by saying what they're not, 'It classifies according to a norm and constructs the excluded as "other" ' (1997: 259). In this way powerful groups construct regimes of representation to reproduce stereotypes which are deployed to keep subordinated and marginalised groups in 'their place'. For example, nationalist rhetoric or colonial discourses of dependency may be invoked as the means by which dominant populations are brought alongside. During the 1980s many black artists began to form collectives in order to develop strategies of resistance to counter this process (Hall 2005). Indeed, under the regimes of the GLC and Arts Council some groups gained state funding, thereby increasing the visibility and, to some extent, the recognition of marginalised voices. Despite the fact that a number of artists, critics and curators involved in this movement are ambivalent about its long-term effects (Araeen; Himid; Khan in Bailey *et al.* 2005), particularly because of the way it perpetuated differences in relation to ethnicity. However, despite the regrets, there are now organisations which do recognise at the level of policy how Britain is, and has always been, an intercultural site where notions of racial purity are ideological fictions deployed for the construction of national, imperial and class identities; in this respect ethnicity is no longer a criteria for inclusion in, for example, the diversity forum. However, in the discourses of the mass media, the street, the school curriculum and the playground many racist stereotypes, and particularly those based on skin colour, are securely embedded and can still be played out in increasingly divisive and insidious ways (Troyna and Hatcher 1993). This is why it is still important to question such stereotypes in schools and to offer a more complex understanding of identity (Skeggs 2004) in which age, class, gender, race and sexuality intersect and overlap to produce the way pupils see themselves, and, importantly, are seen by others; the work of artists can be one source for this counter-project.

The inIVA *Portrait: Education Pack* (Malik 1996) provides an accessible resource to introduce a range of strategies employed by artists to counter these stereotypes; I refer to a number of the reproductions from this pack in the categories below but add to them to meet both local interests and to ensure examples are current. These strategies include:

a) self-representation

For example Sonia Boyce *She Ain't Holding Them Up, She's Holdin On (Some English Rose)* (1986). In this well-known pastel, at one time much imitated in schools, Boyce represents herself and family members in a collage-like composition, juxtaposing patterned, photographic and mirrored elements to produce what at first appears a confident self-portrait. Further looking reveals ambiguities and tensions, already evident in the title, so that signs of family and nationhood, gender and sexuality tell of both belonging and rejection; as Malik suggests: 'this English rose is black and she is staking a claim to a British identity from which she has been excluded' (1996: slide note 22). Likewise your pupils' experience is unlikely to be uniform and image making can provide a vehicle for defining and exploring aspects of identity using material and symbolic means and thus avoiding overtly confessional idioms.

b) subverting stereotypes

For example Chila Kumari Burman's *Walk Tall – Self Portrait* (1995) questions the notion of the passive Asian woman by layering multiple signs: a demure facial portrait, a martial arts' kick, graffiti, some of which may be seen as contradictory, conflicting or perhaps complementary? As well as countering stereotypes such images suggest the possibility of multiple identities, a phenomenon which has been called 'hybridity' (Hall 1990).

An alternative subversive strategy is to negate the power of stereotypes through ironic appropriation (see Plate 14). Chris Ofili's *Captain Shit* series (1997) presents the hero (anti?) of 1970s 'blacksploitation' movies as both a figure of uncritical adoration and absurd macho preening.

The US artist Kara Walker looks to the past for her representational resources appropriating eighteenth- and nineteenth-century silhouettes to produce mural-sized installations depicting the history of slavery and intercultural relations in the southern states, e.g. *Insurrection* 2002. Because she makes use of degrading stereotypes her work has been roundly criticised by some for sensationalising a painful past. Yet her narratives play on assumptions and prejudices, enticing viewers into a seemingly picturesque world of pastoral dalliance and childhood illustration only to subvert expectations by revealing a tableau of corporeal and sexual abuse. As she says: 'Working with such loaded material as race, gender, sex, it's easy for it to become ugly. I really wanted to find a way to make work that could lure viewers out of themselves and into this fantasy. Seduction and embarrassment and humor all merge at a similar point in the psyche, vulnerability' (Sheets 2002). Although it is mostly figures from the 1960s' Black Civil Rights movement who have found her work offensive, it is important, as hooks argues, to include 'white interventions' within Black history, although, and perhaps because, whites (and others) find this threatening:

> To name that whiteness in the black imagination is often a representation of terror: one must make a palimpsest of written histories that erase and deny,

that reinvent the past to make the present vision of racial harmony and pluralism more plausible.

(hooks 1992: 338)

c) deconstructing stereotypes

Deconstruction is a more analytical process which enables you to seek the ideological structures which support systems of power, in this instance, racism. The expanding territories and methods of art history can demonstrate how to deconstruct colonial texts as in Coombes' exacting exposé of the racist theory of degeneracy applied to the material culture of Benin by nineteenth-century anthropologists and scholars (Coombes 1994: 43–62). Whole areas of history can be ignored to reinforce a racist ideology in which the fine arts are written up as the preserve of a dominant people who define their power in terms of race; the exclusion of Afro-Caribbeans from the development of art in the USA is a case in point (Boime 1990). Exposing these structures allows you to recognise the methods and devices used to construct difference. One way in which these structures of difference can be theorised is as a form of projection, in which the secret or repressed desires of whites are played out in their representation of others (Said 1980). The endless portrayals of Harem scenes in nineteenth-century painting are a case in point (Smith 2001) and the stereotypes of the 'lascivious and indolent oriental' and the 'insatiable black' are repeated in the cinema, literature and advertising to this day (Addison 2002).

The historical turn required for an investigation of stereotypes feeds into the next category in which the hidden or repressed experiences of subjugated peoples are revisited and re-presented in order to reconfigure western metanarratives and rewrite history.

d) re-configuring/rewriting histories

e.g. Keith Piper's *Trade Winds* (1992; Plate 17), now included within a CD-ROM (Piper 1997), is a multimedia installation, comprising a dozen makeshift crates containing monitors showing images of imperialism and slavery. It explores the concept of measurement as a system and tool of control. Through image manipulation Piper is able to intervene, juxtapose, refashion and transform the representations of the past in such a way that their fictions are exposed and the viewer is invited to 'rewrite history'.

Task 11.3.2 Countering stereotypes

Devise a SoW that draws on some or all of the strategies outlined above.

How can you introduce these strategies so as to avoid the polarity between guilt and victimisation which could be the unintentional result of confrontational methods?

Further reading

Hall, S. (1997) *Representation: Cultural Representations and Signifying Practices*, London: Sage in association with the Open University.

Ladson-Billings, G. and Gillborn, D. (eds) (2004) *The RoutledgeFalmer Reader in Multicultural Education*, London: RoutledgeFalmer.

Powell, R. (1997) *Black Art and Culture in the 20th Century*, London: Thames and Hudson.

Interculturalism and hybridity

Increasingly, the difficulties faced by teachers in navigating such sensitive territory and the sometimes inhibiting effect of political correctness have encouraged commentators to develop a new approach, which, while acknowledging difference, seeks to focus on the interaction of cultures. For Maharaj (2000) 'interculturalism is the scene of translation and transmission': it constitutes the perpetual condition of diaspora that has produced and transformed cultures (see Plate 22). In historical terms such inquiry would have to focus on migrations and conquests, trade routes and monopolies, and, sure enough, investigations into the transference and transformation of cultural forms through acquisition, appropriation, imposition and diasporic cross-fertilisation do exist. For example Riegl examines the so called 'barbarisation' of the Roman Empire and the development of its decoration by Byzantine and Islamic artists in *Problems of Style* (1893) mentioned in his *Late Roman Antiquity* (1901) (in Fernie 1995: 120–126) and, more recently, Jardine (1996) argues that the trade between east and west during the supposedly western and Christian Renaissance, was in fact initiated and fuelled by intercultural exchange, particularly with Islam. However, it is perhaps the contemporary situation of post-colonialism and the transcultural phenomenon of hybridity that should engage your attention here.

Hybridity is the term most often used to define a condition of 'inbetweeness', the coexistence and cohabitation, not the integration, of difference; a paradoxical unity in diversity. The construction of hybrid forms has often been used as a subversive or transgressive strategy by artists wishing to challenge aesthetic and political authority. Unlikely juxtapositions, graftings and confrontations are particularly evident in the work of artists from Dada and Surrealism, such as Ernst's defamiliarisation of the familiar, through to the present day with Orlan's reconstruction of her own body. However, hybridity has become one of the central tenets of postmodern pluralism and has infiltrated most areas of cultural discourse. For example the ethnographic museum has become the site for cultural exchange in such exhibitions as *Lost Magic Kingdoms and Six Paper Moons from Nahuatl* (1986):

> Crucially, the deliberate focus on transculturated or 'hybrid' material culture has also been promoted as the sign of a mutually productive culture contact – an exchange on equal terms between the western centres and those groups on the so called 'periphery' . . . hybridity has often been an important cultural strategy for the political project of decolonisation. Additionally the

> many manifestations of creative transculturation by those assigned to the
> margins do potentially provide productive interruptions to the West's
> complacent assurance of the universality of its own cultural values. And
> certainly, the celebration of hybridity also implies an acknowledgement of
> the ways in which western culture has been, and continues to be, enriched
> by the heterogeneous experience of living in a multi-ethnic society.
>
> (Coombes 1994: 217)

Often the site of this hybridity has been identified as the western, cosmopolitan
city, the Modernist 'centre' and its migrations: Paris, New York, etc. This has been
complicated into the notion of centres: Berlin, Mexico, Tokyo and so on, but these
revisions merely multiply the centres of power rather than question the city as the
determinant of modernity. A new separation is ensured: only the citizens of this
global and international community can produce art that is 'cutting edge', that pushes
the boundaries. This still leaves peripheral territories whose inhabitants cannot be
expected to participate in international discourse except as outsider guests:

> While the European artist is allowed to investigate other cultures and
> enrich their own work and perspective, it is expected that the artist from
> another culture only works in the background and with the artistic
> traditions connected to his or her place of origin . . . If the foreign artist
> does not conform to this separation, he is considered inauthentic,
> westernised, and an imitator copyist of 'what we do' the universal is
> 'ours, the local is yours'.
>
> (Garcia Canclini in Preziosi 1998: 506)

In schools this can manifest itself in expectations about the behaviour and aesthetic
practices of pupils from ethnic minorities who may be encouraged to conform to
stereotypical, essentialist notions of their family's culture. British and US artists
have explored this issue, notably Rasheed Araeen, Lyle Ashton-Harris, Sonia Boyce,
Mona Hatoum, Lubaina Himid, Issac Julien, Steve McQueen, Chris Ofili, Adrian
Piper, Donald Rodney, Yinka Shonibare and Maud Sulter, whose work questions (or
celebrates ironically?) dominant representations of identity.

Task 11.3.3 Pupils: a multicultural resource

Discuss in tutor groups:

- how to encourage pupils to investigate, value and represent the diversity of their cultural backgrounds;
- how to include pupils' knowledge of additional languages and traditions in the art and design curriculum.

Further reading

Araeen, R. (1989) *The Other Story*, London: Hayward Gallery.

Bailey, D. A., Baucom, I. and Boyce, S. (eds) (2005) *Shades of Black*, Durham, NC: Duke University Press in association with inIVA and Aavaa.

Cohen, P. (1998) 'Tricks of the trade: on teaching arts and "race" in the classroom', in D. Buckingham (ed.) *Teaching Popular Culture*, London: UCL Press.

Hiller, S. (1991) *Myths of Primitivism*, London: Routledge

Third Text is a quarterly journal presenting perspectives on contemporary art from the developing world.

On gender

When you critically investigate art, craft and design with your pupils and, more broadly, the social and cultural contexts within which they have been produced, it is not only questions of race that arise but also those of class and gender. The same concern with visibility and the rewriting of histories became central to feminist discussion of the arts in the 1970s. Nochlin's *Women, Art and Power* (1991) and Chadwick's *Women, Art and Society* (1990) examine the barriers inhibiting female participation in western fine arts. They also celebrate those women who managed to contribute to this history despite the dominance of patriarchal discourses and institutions. The privileged status that has accrued to the fine arts in western culture is questioned in *Old Mistresses* (1981) in which Parker and Pollock argue that certain forms of craft activity constitute a more representative example of women's cultural production and the exchange of meaning. By questioning the universal connotations of a hierarchical, male aesthetic dependent on the notion of genius they also add ammunition to those anthropological and Marxist theories which already proposed more inclusive definitions of the arts in an attempt to undermine exclusively western and bourgeois perspectives (Clifford and Marcus 1986; Geertz 2000; Hauser 1951 [new edition 1999]; Bourdieu 1993). However, because dominant practices are seen as determined by male ambition, the notion of negotiated practice has been espoused as a pragmatic measure. Pollock (1996c) quotes Susan Hiller:

> The idea was that women were creative and their creativity came out in unrecognised ways. And that this was the way of the future. This should be encouraged and one should withdraw from male notions of exhibitions, careers, vast projects and goal oriented work. Any of these sorts of things were seen as wrong. In the United States the women who made it are still seen as having sold out. Of all the political movements feminism, above all other, demands that we live our ideal feminism in a society that is not feminist in a way that socialists are not required to live their socialism in a non-socialist society, to the same extent. Yet on the other hand we agree that our actions now must pressage future social relations. Now we face the

situation of having to live with the struggles and make some peace with them.

(p. 57)

One way of avoiding compromise is through forms of critical practice, or what Pollock (1988) has termed feminist 'interventions', an approach that is methodologically eclectic using the critical tools of complementary disciplines, in particular, sociology, semiotics, anthropology and psychoanalysis. This emphasis on interdisciplinary alliances has had the effect of forming what constitutes a new subject, namely, Visual Culture (Bird *et al.* 1996; Mirzoeff 2002; Elkins 2003). Such alliances have not however, until recently, characterised critical studies in schools which, although materialising alongside these developments, tended to use a language rooted in more conventional and supposedly neutral discourses (Eisner 1972; Taylor, R. 1989).

Task 11.3.4 Patterns of gender

In tutor groups discuss the gender make-up of both the students and the teaching staff when you were at college and professionals you have witnessed in 'industry'.

Can you discern any patterns that suggest gendered divisions of practice?

If so, what are the reasons for these divisions?

How can you question these divisions in the way you present the work of others to pupils?

Review some of the advisory comments you have written for pupils. Can you see any differences in the type of advice you write for girls as opposed to boys?

Further reading

Chadwick, W. (1990) *Women, Art and Society*, London: Thames and Hudson.

Dalton, P. (2001) *The Gendering of Art Education: Modernism, Identity and Critical Feminism*, Buckingham: Open University Press.

Meskimmon, M. (2003) *Women Making Art*, London: Routledge.

Parker, R. and Pollock, G. (1981) *Old Mistresses*, London: Pandora.

On class

Art in Britain is frequently perceived as a special and privileged domain, with its objects made for and by a privileged minority. In order to question this perception:

Task 11.3.5 Grounded aesthetics, grounded taste

Ask each pupil to identify and define art, craft and design, suggesting they investigate the origins of their beliefs by seeking the perceptions of their family and friends and by collecting evidence of how art, craft and design is reported, received, consumed, and produced in their home and local community.

In small groups ask pupils to present, discuss and record their findings followed by a whole class debate in which their evidence is examined for consensus and difference.

To ensure pupils consider their perceptions in relation to broader contexts you should provide them with social and historical information to help them locate the conditions which determine or make possible such perceptions.

Additionally you can ask them to research the interplay of supposedly antithetical class-based phenomena, for example by resourcing an investigation into the appropriation of popular forms by high culture and vice versa: commercial imagery in Pop, fine art in advertising.

Popular culture

Many teachers enter the profession in their twenties, leaving little distance between their own age and that of the older pupils. Some of you may still be actively engaged in the production and reception of 'youth' culture, the aspect of popular culture with which your pupils are likely to be most familiar. Others may feel a need to familiarise themselves with it in order to get to know a significant area of pupils' lives. Whatever the motivation, it is a good idea to develop ways of allowing pupils to reference popular culture because for them it may be extraordinarily potent, a world with which they can readily identify. Hebdidge (1988) and McRobbie (1994) record changes in youth styles and develop modes of writing that, while remaining critical, draw on autobiography and popular culture. Willis (1990b) demonstrates how exclusive attention to 'high' culture can have an alienating effect on pupils and suggests ways in which the presentation of popular culture can aid self-esteem and allow pupils to engage in the same learning skills as they would if the subject were more canonical. It is vital that at some point you consider your pupils' own social contexts, not just in relation to how you teach them but in what you teach as well. In considering 'where they are coming from' you can pursue so called 'mind to world' strategies (Cunliffe 1996: 315–318) but compare these to the 'world to mind' strategies (Cunliffe 1996: 318–325).

Many teachers despair when pupils turn to stereotypes and popular logos. In the last unit you have seen how such forms can be questioned, but is it necessarily a negative phenomenon when pupils have recourse to familiar and dominant signs? What if these signs become the object of parody, a process akin to the ironic appropriation mentioned in the previous section:

> Of course, parody necessarily entails imitation, although imitation does not have to be parodic. Yet when it comes to discussions of popular culture, this distinction is also highly pejorative. Imitation is seen here as an essentially

unthinking process, in which particular behaviours, values and ideologies are simply reproduced – and therefore reinforced. Parody on the other hand, is generally seen to be a matter of conscious deliberation. While parody does not necessarily involve a rejection of that which it parodies, it must involve a form of critical distance from it – however affectionate it may be.

(Buckingham 1998: 68)

Task 11.3.6 Parody as a critical strategy

Discuss with other students how to encourage pupils to parody popular forms rather than imitate them.
  What other strategies can you think of to enable pupils to use popular forms constructively and/or critically?

When giving attention to popular culture it is worth considering the possible differences between 'youth', 'commercial' and 'home' cultures, all of which might be seen as belonging to its orbit. The popular and the commercial (mass media and consumer products) are sometimes presented as synonymous, but youth and home cultures often draw on activities outside the screen or the shopping and leisure centre, from decorating a personal space to caring for relatives; from energetic military training to profound religious devotion. If the classroom environment is a secure one, the things closest to your pupils are likely to prove motivating.

Task 11.3.7 Building on pupils' interests

Discuss with other students:

- how to discover pupils' interests;
- how to build pupils' confidence so that they can present and discuss their home culture with others.

Devise a SoW that starts from pupils' interests and moves out to investigate them in relation to historical precedents.
  How could Task 11.3.5 be presented visually?

Further reading

Buckingham, D. (ed.) (1998) *Teaching Popular Culture*, London: University College London Press.

Fiske, J. (1989) *Understanding Popular Culture*, London: Routledge

McRobbie, A. (1994) *Postmodernism and Popular Culture*, London: Routledge.

Willis, P. (1990a) *Moving Culture*, London: Calouste Gulbenkian.

Sexuality and art and design

It is in the context of a discussion of popular culture (for example) that you might introduce the issue of sexual orientation, in particular non–heteronormative sexualities such as homosexuality. Modern artists have frequently questioned the dominant, bourgeois moralities that have produced and naturalised heteronormative values in western culture, although it is not until recently that non–heteronormative groups have been afforded legal rights and a degree of equality as citizens (Stanley 2007). In ensuring that aspects of citizenship are integrated into teaching across the curriculum QCA (1998b) point out:

> We must recognise that teaching about citizenship necessarily involves discussing controversial issues. After all, open and informed debate is vital for a healthy democracy . . . Teachers are aware of the potential problems and are professionally trained to seek for balance, fairness and objectivity.
>
> (p. 1.9)

The investigative questions provided in *Working with Modern British Art* (Adams *et al.* 1998: 46) direct pupils' attention to a painting by Hockney, *Third Love Painting* (1960) which represents, partly covertly, aspects of homosexual desire. It was painted at a time when homosexuality itself was illegal in Britain. As with many artists of the 'Pop' generation, Hockney makes reference to an eclectic and culturally diverse range of sources, from Whitman to Cliff Richard, Abstract Expressionism to graffiti, addressing the theme of same-sex love with humour and from the point of view of subjective experience. This, and similar work, is a useful vehicle for introducing sexuality at a time when pupils are likely to be exploring and defining their own sexual orientation. This is very important when you consider the divergent ways through which pupils are introduced to discourses on sex and sexuality, particularly the contrast between a popular Media saturated with sexual imagery and school, sex and relationship educa-tion, where sexuality appears in relation to health and morality. This disjuncture produces an environment in which pupils discuss sex and desire amongst themselves but rarely in contexts informed by experienced adults (Allen 2001). This can result in misconceptions and stereotyping, especially as engendered by the bigotry and/or prurience of sections of the Media and other reactionary discourses. A discussion of desire and eroticism through works of art can provide pupils with the distance to discuss a taboo subject in ways that avoid the confessional practices so prevalent in TV and on the Internet (Addison 2002; 2006; 2007).

Issues surrounding gender and sexuality are central to the work of many con-temporary artists; you are advised to choose examples with great care, paying atten-tion to cultural and religious perspectives and the wishes of parents and governors, but without being intimidated into rejecting an issue that is of vital significance in pupils' understanding of human relationships both historically and as manifest in contemporary and particularly popular culture (Addison 2005).

Further reading

Addison, N. (2005) 'Expressing the not-said: art and design and the formation of sexual identities' *International Journal of Art and Design Education*, 24, 1: 20-30.

Stanley, N. (ed.) (2007) Special Edition: Lesbian and Gay Issues in Art & Design Education, *International Journal of Art and Design Education*, 26, 1.

A critical approach to a plural art and design curriculum

A critical approach to given bodies of knowledge is characteristic of changes in education since the 1970s. As you have seen, the various canons relating to the visual arts are established ideologically and reinforced by available resources. But they can be the target of critical investigation as well as the basis of cultural indoctrination (Gretton 2003). It is important to remember, however, that the critical procedures deployed to question the validity of such canons can themselves become the basis of new orthodoxies. The anti-aesthetic procedures of Dadaists; chance, transience, mixed or multiple modes, were partly introduced to undermine the fine arts and its relation-ship to capitalism, yet both conservative and radical critics would argue that they have become the established means guaranteeing access to the Art Market at the turn of the century (Bourdieu 1993; Stallabrass 1999). Foster (1996) has argued that the 'retroversive' power of the Right in the USA and its populist interventions into critical debate on the arts are pernicious and dangerous: to what extent is the situation in Britain different? He warns against any diminution in theoretical rigour or dilution in oppositional and transgressive practice, although he notes a move from 'grand oppositions' to 'subtle displacements'. He suggests that the Right has bestowed advanced art with such symbolic significance that it can still be an effective means of cultural and social change, and can therefore be the means to pursue critical transformations.

It is important then to apply a critical approach to all phenomena, and, as would a nurse or doctor, keep abreast of current theoretical developments as they pertain to your practice. There will be periods when you need to 'rest', to consolidate your teaching, but you should be wary of formulae and outmoded practice. Additionally, Giroux argues that teachers need to counter the privitisation and fragmentation of culture (2003).

The critical approach demands that nothing should be taken at face value (Price 1989). Not only does it enable pupils to review and modify their own work but it enables them to question the values embodied in the objects of visual culture, the perpetuation of canons and their own place within, or relationship to, particular traditions. Ultimately a critical approach enables pupils to understand that art is not limited to the making of beautiful, useful or imitative things alone, it is also a form of communication, an opportunity for them to find, construct and share meanings, and make sense of experiences both from within and outside the curriculum, In this sense critical approaches enable an inclusive and plural education. If pupils are to negotiate their place in visual and material culture they need to understand how art is value laden and consists of a changing series of systems open to analysis and interpretation, their own no less than others.

Further reading

Dawtrey, L. *et al.* (eds) (1996) *Critical Studies and Modern Art* and *Investigating Modern Art*, Milton Keynes: Open University, is a useful dual publication, where many of the issues that can only be touched on in this chapter are examined in greater detail.

Foster, H. (1996) *Return of the Real*, Cambridge, MA: MIT Press.

Giroux, H. and McLaren, P. (eds) (1994) *Between Borders*, London: Routledge.

12 Values in Art and Design: Addressing the Cultural, Moral, Political, Social and Spiritual Dimensions of Art Education

Nicholas Addison

What and whose values are being promoted in the art and design curriculum?

Should the art and design curriculum sustain prevailing normative values?

If so, how are minority and marginal views to be represented?

How can you select and represent values other than your own, including different cultural, social, political and religious attitudes, beliefs and practices?

How can an inclusive and plural curriculum, one that attempts to represent different traditions and faiths, reconcile conflicting moral positions and beliefs?

Objectives

This chapter should help you to:

Unit 12.1	identify the values which have informed your own practice in and through different traditions of art, craft and design, and their possible influence on your teaching;
Unit 12.2	identify and differentiate the intrinsic and extrinsic values embedded in art education;
Unit 12.3	consider your knowledge and understanding of your own and other value systems, and their interactions;
Unit 12.4	investigate the implications of teaching art and design in a plural society;
Unit 12.5	question and begin to resolve differences in value.

This chapter identifies some significant ethical questions arising from teaching. It helps you to investigate your own position within a plural society and to navigate a course for yourself and others through its complexities, its histories and possible futures. As a part of this investigation it is important to acknowledge pupils' cultural, moral, political, social and spiritual values, for it is from these that they derive their identity and worth. To neglect such values is therefore to limit the status and potential of the subject you are teaching, the pupils and yourself.

It would be highly unusual if you were not confronted by moral paradoxes and dilemmas of conscience during a course of initial teacher education (ITE). Your expectations may not meet with the reality you find: pupils may be unmotivated by the beliefs and issues you hold most dear, colleagues may teach in ways that conflict with your ideals, you may be required to teach Schemes of Work (SoW) that hold little interest or sense of worth for you. In this way personal and institutional values may conflict and you will be compelled to make decisions of conscience; to what extent are you prepared to negotiate, compromise or conform? On the one hand you may find yourself in complete agreement with the values inherent in the way art and design is taught in one placement school, seeing no need to question the status-quo. On the other hand you may have to adjust your strategies radically when you move schools; a different context requires a different approach. These differences can be very unsettling and you may start to pathologise either your own teaching or pupils' responses to it in unhelpful ways (Atkinson 2003). A constructive way to make sense of these differences is to reflect continuously on your experience and discuss with others working in different schools what is, and what is not, context specific. Such reflective practice is a prerequisite for developing a critical pedagogy, one rooted in the lives of young people which can help you to help them transcend the limitations of their social situation, whether these are considered advantageous or inhibitive. It is also vital that you examine the traditions and discourses within which your own beliefs and practices are embedded, for it is within these that your understandings of art have been formed. You are therefore encouraged to reflect on your current experience in schools in relation to these historical and contemporaneous discourses.

UNIT 12.1 THE VALUES OF THE AESTHETIC

By the end of this unit you should be able to:

- identify and reflect on the values that have informed your understanding of, and practice in, art, craft and design;
- investigate the origins of belief systems in art and design;
- consider the moral and spiritual dimensions of different types of art practice.

Student teachers arrive on an art and design PGCE course from a range of specialisms each with its own values and attitudes. The way you value and make judgements about your own and others' practice is informed by this experience. For instance, some fine art graduates may believe that art is an autonomous endeavour and may not have considered the reception of their work by non-specialist audiences. In contrast, for most design graduates, decisions are conditioned by functional and commercial factors such as the needs of target groups and potential markets. Some students work in isolation, others collaboratively, some from within existing cultural traditions, others from emerging ones. On the one hand you may see aesthetic education as an introduction to dominant cultural norms both conceptual and technological, on the other, you may see it as a challenge to the prevailing orthodoxies of the mass media and the new technologies (Adorno and Horkheimer 1999) what the revisionist critic Peter Fuller called the 'anaesthetic' world of the 'mega-visual tradition':

> The underlying struggles . . . are between those who are basically 'collaborationist' in outlook towards the existing culture, and those who perceive that the pursuit of 'the aesthetic dimension' involves a rupture with, and refusal of, the means of production and reproduction peculiar to that culture.
>
> (Fuller 1983: 23)

For a different view you have only to turn to Raymond Williams, who, writing in the early 1960s, saw the then prevailing 'aesthetic' orthodoxy as uncritical and moribund, as totally out of touch with the realities of people's lives:

> It is a meagre response to our cultural tradition and problems to teach, outside literature, little more than practical drawing and music with hardly any attempt to begin either the history and criticism of music and the visual art forms, or the criticism of those forms of film, televised drama, and jazz to which every child will go home . . .
>
> (Williams 1971: 172)

Fuller suggests that aesthetic practice is, in itself, a moral activity, a critique of alienating contemporary systems of communication and organisation. For him, those who advocate the virtual means of cyberspace or hyper-reality are mistaken if they think they are engaging in aesthetic practice. In relation to Williams (1979), this perception is deterministically pessimistic. He addresses a key moral question; should education provide people with the critical tools necessary for participation in democracy? In his discussion of television he implies that only through a critical approach to developing technologies can people discriminate between what does and does not answer a need. Equipped with the necessary tools they can better argue their case and effect change through the democratic processes of demonstration and negotiation.

Why is the aesthetic domain such a contested area?

Art education has a long history, but your position as a teacher of art and design within universal schooling is a product of modernist practices and it is important that you examine the ways that your beliefs and practices have been formed through its discourses (Dalton 2001). I therefore discuss here what can only be an introduction to the complex history of the development of aesthetics as a branch of philosophy in western culture, a discourse that has placed the arts, and the visual arts in particular, as a central indicator of human development, and thus a vehicle for education.

Aesthetics as a discipline

There is a major western tradition stemming from the Enlightenment, which posits the aesthetic as a form of knowing that is different, but in no way inferior, to logic. Until the eighteenth century, 'knowing through the senses' was considered a pre-liminary stage in a process leading to higher cerebral activities such as analysis and synthesis. The new philosophical study of aesthetics, or the inquiry into the nature of the beautiful and the sublime, was initially part of a moral quest to find ways of judging how things ought to be, a complement to the scientific quest to understand how things are. For some, given this moral aim, aesthetics became an alternative or surrogate religion, for others it became an end in itself. In the latter instance, all other interests, for example those of appetite, moral consequence, utility, had to be rejected so that nothing should impede the pure and disinterested act of judgement. In Kant's 'Critique of judgement' (1790) and unlike the tradition since antiquity, beauty is no longer the outward sign of inner-goodness, nor is the sublime dependent on divine revelation or the actions of great and moral people. The beautiful and the sublime are theorised as a subjective response to a phenomenon, cultural or natural, the goodness of which is judged through the faculty of taste. It is not beauty itself that is universal but the ability of people to make critical judgements: '*taste*, is the faculty of estimating an object or a mode of representation by means of a delight or aversion *apart from any interest*. The object of such delight is called *beautiful*' (Kant in Preziosi 1998: 84). In this sense it could be said that 'beauty is in the eye of the beholder'. However, Kant does refer to the notion of a 'general validity', an idea that implies some sort of universality, and emphasises that aesthetic and moral judgement are analogous activities.

Kant's followers and apologists have frequently avoided this aspect of his phil-osophy and also the fact that his theory is concerned with aesthetic experience, not specifically with art. This avoidance has produced a tradition positing art as a transcendental and autonomous entity, the pursuit of which is, in and of itself, worth-while. In this tradition the practice and appreciation of art needs no explanation, need serve no purpose, social, moral, political or otherwise. Walter Pater, the advocate of aestheticism in late Victorian England, wrote in the conclusion to his influential essay, 'The Renaissance' (1873):

> Of such wisdom, the poetic passion, the desire for beauty, the love of art for its own sake, has most. For art comes to you proposing frankly to give nothing but the highest quality to your moments as they pass, and simply for those moments' sake.
>
> (in Golby 1986: 221)

The pursuit of art for art's sake towards the close of the nineteenth century was immediately countered by a succession of socially motivated cultural initiatives, from the Arts and Crafts and Secessionist movements to the Weimar Bauhaus and early Soviet experiments in the 1920s. However, the idea that art is somehow separate from mainstream social concerns, a realm of higher feelings and sensibilities, has led some people to imagine art and design in negative terms. For example its products may be seen as a luxury or as an adjunct of power, wealth and education and thus a mark of distinction (Bourdieu 1984). Alternatively, when thought of as a set of practices, they may be seen as a type of recreation, a mode of production that is enjoyed once 'real' work is over. Many people still hold these attitudes, but in doing so they overlook the ways in which artists, craftspeople and designers produce the visual environment in which they live, from the built environment to the landscape of images that circulates in print and on the screen, a saturation that is a result of modernist technologies and philosophies. But before examining the social and cultural ambitions of modernism it is important to introduce a set of ideas that were parallel to Kant's but no less influential.

Hegel, a generation younger than Kant, argued that art is the most complete manifestation of the 'spirit' or 'essence' of a people, indeed he countered Kant's speculations in his *Philosophy of Fine Art* (1835–38) by theorising art as a secondary process, a sensuous vehicle for the representation of the 'primary Idea'. If Kant's theories had suggested that humans could only get a sense of their universality through aesthetic experience Hegel argued that Art (not aesthetic experience) is the most concentrated and representative manifestation of the 'Zeitgeist', the character or spirit of the age as revealed in a particular culture. For him, seeking to understand art is therefore:

> . . . an attempt to understand the entirety of history as both a system and a *process* – a process of the unfolding of the Divine Idea in the (sensory and hence illusory) temporality of artistic change. The Divine Idea is unchanging and immutable: its changing representations over time are but confused ways in which mortal beings attempt to grasp the unchanging and singular Divine perfection.
>
> (Hegel in Preziosi 1998: 67)

It was this aspect of Hegel's thinking that ultimately secured a place for the history of art as a significant academic discipline, for it is with such study that the inquirer is able to construct a history of culture using concrete evidence to record the unfolding of the 'human spirit'. However, the danger with this process is that the visual and plastic arts come to illustrate ideas that pre-exist their expression, a formulation that sets up a temporal sequence and hierarchical relation between thought and other forms of semiotic practice. As many artists, craftspeople and designers attest, an idea does not necessarily come before its expression, rather it is formed through the process of making itself; the idea and the form are indivisible and concurrent, what Crowther (1993) terms a peculiarly condensed, 'sensuous and conceptual manifold'. Danto (1986) argues that the development of conceptual practices in the arts from the 1960s to this day is as a direct result of philosophical thinking, and in particular Hegelian philosophy. But Hegel's two legacies to the field: the history of culture examined through the arts and conceptual, dematerialised practices, are notably absent in secondary schools where practice is rooted in theories of technical proficiency, accuracy and/or self-expression (Atkinson 2002). Nonetheless, critical, historical and conceptual practices are the bedrock of study in Higher Education, and in this disjunction can be found one reason for the differences in pedagogical practices across the sectors (Robins 2003).

The legacies of Kant's and Hegel's philosophies impact on art education in yet further ways: Kant's speculations allow for coexistent, equivalent aesthetic systems, differences in cultural outcomes that are not hierarchical but determined by such variants as geography and climate, whereas, for Hegel, western and Christian forms are more fully developed than any other culture or religion because they are nearer to the 'Divine Ideal'. This tradition posits art as a cultural phenomenon representing a people's collective or social consciousness and thus their position on an evolutionary scale towards ultimate perfectibility, a teleological notion which Hegel's most ardent, if critical, student, Marx, was to apply to a theory of economics. These theoretical positions may seem very distant from teaching young people to engage with art in twenty-first-century British schools, but such ideas are reinforced and sedimented in practice so that they still manifest themselves within the curriculum. For example, look at the *National Curriculum Order for Art and design* (NC) (DfEE 1999) where it states pupils should be taught about:

> ... continuity and change in the purposes and audiences of artists, craftspeople and designers from Western Europe and the wider world (for example, differences in the roles and functions of art in contemporary life, mediaeval, Renaissance and post-Renaissance periods in Western Europe, and in different cultures such as Aboriginal, African, Islamic and Native American).

(DfEE 1999: 20)

Task 12.1.1 Egalitarian and/or pragmatic principles

Discuss the following questions within your tutor group.

If the NC implies an evolutionary western tradition, does such a progression correspond in any way to Hegelian notions of cultural superiority, or is it just a matter of knowing 'Western European' culture within the context of different cultures?

What evidence is there to suggest that recommendations to investigate 'different' cultures relate to: i) egalitarian principles, ii) colonial legacies, iii) pragmatism (responding to the postcolonial, multicultural context of today's classroom)?

The spiritual, the liminal and the unconscious

There is then, a modern western tradition from Kant to Greenberg (1965) which posits the aesthetic, or the contemplation of sensate experience, as transcendental, a phenomenon that through the imagination goes beyond the physiological into liminal experience, that is a disengaged state of heightened consciousness. Thus a sort of metaphysical or spiritual domain is suggested that might be considered 'religious' but without being associated directly with any specific religion. Are there any vestiges of this notion in beliefs about the way art affects people? Is this notion one with which you identify? If so, in what context does this happen to you? If you read Taylor (1986) or Hargreaves (1983) you will see both arguing that art is able to engage pupils in transcendental ways, for the former through 'the illuminating experience' and for the latter through a 'traumatic conversive experience'. These claims are sustained with reference to specific cases, but they are far from typical and organising learning situations designed to engender such liminal states is a hapless task. It is like insisting to pupils: 'you *will* find this moving'. This is not to suggest that such experiences do not happen, but the public forum of a school trip to a gallery, museum or studio, or the contemplation of art works reproduced on page or screen do not provide the ideal contexts for such encounters. However, encouraging pupils to look, ask questions, develop interest is another matter.

For many it is in the context of religious practice that spirituality is most readily manifest and it is worth evaluating the part played by the material and visual environment in which these experiences happen, whether Church, Mosque, Synagogue, Temple or other religious site. The major world religions are self-evidently the most representative vehicles for addressing spiritual needs: but in a multi-faith society it is dangerous for you to attempt to engage with religion in the context of the classroom through any one system of aesthetic practices, one person's icons are another person's idols; this is an area of real difference. It does not mean, however, that you should consider religious art out of bounds: it can be viewed, discussed and provide individual pupils with stimulus for their own work. You are merely advised that whole classes should not be asked to make work in ways or with content that may be antithetical to their religious or philosophical beliefs. This advice also indicates some of the moral values associated with different types of educational activity: inquiry and comparative study, which are investigative processes that challenge but do not deny beliefs, and representational action, where the process is in itself a contamination of those beliefs.

> In places like universities, where everyone talks too rationally, it is necessary for a kind of enchanter to appear.
>
> (Beuys in Gablik 1991: 41)

Since the 1960s there has been a renewed interest in the practices of the shaman and her/his ability to contact and move through the world of the spirits. This desire to understand alternative forms of knowledge, initially driven by anthropological and psychoanalytical inquiry at the turn of the twentieth century, was pursued in western aesthetic experimentation as an anti-materialist, anti-bourgeois strategy. Artists as diverse as Picasso and Beuys have adopted the ritual objects and performance of animistic cultures. The Surrealist quest for altered states of consciousness found in the shaman the priestly other on which they could build their own artistic identities, the holy fool, at once foolish and wise, transgressive and sacred (Ades *et al.* 2006). Performance artists such as Fern Shaffer (Gablik 1991: 42–45) and Carolee Schneeman (Broude and Garrard 1994: 161) have also appropriated something of the ritualised activity of the shaman, incorporating gesture, dance, chanting, imaging in a multi-sensory and often hypnotic totality (see Plate 25). Susan Hiller, who before turning to art was a practising anthropologist, co-ordinated a group work, *Dream Mapping* (1976) in which participants gathered at a ritual site to record their dreams in visual form (Hiller 1996: 129–130).

When discussing shamanism it is important not to compound all its cultural manifestations into one homogeneous practice. Unless you approach such processes from a position of knowledge and understanding you are in danger of reinforcing myths of the primitive, 'others' as irrevocably irrational. If postmodern performance has appropriated aspects of shamanistic practice for ethical and political reasons, for example, its ephemeral nature makes it a difficult phenomenon to market, it is also open to sensationalist exploitation, the cult of spectacle, to 'shamism'. As Hiller herself points out in response to the question:

'Post?modern?ism IS IT WORTH DISCUSSING?'

NO	YES
if it means falling back on a superficial and ideologically-constructed 'primitivism'	if it represents a real desire to break the enforced superficiality of some late-modernist/conceptual practices
if jokey historical references are used to disguise real contemporary recognition of new struggles, practices, marginalised groups, radical attitudes.	if it means we are closing the gap between 'experience' (ours) and 'reality' (theirs)

(Hiller 1996: 141–142)

For Hiller, such practice is integral to the ongoing critique of dominant forms of western knowledge and representation and is profoundly political in intent. Other more universalising tendencies attempt to theorise territories that are hidden to the conscious mind, the mythopoetic and visionary as manifest in art. Building on Freud's notion of the 'Unconscious' Jung's theory of universal archetypes (1964) suggests the underlying commonality of the human psyche:

> What we properly call instincts are physiological urges, and are perceived by the senses. But at the same time, they also manifest themselves in fantasies and often reveal their presence only by symbolic images. These manifestations are what I call the archetypes. They are without known origin; and they reproduce themselves in any time or in any part of the world – even where transmission by direct descent or 'cross-fertilization' through migration must be ruled out.
>
> (Jung 1964: 69)

Rolland's 'oceanic experience' suggests a condition in which people feel at one with their environment, a condition Freud saw as 'a regression "to an early phase of EGO-feeling" [which] revives the experience of the infant at the BREAST' (Rycroft 1968: 105). Freud considered this sense of overwhelming benevolence as something akin to religious experience, (for him, an equally regressive desire for parental protection) feelings which others locate in the contemplation of nature and art. In this sense the practice of art becomes a form of wish fulfilment, fulfilling a therapeutic rather than educational purpose.

Kandinsky's 'Concerning the Spiritual in Art' (1911) proposes methods whereby 'the psychological power of colour becomes apparent, calling forth a vibration from the soul. Its primary, elementary physical power becomes simply the path by which colour reaches the soul' (in Harrison and Wood 1992: 92). Kandinsky was only one of many artists who conceived a universal, spiritual aesthetic which, in its historical context, was not only an attempt to reject academic, bourgeois and scientific values but also an effort at healing deep historical divisions (Smolik 2006). However, today such universalising rhetoric tends to appeal to those who uncritically appropriate all things mystical in the production of a new-age utopia. It might be more educationally useful to look at the concept of synaesthesia, of equal interest to Kandinsky, as outlined in Unit 3.7.

The early modernist attempts to merge religious difference in universalising theories of the spirit around the beginning of the twentieth century were concurrent with the development of psychoanalysis. Whereas psychologists had attempted to explain the spiritual in terms of potent myth-making, it could be argued that psychoanalysis provided analysts with the interpretative tools with which to limit difference in the name of phantasy. However, in recent years, psychoanalytical approaches to understanding art and visual culture have greatly extended the scope and significance of interpretative practices (Kristeva 1982; Zizek 1991; Adams 1996), but such analysis is very difficult, and although it can undoubtedly develop your own understandings and those of post-16 art and design students, its direct application to KS3 and GCSE is questionable. Nonetheless, a psychoanalytic language has begun to enter popular discourses so that such terms as compulsion, denial, ego, fetish sometimes enter discussions and it may be that you can encourage pupils who use them to research their meaning and to use them with more discrimination.

Modernism

These spiritual and/or non-conscious notions about the nature and purpose of art are somewhat at odds with the progressive trajectory of modernism in which design combines with art not only to suggest what life ought to be, but what it can be. The modernist project to improve the material environment in which people live is sometimes overlooked in art and design teaching in favour of those aspects that valorise self-expression, the authentic and the autonomous. Many teachers pay lip service to the idea of self-expression while devising SoW with clearly defined technical procedures and fixed outcomes. Remember that modernist artists often chose to work at the limits of the historical traditions within which they were embedded (whether related to representational resources or material practices) and, within the limits of this discourse, sought to overturn outmoded and moribund conventions. This was, and still is, often achieved by the 'shock of the new', an attempt through aesthetic reworkings to jolt people out of the complacency of naturalised assumptions, the easy and gratifying ways of seeing and engaging with the world. It is unlikely that many of your pupils will be directly contributing to this project, however, it is vital that they are aware of the attempt and in what ways it has succeeded and how it still continues. T.J. Clark states that 'Modernism is our antiquity' (1999: 3) by which he seems to be suggesting that the project of modernism can serve as the foundation for future development just as the philosophies of antiquity served the development of Islam and Christendom in the middle ages and beyond. It is therefore important to revisit this project and not assume that postmodernism has somehow overturned its values (Meecham and Sheldon 2004). As Harris (1996) argues:

> The culture of 19th- and 20th-century modernism *itself* offers much, now on the historical and cultural margins, in the way of critiques of actual society and views of projected kinds of society radically different from that within industrial capitalism. In this sense, once again, the 'modern' and 'postmodern' are inseparable, despite arguments to the contrary.
>
> (p. 185)

Materialism and embodiment

What you find in the modernist experiment if you look closely is far from an homogeneous enterprise: it is neither a monolithic metanarrative of western progress nor a universal theory of creativity, nor is it just a quest to push semiotic practices to the limits of their understanding; rather it is a heterogeneous collection of social and cultural practices through which artists, craftspeople and designers attempt to reflect on, inform and ultimately transform the conditions of life in specific historical and geopolitical contexts. Nonetheless, in terms of universal theories there is a materialist alternative to the spiritualist notions outlined above in which art can supposedly transcend the specificities of its social and historical contexts, including religions and politics, by stressing the moral concept of a common humanity rooted in the body as a material entity. Should it be this area of commonality, if it can be defined, and this alone, that forms the basis of a curriculum in a secular democracy?

Dissanayake (1992) suggests that the biological basis of art, being universal, is its most ethical dimension. She bemoans the western tradition that exalts individuality and the cult of genius contending that art is essentially a biologically determined social activity:

> . . . elements of what we today call the arts (e.g., pattern, vividness) would have existed first in nonaesthetic contexts. But because these elements were inherently gratifying (perceptually, emotionally, cognitively) to humans, humans who had an inherent proclivity for making special would use them – not for their own sake, but, instead, in ecological terms, as 'enabling mechanisms' – in the performance of other selectively valuable behaviours.
>
> (p. 51)

However, any easy reduction of the body to a biological essence is contested in much aesthetic theory. For example, Crowther (1993), like Disanayake, also looks at works of art as a special instance of human activity, but he argues that through making art people do not simply react or adapt to a given environment but construct alternative if related environments in which their existence as sensible beings, that is perceiving/ thinking/doing beings, is given its most concentrated expression. Not only is the old

Cartesian split between mind and body refuted in this definition (a split that is reinforced within the logocentric, school curriculum) it is collapsed since Crowther goes on to suggest that works of art are analogous to human beings themselves, an inseparable cohabitation of physical and mental energies. Further to this, the social activity insisted on by Disanayake is not determined by the body as biological fact for, as Betterton asserts 'the body itself becomes the site of social and political inscription rather than a given biological truth' (1996: 15). As Meskimmon (2003) points out: 'Embodiment refutes the division of 'mind' from 'body', arguing that what we call subjectivity is the effect of human, corporeal existence in the world . . . Moreover, embodied subjectivities are relational, formed through corporeal encounters with other bodies in the world' (p. 76). Therefore, as the National Curriculum insists, making works of art, craft and design provide not only 'visual, tactile and sensory experiences and a unique way of understanding and responding to the world [but also] 'communication' and a little later 'becoming actively involved in shaping environments' (DfEE 1999: 14) marking out art and design as an interactive social practice as well as a personally transformative one. In this way the aesthetic potential of art and design not only enables young people to express themselves as embodied beings and helps them forge personal identities, but it also helps them to develop agency within the larger, social and cultural structures that at first appear so determining.

Dialogical aesthetics

The role of the body is evidently central to making practices in art and design but so too is the body collective, making as social action. In recent years a number of artists have moved away from an authored practice towards public forms of art in which they engage with and join communities to explore their situated aesthetic, environmental and political concerns. Kester (2004) argues that such practice has its roots in the public events and performances of the 1960s (although he might have looked further back at Futurist and Soviet strategies for engaging the public). But he rightly observes that what is different about 'dialogical aesthetics' is that these engaged practices:

> . . . expand that tradition, [which is] often focused on an internal critique of the work of art, into a set of positive practices directed towards the world beyond the gallery walls, linking new forms of intersubjective experience with social or political activism. Important transition figures here include Stephen Willats and the Artists Placement Group (John Latham and Barbara Steveni) in the United Kingdom and Suzanne Lacy and Helen and Newton Harrison in the United States.

(p. 9)

Participation and collaboration are central to the ways these artists work and always in the context of real community situations. The projects therefore emerge in dialogue with participants and are negotiated and carried out in collaboration. Examples are numerous but include a project to enable tenants in public housing to self-organise (pp. 91–94) or as with the 'Roof is on Fire' (1994) to provide a platform for young 'people of color' to discuss the problems they face with officials (pp. 3–5). What is particularly significant here is the way *young* people have been involved in such projects and it may be that you can work with galleries and their network of artists to propose and carry out locally specific collaborations. Kester admits that the criteria by which modern art tends to be evaluated is redundant in these cases and he provides a theoretical framework through which they can be assessed (pp. 125–151). It is worth looking at this to consider how you might relate his framework to the assessment of individual pupils and their contributions so that you can meet statutory expectations over the monitoring and assessment of learning whilst also meeting the core skills of working with others and reflecting on own practice. This approach can be very liberating for all involved, providing pupils with a real degree of agency and, as Kester notes, an opportunity for reciprocal learning: 'Active listening and intersubjective vulnerability play a more central role in these projects, as the artist does not always occupy a position of pedagogical or creative mastery' (p. 151).

Task 12.1.2 Evaluating values

In tutor groups, write down what you understand by the terms: cultural, moral, political, social and spiritual. Argue your explanations and compare them with dictionary definitions.

In relation to these terms, reflect on your own experience and practice of art and design by listing the main factors that have influenced and shaped it.

To what extent do you feel that your experience has been conditioned by any of the traditions or systems of value outlined in this unit?

Identify and discuss the significance of any other traditions that have helped form your values.

Further reading

Gablik, S. (1991) *The Reenchantment of Art*, London: Thames and Hudson.

Kester, G. H. (2004) *Conversation Pieces: Community + Communication in Modern Art*, Berkeley and LA, University of California Press.

Meecham, P. and Sheldon, J. (2004) (2nd Edition) *Modern Art: A Critical Introduction*, London: Routledge.

Preziosi, D. (1998) *The Art of Art History: A Critical Anthology*, Oxford: Oxford University Press.

Williams, R. (1971) *The Long Revolution*, Harmondsworth: Pelican.

UNIT 12.2 INTRINSIC AND EXTRINSIC VALUES IN ART, CRAFT AND DESIGN EDUCATION

By the end of this unit you should be able to:

- identify and differentiate between the intrinsic and extrinsic values embedded in art education;
- consider the social and cultural functions of art, craft and design;
- examine ways to question perceived conflicts between different values.

In historical terms art education has served a variety of purposes, from training an elite group with specialised skills to producing a skilled working population; from providing a leisured class with cultivated pursuits to contributing to the fundamental biological and spiritual development of all. Clearly, these purposes relate to systems of value formed through specific social and historical contexts, but traces of these and other values still saturate the art and design classroom of today as well as the institutions of further and higher education. It is important that you identify these values and examine their present function and validity.

The management of art and design education by the state in Britain has a history that was, and still is, closely entwined with its history as an industrial and colonial power. For example from 1852, under the direction of Henry Cole, mandatory art instruction at elementary level focused on drawing and the ability of pupils to replicate designated models, usually of an ornamental nature (Swift 1995). This education was structured to produce a designer workforce capable of competing with Britain's industrial neighbours as well as producing goods for sale in the new markets of empire (MacDonald 1970). It involved copying with great exactitude, ensuring technical competence and a workforce that was diligent, patient and compliant, a process disciplining the working person into what Foucault termed 'the docile body' (1976). The fine arts were taught separately along the lines established by academic practice. Access to this education was largely determined by class (Dalton 2001). You can characterise these kinds of education, or training, as *instrumental* and *utilitarian*: they have a clear economic and social basis designed to develop 'useful' skills and behaviour.

The Victorian critic John Ruskin thought that such forms of education were denigrating and immoral. His writings forcefully castigate the whole basis of industrial production defining its division of labour as soulless and dehumanising, and the second-hand formulae perpetuated in the academies as tainted and at worst poisonous and corrupting. He also bemoaned practice aimed at 'art for art's sake', where pleasure in making and looking is the primary goal, for it 'has an influence of the most fatal kind on the brain and heart, and it issues . . . in the *destruction both of intellectual power* and *moral principle*' (in Clark 1985: 146). He proposed an art education in which nature, not art, is the true model, reflecting the glory of God's craftwork, one in which students should be trained to work with their hands, perceive and think for themselves and develop the capacity for critical judgement: in 1852 he wrote in *The Stones of Venice*:

Understand this clearly: You can teach a man to draw a straight line and to cut one; to strike a curved line and to carve it; and to copy and carve any number of given lines or forms, with admirable speed and perfect precision; and you find his work perfect of its kind but if you ask him to think about any of those forms, to consider if he cannot find any better in his own head, he stops; his execution becomes hesitating; he thinks, and ten to one he thinks wrong; ten to one he makes a mistake in the first touch he gives to his work as a thinking being. But you have made a man of him for all that. He was only a machine before, an animated tool.

(in Clark 1985: 281–282)

Ruskin was enormously influential in nineteenth-century Europe and the USA advocating the centrality of education in and through the visual arts for the social, moral and spiritual health of a culture. In the writing of his disciple, William Morris, and the development of the Arts and Crafts Movement, you find principles which value the visual and material arts for their social significance both in work and leisure:

The chief source of art is man's [sic] pleasure in his daily necessary work, which expresses itself and is embodied in that work itself; nothing else can make the common surroundings of life beautiful, and whenever they are beautiful it is a sign that men's work has pleasure in it, however they may suffer otherwise . . .

. . . leisure and desire are sure to produce art, and without them nothing but sham art, void of life or reason for existence, can be produced: therefore not only the worker, but the world in general, will have no share in art till our present commercial society gives place to real society – to Socialism.

(in Denvir 1986: 71–74)

Task 12.2.1 Art education, a spiritual and moral force?
You can deduce from these differing views from the Victorian era that art education was seen as central to the material and moral, physical and spiritual well-being of the nation. In your tutor group discuss: • whether art education holds a similar position in Britain today; • what instrumental arguments are made for its inclusion in the curriculum at secondary level; • what evidence you can find to suggest that people believe the practice of art will have a spiritual or moral effect.

Intrinsic and extrinsic values in art and design

Instrumental or utilitarian objectives serve extrinsic educational ends, objectives designed to meet economic and social imperatives rather than the specific needs of an individual or the peculiar qualities of a curriculum subject, although they may make possible that individual's or subject's public role in society. Arguments for the extrinsic value of art education often talk about the transferability of skills (Harland *et al.* 1998). This establishes a hierarchy of subjects and related skills where those considered secondary or peripheral serve primary objectives, namely: literacy, numeracy and ICT. In addition, the core, now key, skills introduced initially for post-16 education reiterate the old hierarchy but add 'improving own learning and performance', 'problem solving' and 'working with others' (http://www.qca.org.uk/603.html). These skills establish objectives which, however commendable, further prescribe the ways in which you are expected to approach your subject. If the extrinsic value of employable skills is applied to art education, apologists cite the share of Gross National Product provided by Britain's creative industries, including the visual arts of fashion, advertising and television (Steers 2003). This perceived need is partly being met in the recent development and proliferation of vocational courses and in the new applied diplomas.

The intrinsic values of art are those which refer to its peculiar character (Wolheim 1983). This is a proudly modernist position where art (rather than design) has been associated with its visual and aesthetic properties, to its status as a phenomenon related to, but different from, ordinary reality, a parallel universe. For art education this has suggested practices in which the furthering and questioning of tradition can be allied to personal development, the artist as self-expressive agent. Consequently, the (western) artist has come to personify agency and thus freedom (Preziosi 2006: 57–59) and as such her/his work provides their nation/culture with a symbolic marker of this ideal. Such an approach can be seen as a strength in that it has defined a unique realm of experience requiring expertise, but is also a limitation in that it has isolated art from both its broader social functions and its related disciplines, aspects that are becoming increasingly significant for contemporary artists (see the earlier sections on dialogical aesthetics).

In philosophy, intrinsic values are differentiated into four basic areas; ethics, logic, aesthetics and religion (goodness, truth, beauty and holiness). Whether art can be associated with only the third of these is open to debate particularly in the context of art education. Today, any definition of art's intrinsic values would also include its worth as a means to interpret and understand experience: art is no longer perceived as a discrete and formalist universe or a means of reflecting the world out there, but as an integral part of interconnecting systems of knowledge that in part produce and thus transform experience. In the sense that these intrinsic values make possible understandings which can be communicated; you might say that the intrinsic has a social value. However, you should remember that the values of art, along with its education, are socially and culturally contingent and thus variable:

> A work of art can have many different kinds of value – a cognitive value, a social value, an educational value, a historical value, a sentimental value, a religious value, an economic value, a therapeutic value; it can possess as many kinds of value as there are points of view from which it can be evaluated.
>
> (Budd 1995: 1)

The intrinsic qualities of art include its material, aesthetic, communicative and trans-formative properties both in terms of production and reception. The additional uses to which these properties can be applied are its extrinsic values. Ethical cases can be put forward for both.

Task 12.2.2 Extrinsic and intrinsic objectives

In your tutor group identify:

- the extrinsic and intrinsic objectives of the art and design curriculum;
- any conflicts between them.

Look for evidence in:

- QCA exemplification and advisory documents;
- the PGCE course documentation;
- the NC;
- the examination syllabuses;
- the Manifesto for Art in Schools (Swift and Steers 1999);
- your placement schools;
- your own practice.

Discuss ways you might resolve any conflicts.

Further reading

Clark, K. (1985 [1964]) *Ruskin Today*, Harmondsworth: Penguin.

Dalton, P. (2001) *The Gendering of Art Education: Modernism, Identity and Critical Feminism*, Buckingham: Open University Press.

Thistlewood, D. (ed.) (1992) *Histories of Art and Design Education*, Burnt Mill, Harlow: Longman.

UNIT 12.3 DIFFERENCE AND COMMONALITY: VALUES AND THEIR INTERACTIONS

By the end of this unit you should be able to:

- question how you come to know your own and other value systems and their interactions;
- examine the benefits and dangers of using canonical exemplars to define cultures.

When attempting to identify cultures most people look first for concrete symbols to represent them. They are attracted to a process of reduction, seeking out key ideas, personalities, artefacts, behaviours, etc. The resulting characteristics provide a sense of certainty and permanence standing in for what is likely to be a complex, paradoxical and diverse reality. There is a danger that the pressures on secondary school teachers to educate pupils within a clearly defined and articulated curriculum invite just such a reductive process. At worst this produces stereotypical representations that permeate discourses to become fixed and naturalised (Hall 1997), particularly in the process of othering by which one group defines who they are by demonstrating, in relation to 'others', what they are not. Within the context of modern history the emerging plural societies formed through colonisation and empire tended to construct a genealogy in which the 'host' culture presented dominant values through myths of purity. In the cultural sphere this usually manifested itself as a canon constituting key or exemplary 'texts'. A traditional reading of the 'West', for example, defines it as the product of Greco/Roman culture modified by Judeo/Christian teaching passed on and developed down the ages through a succession of European heirs, and, more latterly, their colonial 'subjects'. Recent scholarship presents a less certain story acknowledging that cultures are neither strictly linear nor fixed phenomena but fluid and interactive.

In the multicultural art and design classroom teachers tend to introduce 'world' cultures by showing a limited number of historical and emblematic reproductions; Africa = central African masks, India = court miniature painting from the sixteenth to eighteenth centuries: the past stands in for the present, contemporary practice is usually ignored. What is seen as 'familiar' or 'our' culture (i.e. Euro/British) has similar, if fuller, representation but is balanced by pupils' lived experience and the tacit knowledge they bring through living in or with a dominant culture. But its complex fabric and multiple signifying systems are distorted when reduced to canonic exemplars. A brief examination of a selection of key western texts demonstrates just how misrepresentative it can be to represent cultures through decontextualised exemplars.

> Thou shalt not make thee *any* graven image, or any likeness of *any thing* that *is* in heaven above or that *is* in the earth beneath, or that *is* in the waters beneath the earth.
> (The Bible, King James' version, Deuteronomy: 5: 8; see also Exodus: 20: 4)

> The imitative art is an inferior who marries an inferior, and has inferior offspring . . . for he (the imitative poet) is like him (the painter) . . . he is the manufacturer of images and is very far removed from the truth.
> (Plato c380 BC: 671–674)

These extracts concerned with the ethics of representation are selected from texts which are perceived as fundamental to the idea of 'Western Culture'. More pertinently, as texts, they lie at the very centre of what the British claim to have achieved in the development of their culture, a synthesis of Judeo/Christian morality and Athenean civic and democratic principles.

Task 12.3.1 Questioning tradition

Discuss in tutor groups:
    To what extent does this notion of an homogenised and linear tradition define your identity? What aspects of your identity fall outside its parameters?
    With what criteria of exclusion do you construct this; for example by age, class, race, gender?
    In planning SoW, how can you plan effectively both within and outside this tradition?

'Us' and 'our' canon?

Returning to the quotations from the Bible and Plato's Republic you recognise that both reject mimetic forms of visual representation. Examine the following parallel texts:

> I [God] have filled him [Bezalel] with the spirit of God, with skill, ability and knowledge in all kinds of crafts to make artistic designs for work in gold, silver and bronze, to cut and get stones, to work in wood and to engage in all kinds of craftmanship . . . And all the men among the workmen made the tabernacle with ten curtains; they were made of fine twined linen and blue and purple and scarlet stuff, with cherubim skilfully worked.
>
> (The Bible, King James' version, Exodus: 31: 3–5 and 35: 8)

> 58. Myron seems to be the first sculptor to have extended the scope of realism; there was more harmony in his art than in Polyclitus', and he exercised more care with regard to proportions . . .
> 59. Pythagorous of Rhegium in Italy outclassed Myron . . . He was the first sculptor to show the sinews and veins and to give a lifelike rendering of the hair.
>
> (Pliny the Elder, first century AD: 315)

Western culture is noted for its visualising tendencies, yet canonic texts provide what seems to be contradictory and conflicting advice. Which texts are indicative of western culture? Both the iconoclasts (sixth century AD) and the puritans of the Reformation (sixteenth century) relied on literal interpretations of Exodus, neglecting the context of its writing and evidence elsewhere in the Bible denoting the multiplicity of images in the Tabernacle (presumably representations of cherubim are a likeness of something in heaven). Historical 'Western' commandments and cautions show great congruence with parallel belief systems, in Hinduism, a warning that mimetic representation (likeness to appearance) is but a veil before the truth:

> That aesthetic *sadrsya* (concomitance of formal and pictorial elements) does not imply naturalism, verisimilitude, illustration, or illusion in any superficial sense is sufficiently shown by the fact that in Indian lists of factors essential to painting it is almost always mentioned with *pramana*, 'criterion of truth',

here 'ideal proportion' . . . *Sadrysa* is then 'similitude', but rather such as is implied by 'simile' than by 'simulacrum'. It is in fact obvious that the likeness between anything and any representation of it cannot be a likeness of nature, but must be analogical or exemplary, or both of these. What the representation imitates is the idea or species of the thing, by which it is known intellectually, rather than the substance of the thing as it is perceived by the senses.

(Coomaraswamy 1956: 12–13)

In Islam (Sura 6.74) there is a clear proscription in relation to the worship of idols: 'O Believers, wine and arrowsmithing, idols and divining arrows are an abomination, some of Satan's work; so avoid it: haply you will prosper' (in Yeomans 1993: 85) and further proscription in relation to representation of any living thing. The thirteenth-century legist, Nawami, stated:

The learned authorities of our school and others hold that the painting of a picture of any living thing is strictly forbidden and is one of the great sins, because it is threatened with the above grievous punishment as mentioned in Traditions, whether it is intended for common domestic use or not. So the making is forbidden under every circumstance, *because it implies a likeness to the creative activity of God* . . . On the other hand, the painting of a tree or of camel saddles and other things that have no life are not forbidden.

(*ibid.*)

These historical rules do not seem at all indicative of the visual landscape of contemporary Britain, nor indeed Pakistan or India. Clearly both the spiritual and temporal domains of western and eastern cultures have by and large come to terms with any dilemmas of conscience over the use of images, and the ubiquity of the photograph, with the distancing effect of its mechanical processes (Sontag 1977), has realised a place for the mimetic in cultures where it was once marginalised if not outlawed.

You may be tempted to define cultural ancestry in terms of ideas embedded in canonical texts (whether word or image); this is the easy answer. If you investigate further you are likely to find them contradictory, replete with compromise, open to antithetical interpretations and embraced by cultures you might assume in all other respects to be different; this is a more demanding task, but it rewards understanding.

Task 12.3.2 To represent, or not to represent?

In tutor groups:

- discuss the possibility of constructing an art and design curriculum that is open to non-representational tasks;
- plan a SoW that has the choice between a non-representational and a representational outcome.

Further reading

Hall, J. (2004) 'Art education and spirituality', in R. Hickman (ed.) (2nd Edition) *Art Education 11–18: Meaning, Purpose and Direction*, London: Continuum.

Yeomans, R. (1993) 'Islam: the abstract expressionism of spiritual values', in D. Starkings (ed.) *Religion and the Arts*, Sevenoaks: Hodder and Stoughton.

UNIT 12.4 TEACHING IN A PLURAL SOCIETY

By the end of this unit you should be able to:

- explore the implications of teaching art and design in a plural society;
- examine how a knowledge of and respect for difference can be ensured while seeking the interrelatedness and interdependence of cultures;
- consider how to develop a democratic education.

Others without?

Tate (1997) (the then head of SCAA, now QCA) advocated the necessity to construct a cultural identity by exposing pupils to canonical exemplars. He admitted that:

> Because identities are multiple, this involves a sense of how they locate themselves within a variety of cultural traditions: above all those of England, Britain and Europe, but also the traditions of those parts of the non-European world with which this country has close and long-established links.
>
> (p. 13)

Where those differences are blurred or indistinct they tend to be written out. Tate warned against the distractions of:

> . . . a pervasive cultural egalitarianism which refuses to recognise that cultures (especially majority ones) are very special to those who belong to them and need to be nurtured and transmitted through careful attention and special treatment. It is in part a result of the prevailing postmodernist intellectual climate with its emphasis on fictions and constructions and its sense that nothing is sufficiently substantial or objective for it to be worth passing on.
>
> (*ibid.*: 12)

However, it must be one task of a critical curriculum to uncover or deconstruct the structures of difference which support a dominant culture, even if that culture feels fragile and threatened by such scrutiny. If it has anything ethically vital or cogent to offer it can only benefit from the experience. To invoke the agenda of one 'postmodernist intellectual':

> Foucault completely upsets our conventional expectations of history as something linear – a chronology of inevitable facts that tells a story which makes sense. Instead, he uncovers the underlayers of what is kept suppressed and unconscious *in* and *throughout* history – the codes and assumptions of order, the structures of exclusion that legitimate the epistemes (systems of knowledge/belief) by which societies achieve their identities.
>
> (Appignanesi and Garratt 1995: 83)

Cultures typically define themselves through a system of difference and exclusion and this process has left postcolonial Britain with a host of minority groups whose identities were once seen as marginal to the grand-narratives of its national identity: this would include groups differentiated on such grounds as race, gender, class, sexuality, age, disability. Empowering these groups with the means to find and disseminate their voices is one of the principles of equal opportunities. But it is entirely counter-productive to lump them all together as if they were some form of homogeneous, disenfranchised 'other' who require uniform and patronising assistance. Equal opportunities does not mean treating everyone in the same way: this would deny the concept of differentiation by need which is at the heart of its strategies. For example, racism based on the grounds of skin colour is still pervasive, and teachers may be tempted to celebrate, for example, 'black arts' by showing examples of limited and usually historical artefacts. But in countering prejudice, in what ways would this benefit their pupils? What messages would it send to the class about 'blackness'? All are likely to have experienced the monolithic effects of undifferentiated prejudice, whether produced or received, but the means to counter racism cannot be achieved through a celebration of an essentialist blackness exemplified by token (mis)representation. What is needed is an understanding of the mechanisms by which colour prejudice has been constructed and perpetuated (see Chapters 5 and 11). At the same time this must be managed by avoiding strategies that engender a culture of victimisation and guilt:

> Otherization is unavoidable, and for every One, the Other is the Heart of Darkness. The West is as much the Heart of Darkness to the Rest as the latter is to the West. Invention and contemplation of the Other is a continuous process evident in all cultures and societies. But in contemplating the Other, it is necessary to exhibit modesty and admit relative handicap since the peripheral location of the contemplator precludes a complete understanding; this ineluctability is the Darkness.
>
> (Oguibe 1993: 3–8)

For many years anthropologists have been demanding that represented 'others' should be allowed to speak for themselves. The ethical way forward is to share 'textural authority with subjects themselves, autobiographical recanting as the only appropriate form for merging cultural experience with the ethnographer's own' (Clifford and Marcus 1986: 168) and museums have made concerted efforts to enable self-representation to take place (Stanley 1998). At one time the ethnographer's remit was to explain 'other' people from *without* their own, usually colonialist, culture, to 'us', an audience from within: increasingly they are exploring people from *within* their own, usually, postcolonial culture, to 'others as us' and 'us as others'.

> One of imperialism's achievements was to bring the world closer together and, although in the process the separation between Europeans and natives was an insidious and fundamentally unjust one, most of us would now regard the historical experience of empire as a common one. The task then is to describe it as pertaining to Indians *and* Britishers, Algerians *and* French, Westerners *and* Africans, Asians, Latin Americans, and Australians despite the horrors, the bloodshed, and the vengeful bitterness.
>
> (Said 1993: xxiv)

Although containing a critique, modernism's appropriation of the exotic and the other is closely bound to the colonialist enterprise, it simultaneously disseminated difference while diffusing it by appropriating it in the guise of universalism (Hiller 1991). Likewise postmodernism is part and parcel of the late capitalist world of international corporatism, exploring and embracing its technologies while deconstructing and attacking its structures. Whatever you may think of global capitalism, one of its by-products is the erosion of cultural, if not economic, difference:

> Global homogenization is more credible than ever before, and though the challenge to discover and represent cultural diversity is strong, doing so in terms of spatio-temporal cultural preserves of otherness seems outmoded. Rather, the strongest forms of difference are now diffused within our own capitalist cultural realm, gender and lifestyle constructs being two prominent fields of representation for exploring cultural difference.
>
> (Clifford and Marcus 1986: 168)

Market incursions are liable both to appropriate the material cultures of indigenous peoples and gradually replace them through a system of imports either imposed or made desirable through targeted advertising. This has the effect of diluting a major basis of indigenous cultural identities. The resulting losses in difference, although welcomed by some, are perceived by many as pernicious and destructive. Postmodern pluralism is often hailed as an egalitarian triumph, all voices have their say, hierarchies are abandoned, the transcendental with the transient, the sublime with the pornographic, all hold hands together and dance the dance of relative values. But, 'doubt is itself in doubt, doubt should be, but isn't, tolerant of others' beliefs' (Appignanesi 1997: audio). For many in the 'Developing World' postmodernism is the bedfellow of capitalism, a cynical manoeuvre to relegate difference to marketable spectacle. For such critics postmodernism possesses no integrity or originality, it is a parasitic maw with an indiscriminate hunger to appropriate difference for mass consumption and it can do so because it has absolute control of the means of production, marketing and distribution. For people who hold fundamental beliefs the erosion of difference signals a major threat and defensive positions are drawn. 'Postmodernism, with its relativism, scepticism and doubt, appropriates the grand-narratives of others' certainties at its peril' (*ibid.*). Religious and political fundamentalisms have thus been theorised not as returns from, but as products of, postmodernism.

Hughes (1993) reminds you that all cultural and spiritual exposition, including that of marginal groups, requires critical scrutiny if its voice is to take on historical, social and collective significance: 'Now the claims of the victim do have to be heard, because they may cast new light on history. But they have to pass the same tests as anyone else's, or debate fails and truth suffers' (p. 146). The tests and truth referred to by Hughes are not those of scientific empiricism probing divine revelation or metaphor, but historical investigation differentiating actual and fictive events. He argues that the endeavour to empower oppressed groups within particular nation states, in this instance the USA, is doomed to failure if it fails to recognise that the myths of the oppressed are as pernicious and divisive as the myths of the dominant culture:

> Cultural separatism within this republic is more a fad than a serious proposal; it is not likely to hold, but if it did, it would be an educational disaster for those it claims to help, the young, the poor and the black. It would be a gesture not of 'empowerment' but of emasculation. Self-esteem comes from doing things well, from discovering how to tell a truth from a lie, and from finding out what unites as well as separates us.
>
> (*ibid.*: 150–151)

While Hughes warns against a misguided and patronising desire to accept at face value all comers in the name of liberal tolerance, Said reminds us that indiscriminate acceptance is a recipe for chaos:

> In our wish to make ourselves heard, we tend to forget that the world is a very crowded place, and that if everyone were to insist on the radical purity or priority of one's own voice, all we would have would be the awful din of unending strife, and a bloody political mess, the true horror of which is beginning to be perceptible here and there in the reemergence of racist politics in Europe, the cacophony over political correctness and identity politics in the United States, and . . . the intolerance of religious prejudice and illusionary promises.
>
> (Said 1993: xxiii)

Giroux (1992) argues that teachers can overcome the increasing divisions in society if they come together to form a radical project:

> This means we must make an attempt to develop a *shared language* around the issue of pedagogy and struggle, develop a *set of relevancies* that can be recognised in each other's work, and articulate a *common political project* that addresses the relationship between pedagogical work and the reconstruction of *oppositional spheres*. Second, we need to form alliances around the issue of *censorship* both in and out of the schools. The question of *representation* is central to issues of pedagogy as a form of cultural politics and cultural politics as practice *related to the struggles of everyday life*. Third we need to articulate these issues in a *public* manner, in which . . . we're really addressing a *variety of cultural workers* and not simply a narrowly defined audience. This points to the need to broaden the definition of culture and political struggle and in

doing so *invite others to participate* in both the purpose and practice central to such tasks.

(p. 159; my italics)

This may at first appear too openly political a project. But all Giroux is outlining here is the basis for developing a democratic education, for it must be remembered that whatever your personal beliefs you are educating young people to engage and contribute to the practice of art and design within the context of a democracy. Democratic principles ensure that you have the right to hold your beliefs and, in most instances, to speak them, but they also hold that you have no right to impose them on others; rather your task is to enable others to find voice.

Task 12.4.1 Interdependencies

Discuss these questions in tutor groups:
  Can you use exemplars in such a way that they will be understood as contingent (on culture and history) rather than fixed or absolute?
  How can you ensure that the range of knowledge encompassed by your pupils can be shared and made productive in your lessons?
  How can you ensure that your teaching recognises the 'interdependencies' of the past and present?

Others within?

One of the central tenets of official Modernism, that 'Art' has the potential to provide transcendental or liminal experiences, is a device for separating the chaff from the wheat. Thus, most applied art, craft, design and particularly decoration, is immediately placed outside the domain of 'Art' and 'Aesthetics' because it serves utilitarian and domestic functions (Addison 1997). Likewise popular traditions, characterised by Greenberg in 1939 as 'Kitsch', serve lesser purposes, particularly sentimentality and ephemeral fashion (Harrison and Wood 1992: 530–541). There is a powerful tradition in British education that dismisses popular forms of cultural production and upholds an exclusive and elitist notion of culture. Leavis asserted in 'Mass Civilisation and Minority Culture' (1930):

> In any period it is upon a very small minority that the discerning appreciation of art and literature depends: it is (apart from the cases of the simple and the familiar) only a few who are capable of unprompted, first hand judgement . . . upon this minority depends our power of profiting by the finest experience of the past; they keep alive the subtlest and most perishable parts of tradition. Upon them depend the implicit standards that order the finer living of an age, the sense that this is worth more than that, this rather than that is the direction in which to go, that the centre is here rather than there.

(in Easthope and McGowan 1992: 210)

This suggests that teachers, as upholders of standards and advocates of quality, should inculcate their pupils with only the very best. But if you reject the popular out of hand, as Leavis implies you should, you renounce territory with which pupils may feel kinship if not outright ownership; you reinforce the sense that they function from outside cultural codes that are valued because they identify with an inferior if majority culture. This is not to propose that you exclusively embrace the popular but that you acknowledge and approach its forms critically, seeking its function and value. Adorno's analysis of popular music (1941 in Easthope and McGowan 1992: 211–223) is scathing in its conclusions but so much more telling than Leavis' patronising certainties.

Williams (1976) reminds you that the meaning of the word 'popular' has multiple and changing meanings:

> **Popular culture** was not identified by *the people* but by others, and it still carries two older senses: inferior kinds of work (compare **popular literature, popular press** as distinguished from *quality press*); and work deliberately setting out to win favour (**popular journalism** as distinguished from *democratic journalism*, or **popular entertainment**); as well as the more modern sense of well-liked by many people, with which of course, in many cases, the earlier senses overlap. The recent sense of **popular culture** as the culture actually made by people for themselves is different from all these; it is often displaced to the past as *folk* <u>culture</u> but it is also an important modern emphasis.
>
> (p. 237)

Task 12.4.2 Attitudes to pluralism

In your tutor group discuss your attitudes to:

- forms of art other than the fine arts;
- forms of popular production;
- forms of production that acknowledge minority interests.

What implications do your attitudes have on what you choose to introduce to the art and design curriculum?

Task 12.4.3 Developing pluralist resources

Using the types of production outlined in this unit, form banks of reproductions which have, on their reverse, contextualising information, bibliographical references and sets of questions for further investigation, e.g. such packs as *Working with Modern British Art: A Guide for Teachers* (Adams et al. 1998).

Further reading

Appignanesi, R. and Garratt, C. (1995) *Postmodernism for Beginners*, Cambridge: Icon Books.

Giroux, H. (1992) *Border Crossings*, New York; London: Routledge.

Said, E. (1993) *Culture and Imperialism*, London: Chatto and Windus.

Williams, R. (1976) *Keywords*, London: Fontana.

UNIT 12.5 QUESTIONING AND RESOLVING DIFFERENCES IN VALUES

By the end of this unit you should be able to:

- explore and attempt to resolve dichotomies of value in educational and other systems.

In the same way that people look for fixed points of reference when identifying cultures, you may feel it necessary to form a particular allegiance to an educational theory and pedagogical method in order to construct a clear identity as a teacher. This is almost irresistible because it seems to answer the pressures on you to resolve ideological conflicts within your own practice. However, it is advisable to remain open in your approach and not be tempted to fit one method to all circumstances (Atkinson 2003).

Educational theories and traditions often pull you in directions that seem to be opposing, for example, while one suggests that you should encourage 'self expression' and the emergence of the individual, another expects you to provide pupils with the symbolic tools to enter a collective and social universe. The following table outlines some of these dichotomies.

Dichotomies

self-actualisation	socialisation
nature	culture
self-expression	cultural conventions
vocational	academic
child-centred	adult-centred
world-to-child	child-to-world
authenticity	normative standards
peer pressure	adult authority
feeling	cognition
intuition	logic
haptic	visual
non-conformity	conformity (behaviour)
being other	belonging
popular	elite
spontaneity	exactitude

risk	safety
certainty	doubt
action	thought

Task 12.5.1 Working with dichotomies

In your tutor group:

Add to this list any dichotomies that you have witnessed or imagine might arise in your teaching.

Consider the extent to which these dichotomous pairs or contrasting aims and values are opposed: for example the French philosopher Baudrillard says: the opposite of knowledge is not ignorance, but deceit and fraud (Appignanesi and Garratt 1995: 136).

Where you find oppositions, either in the given list or in your own additions and amendments, is it necessary to reconcile them, and if so, how?

Might they be argued as parallel or complementary?

To which of these values do you feel any allegiance? Do any patterns emerge? Where do you see a problem in applying these values to your own teaching?

AFTERWORD

In this chapter you have seen the dangers of stereotypical representation, canonical exemplification and misdirected sympathy. In suggesting that you look for contradictions where consistency is normally posed, for commonality where difference is reinforced, and by questioning 'universal' criteria used to belittle what is actually different, you help pupils to see the structures of value by which cultures choose to identify themselves. The key to effective artistic and cultural study is critical inquiry. Before you can expect pupils to adopt such a position this requires continual critical effort and demonstration by you which should not end in your student years. Without it your representations of cultures are liable to be forms of misrepresentation and mis-evaluation. In Chapter 11 it was recommended that in pursuing a multicultural curriculum you investigate the intercultural dimension of art and design. It is educationally desirable to apply its defining principle, inquiry at the point of translation and transmission, to include and question other categories of difference: elite/popular, male/female, heterosexual/homosexual, young/old. In this way you seek out and engage with those points of contact and interaction where difference is negotiated, and transformation, appropriation, compromise and rejection are put into play.

Bibliography

AAVVAA (African & Asian Visual Artists Archive) Online. Available HTTP: http://www.aavaa.org.uk (accessed 30 October 2006).

Abbs, P. (ed.) (1987) *Living Powers: the Arts in Education*, London: The Falmer Press.

—— (1989) *A is for Aesthetic*, London: Falmer Press.

Adams, E. (1989) 'Learning to see', *Journal of Art and Design Education* 8, 2: 183–196.

—— (1997) 'Public Art and Art & Design Education in Schools', *Journal of Art and Design Education* 16, 3.

—— (2002) 'Power drawing', *International Journal of Art and Design Education* 21, 3: 220–233.

Adams, J. (2000) 'Working out comics', *International Journal of Art & Design Education*, 19, 3: 304–312.

—— (2005) 'Room 13 and the contemporary practice of artist-learners' *Studies in Art Education*, 47, 1, Fall 2005, National Art Education Association.

Adams, J., Meecham, P., and Orbach, C. (1998) *Working with Modern British Art: A Guide for Teachers*, Liverpool: Tate Gallery and John Moores University.

Adams, P. (1996) *The Emptiness of the Image: Psychoanalysis and Sexual Differences*, London: Routledge.

Addison, N. (1997) *In Defence of Decoration*, London: Institute of Education.

—— (2002) 'What are Little Boys made of, made of? Victorian constructions of Gender', *Sex Education*, 2, 1.

—— (2005) 'Expressing the not-said: art and design and the formation of sexual identities', *International Journal of Art and Design Education*, 24, 1: 20–30.

—— (2006) 'Acknowledging the gap between sex education and the lived experiences of young people: a discussion of Paula Rego's *The Pillowman* (2004) and other cautionary tales', *Sex Education: Sexuality, Society and Learning*, 6, 4.

—— (2007) 'Identity politics: queering art and design and the confessional route to salvation', Special Edition: Lesbian and Gay Issues in Art & Design Education, *International Journal of Art and Design Education*, 26, 1.

Addison, N. and Allen, E. (1997) 'Art Histories in Action', *Engage 8*, London: Engage.

Addison, N. (2003) 'Productive tensions: residencies in research', in N. Addison and L. Burgess (eds), *Issues in Art and Design Teaching*, London: RoutledgeFalmer.

Addison, N. and Burgess, L. (2003) *Issues in Art and Design Teaching*, London: RoutledgeFalmer.

—— (2006) 'London cluster research report', *En-quire: Inspiring Learning in Galleries*, London: Engage.

—— (forthcoming) *Teaching Art and Design*, London: RoutledgeFalmer.

Ades, D. (1976) *Photomontage*, London: Thames and Hudson.

Ades, D., Baker, S. and Bradley, F. (2006) *Undercover Surrealism: Georges Bataille and Documents*, London: Hayward Gallery in association with MIT Press.

Adorno, T. (1941) 'On popular music', in A. Easthope and K. McGowan (eds) (1992) *A Critical and Cultural Theory Reader*, Buckingham: Open University.

Adorno, T and Horkheimer, M. (1999) 'The Culture industry: enlightenment as mass deception', in S. During (ed.) *The Cultural Studies Reader*, London: Routledge.

Allan, J. (1999) *Actively Seeking Inclusion: Pupils with Special Needs in Mainstream Schools*, London: Falmer Press.

Allen, D. (1996) 'As well as painting', in L. Dawtrey *et al.* (eds) *Critical Studies and Modern Art*, Milton Keynes: Open University.

Allison, B. (1974) 'Professional art education', *Journal of the NSAE* 1, 1: 3–9.

—— (1997) 'Professional attitudes to research in art education', *Journal of Art and Design Education* 16, 30: 211–215.

Anderson, D. (1997) *A Common Wealth: Museums & Learning in the UK*, London: Department of National Heritage.

Appignanesi, R. (1997) *Postmodernism for Beginners*, Cambridge: Icon, Audio.

Appignanesi, R. and Garratt, C. (1995) *Postmodernism for Beginners*, Cambridge: Icon, Books.

Applied Diplomas information from the DfES. Online. Available HTTP: http://www.dfes.gov.uk/14–19/index.cfm?sid=3 (accessed 13 December 2006).

Araeen, R. (1989) *The Other Story*, London: Hayward Gallery.

Archer, M. (1996) *Installation Art*, London: Thames and Hudson.

Architecture Centre Network. Online. Available HTTP: http://www.architecturecentre.net (accessed 30 October 2006).

Assessment Reform Group (2002) 'Assessment for Learning: 10 principles'. Online. Available HTTP: http://arg.educ.cam.ac.uk/CIE3.pdf (accessed 24 August 2006).

Ash, A. (2002) 'Forms of Experience: Engaging with 20th Century Sculpture', *NSEAD annual conference*, convened by NSEAD at Tate Britain.

—— (2004) 'Bite the ICT bullet: using the World Wide Web in art education', in R. Hickman (ed.) (2nd Edition) *Art Education 11–18: Meaning, Purpose and Direction*, London, New York: Continuum.

Association of Advisers and Inspectors in Art and Design (AAIAD: Midland Group) (1994) *Sketchbooks*, Bromsgrove: AAIAD.

Association of Advisers and Inspectors in Art and Design (AAIAD) (1996) *Art: A Review of Inspection Findings 1995–96*, AAIAD.

Attfield, J. and Kirkham, P. (eds) (1995) *A View from the Interior: Women and Design*, London: Women's Press.

Atkinson, D. (2001) 'Assessment in educational practice: forming pedagogised identities in the art curriculum', *International Journal of Art & Design Education*, 20, 1: 96–108.

—— (2002) *Art in Education: Identity and Practice*, Dordrecht, Boston and London: Kluwer Academic Publishers.

—— (2003) 'Forming teacher identities in initial teacher education', in N. Addison and L. Burgess (eds) *Issues in Art and Design Teaching*, London: RoutledgeFalmer.

—— (2006) 'School art education: mourning the past and opening a future', *International Journal of Art and Design Education*, 25, 1: 16–27.

Aubrey, C. (ed.) (1994) *The Role of Subject Knowledge in the Early Years of Schooling*, London: Falmer Press.

Audit Commission (2002) *SEN: A Mainstream Issue*, Audit Commission Publications.

Avery, H. (1994) 'Feminist issues in built environment education', *Journal of Art and Design Education* 13, 1: 65–69.

Baddeley, O. (ed.) (1994) *New Art from Latin America*, London: Art and Design.

Bailey, D. A., Baucom, I., and Boyce, S. (eds) (2005) *Shades of Black*, Durham, NC Duke University Press in association with inIVA and Aavaa.

Ball, L. (1998) *Crafts 2000: A Future in the Making*, London: Crafts Council.

Bancroft, A. (1995) 'What do Dragons think about their Dark Lonely Caves?', *Journal of Art and Design Education* 14, 1.

Barrett, M. (1982) *Art Education: A Strategy for Course Design*, London: Heinemann.

Barrow (Dame) J. et al. (2005) Delivering Shared Heritage: The Mayor's Commission on African and Asian Heritage, Mayor of London.

Barthes, R. (1957) *Mythologies*, trans. A. Lavers and C. Smith (edition 1990), London: Jonathan Cape and New York: Hill and Wang.

—— (1965) *Elements of Semiology*, trans. A. Lavers and C. Smith (edition 1975), London: Jonathan Cape and New York: Hill and Wang.

Barton, L. (1998) *The Politics of Special Educational Needs*, Lewes: Falmer Press.

Bastick, T. (1982) *Intuition: how we think and act?*, Chichester: Wiley.

Baynes, K. (1982) 'Beyond design education', *Journal of Art and Design Education*, 1, 1: 105–114.

—— (1990) 'Defining the design dimension of the curriculum', in D. Thistlewood (ed.) *Issues in Design Education*, London: Longman.

Beckett, S. (1979) *The Beckett Trilogy: Molloy, Malone Dies, The Unnameable*, London: Picador.

BECTA for DfEE *Choosing and Using CD-ROMs for Art & Design*, London: HMSO.

Belting, H. (1995) *The End of the History of Art?* Chicago, London: University of Chicago Press.

—— (2003) *Art History after Modernism*, Chicago, London: The University of Chicago Press.

Benjamin, W. (2004) *Selected Writings Volume 1 (1913–1926)*, Cambridge, MA.: The Belknap Press of Harvard University Press.

Bergstrom, K. (2005) 'Paving the way forward', *Royal Society of Arts Journal*, October 2005: 50–55.

Berkeley, A. (1987) *Academic Inflation*, London: Crafts Council.

Bermingham, A. (2000) *Learning to Draw: Studies in the Cultural History of a Polite and Useful Art*, New Haven and London: Yale University Press.

Bernal, M. (1987) *Black Athena: The Afroasiatic Roots of Classical Civilization*, New Brunswick: Rutgers University Press.

Betterton, R. (1996) *An Intimate Distance: Women, Artists and the Body*, London: Routledge.

(The) Big Draw. Online. Available HTTP: http://www.thebigdraw.org.uk/ (accessed 28 October 2006).

Binch, N. (1994) 'The implications of the National Curriculum Orders for Art', *Journal of Art & Design Education*, 13, 2, Oxford: Blackwell.

—— (1997) 'Foreword' in D. Allen and A. Rogers, *What's happening to Photography?* London: Arts Council of England.

Binch, N. and Robertson, E. (1994) *Resourcing and Assessing Art, Craft and Design – Critical Studies in Art at Key Stage 4*, Corsham: NSEAD.

Blackburn, J. (1998a) 'It's not all looms and sad crochet', *The Observer* 25 Jan p. 27.

—— (1998b) 'Arts and crafty', *Metro*, 4 July 1998.

Blair, T. (1998) 'Foreword' in the Green Paper *Teachers Meeting the Challenge of Change*, London: DfEE, HMSO.

Blazwick, I. (ed.) (2001) *Century City: Art and Culture in the Modern Metropolis*, London: Tate Publishing.

Blurring the Boundaries: Installation Art 1969–1996, San Diego: Museum of Contemporary Art.

Boime, A. (1990) *The Art of Exclusion, Representing Blacks in the Nineteenth Century*, London: Thames and Hudson.

de Boisbaudran, H. L. (1862) *The Training of Memory in Art*, trans. L. D. Luard (edition 1914), London: Macmillan.

Bonami, F. (2005) *Universal Experience: Art, Life and the Tourist's Eye*, Chicago: Museum of Contemporary Art.

Booth, T., Aiscow, M., Black-Hawkins, K., Vaughan, M. and Shaw, L. (2002) *Index for Inclusion*, Bristol: CSIE.

Boughton, D. (1995) 'Six myths of National Curriculum reforms in art education', *Journal of Art & Design Education*, 14, 2, Oxford: Blackwell.

Boughton, D., Eisner, E., and Ligtvoet (eds) (1996) *Evaluating and Assessing the Visual Arts*, New York: Teachers College Press.

Boughton, D. and Mason, R. (1999) *Beyond Multicultural Art Education: International Perspectives*, New York, Berlin: Waxman Munster.

Bourdieu, P. (1984) *Distinction: A Social Critique of the Judgement of Taste*, trans. R. Nice, London: Routledge and Kegan Paul.

—— (1993) *The Field of Cultural Reproduction*, Cambridge: Polity Press.

Brett, G, (1986) *Through our Eyes*, London: Heretic Books.

British Library Website (2006) online. Available HTTP: http://www.bllearning.co.uk/live/sacredbook/object/SacredBook (accessed 27 October 2006).

Britton, A. (1991) 'The manipulation of skill and their outer limits of function', in *Beyond the Dovetail: Craft, Skill and Imagination*, London: Crafts Council.

Brophy, J. (ed.) (1991) *Introduction to Vol 2 Advances in Research on Teaching*, Greenwich: CTJAI Press.

Broude, N. and Garrard, M. D. (1994) *The Power of Feminist Art*, New York: Harry N Abrams Incorporated.

Bruner, J. (1960) (edition 1977) *The Process of Education*, Cambridge, MA and London: Harvard University Press.

Bryson, N. (1990) *Looking at the Overlooked: Four Essays on Still Life Painting*, London: Reaktion.

—— (1991) 'Semiology and visual interpretation', in N. Bryson *et al.* (eds) *Visual Theory*, Cambridge: Polity Press.

Buchanan, M. (1990) 'Design & Technology: issues for implementation and the role of Art and Design', in *Design & Technology Teaching*, Stoke-on-Trent: Trentham Books.

—— (1993) MA Lecture, Institute of Education.

—— (1995) 'Making art and critical literacy: a reciprocal relationship', in R. Prentice (ed.) *Teaching Art and Design: Addressing Issues and Identifying Directions*, London: Cassell.

Buck, L. (1993) *On the Edge: Art meets Craft*, London: Crafts Council.

Buckingham, D. (ed.) (1998) *Teaching Popular Culture*, London: University College Press.

—— (2005) *Schooling the Digital Generation: Popular Culture, New Media and the Future of Education*, London: Institute of Education, University of London.

Buckingham, D. and Sefton-Green, J. (1994) *Cultural Studies goes to School*, London: Taylor and Francis.

Budd, M. (1995) *Values in Art*, London: Penguin.

Building Schools for the Future. Online. Available HTTP: http://www.bsf.gov.uk/bsf/ (accessed 30 October 2006).

Burgess, L. (1995) 'Human Resources: artists, craftspersons, designers', in R. Prentice (ed.) *Teaching Art and Design: Addressing Issues and Identifying Directions*, London: Cassell.

—— (2003) 'Monsters in the playground', in N. Addison and L. Burgess (eds) *Issues in Art and Design Teaching*, London: RoutledgeFalmer.

Burgess, L. and Addison, N. (2004) 'Contemporary Art in Schools: Why Bother?', in R. Hickman (ed.) (2nd Edition) *Art Education 11–18: Meaning, Purpose and Direction*, London and New York: Continuum.

Burgess, L. and Holman, V. (1993) 'Live Art', *Times Educational Supplement* 2 June.

Burgess, L. and Reay, D. (2006) 'Postmodernism feminisms: problematic paradigms', in T. Hardy (ed.) *Art Education in a Postmodern World*, Bristol: Intellect Books.

Burgess, L. and Schofield, K. (1998) 'Shorting the circuit', in P. Johnson (ed.) *Ideas in the Making*, London: Crafts Council.

Button, V. (2005) (4th Edition) *The Turner Prize*, London: Tate Publishing.

Buzan, T. (1982) *Use Your Head*, London: Ariel Books, BBC.

Cahan, S. and Kocur, Z. (eds) (1996) *Contemporary Art and Multicultural Education*, New York, London: Routledge.

Callow, P. (2001) 'ICT in art', *International Journal of Art and Design Education*, 20, 1: 41–48.

(The) Campaign For Drawing (2006) online. Available HTTP: http://www.drawing-power.org.uk/menu2.htm (accessed 11 April 2006).

Capel, S., Leask, M., and Turner, T. (2005) *Learning to Teach in the Secondary School – A Companion to School Experience*, London: Routledge.

Caruana, W. (1993) *Aboriginal Art*, London: Thames and Hudson.

Causey, A. (1998) *Sculpture since 1945*, Oxford: Oxford University Press.

Chadwick, W. (1990) *Women, Art and Society*, London: Thames and Hudson.

—— (1991) *Women Artists and the Surrealist Movement*, London: Thames and Hudson.

Chalmers, G. (1996) *Celebrating Pluralism Art, Education, and Cultural Diversity*, Los Angeles: Paul Getty Trust.

Chandler, D. (2001) *Semiotics: The Basics*, London: Routledge.

Charman, H., Rose, K. and Wilson, G. (eds) (2006) *The Art Gallery Handbook: A Resource for Teachers*, London: Tate Publishing.

Cheng, F. (1994) *Empty and Full: The Language of Chinese Painting*, London: Shambahla.

Chicago, J. (1996) *The Dinner Party*, New York and London: Penguin.

(The) Children Act (2004) online. Available HTTP: http://www.dfes.gov.uk/publications/childrenactreport/ (accessed 30 October 2006).

Chipp, H. (1968) *Theories of Modern Art*, Berkley; Los Angeles; London: University of California Press.

Clark, G. and Zimmerman, E. (1984) *Educating Artistically Talented Students*, Syracuse, NY: Syracuse University Press.

Clark, K. (1985) *Ruskin Today*, Harmondsworth, Middlesex: Penguin.

Clark, S. (2003) *Beaded Prayer Project* catalogue courtesy of Baltimore Museum of Art.

Clark, T. J. (1999) *Farewell to an Idea: Episodes from a History of Modernism*, New Haven, CT, and London: Yale University Press.

Clement, R. (1986) *The Art Teacher's Handbook*, London: Hutchinson Education.

—— (1993) (Second Edition) *The Art Teacher's Handbook*, Cheltenham: Stanley Thornes.

Clifford, J. and Marcus, G. E. (1986) (eds) *Writing Culture: The Poetics and Politics of Ethnography*, Berkeley; London: University of California Press.

Clive, S. and Binch, N. (1994) *Close Collaborations: Art in Schools and the Wider Environment*, London: ACE.

Clive, S. and Geggie, P. (1998) *Unpacking Teachers' Packs*, London: ENGAGE.

Clough, P. and Corbett, J. (2000) *Theories of Inclusive Education: A Student's Guide*, London: Paul Chapman Publishing.

Cohen, P. (1998) 'Tricks of the trade: on teaching arts and "race" in the classroom', in D. Buckingham (ed.) *Teaching Popular Culture*, London: University College Press.

Cole, A. (1992) *Perspective*, London: Dorling Kindersley and National Gallery.

(The) Commission for Architecture & the Built Environment. Online. Available HTTP: http://www.cabe.org (accessed 30 October 2006).

(*The*) *Concise Oxford Dictionary* (1985), Oxford: Oxford University Press.

Conran, T. (1993) in *Sunday Express Magazine*, 3 October.

Coomaraswamy, A. K. (1934) (1956 edition) *The Transformation of Nature in Art*, New York: Dover.

Coombes, A. (1991) 'Ethnography and the formation of national and cultural identities', in S. Hiller (ed.) *The Myth of Primitivism*, London: Routledge.

—— (1994) *Reinventing Africa*, New Haven, CT, and London: Yale University Press.

—— (1998) in Preziosi (ed.) *The Art of Art History*, Oxford: Oxford University Press.

Corbett, J. (1997) 'Include/exclude: redefining the boundaries', *International Journal of Inclusive Education*, 1, 1.

—— (1999) 'Inclusive education and school culture', *International Journal of Inclusive Education*, 3, 1.

—— (2001) *Supporting Inclusive Education: A Connective Pedagogy*, London: Routledge/Falmer.

Courtney, C. (1995) *Private Views and Other Containers: Artists' Books Reviewed by Cathy Courtney for* Art Monthly *1983–1995*, London: Estamp Editions.

Crafts Council (1982) *Crafts Conference for Teachers*, London: Crafts Council.

Crafts Council (1996) (Exhibition Catalogue) *Recycling: Focus for the Next Century – Austerity for Posterity*, London: Crafts Council.

Crafts Council (1998) Definitions of Craft as provided by the Crafts Council.

Crafts Council Survey (1995) *Pupils As Makers, Part 1*, London: Crafts Council.

Crafts Council Survey (1998) *Pupils As Makers, Part 2*, London: Crafts Council.

Crafts: the decorative and applied arts magazine. Published bi-monthly.

Creative Partnerships (2006) Online. Available HTTP: http://www.creative-partnerships.com (accessed 30 October 2006).

Creber, P. (1990) *Thinking Through English*, Milton Keynes: Open University Press.

Crowther, P. (1993) *Art and Embodiment*, Oxford: Clarendon Press.

—— (1996) *Critical Aesthetics and Post Modernism*, London: Clarendon Press.

Cunliffe, L. (1996) 'Art and world view: escaping the formalist labyrinth', *Journal of Art and Design Education* 15, 3.

Dahl, D. (1990) *Residencies in Education*, Sunderland: AN Publications.

Dalton, P. (1987) 'Housewives, leisure crafts and ideology: deskilling in consumer craft', in G. Elinor *et al.*, *Women and Craft*, London: Virago.

—— (2001) *The Gendering of Art Education: Modernism, Identity and Critical feminism*, Buckingham: Open University Press.

Daniels, H. (2001) *Vygotsky and Pedagogy*, London: RoutledgeFalmer.

Danto, A. (1986) *The Philosophical Disenfranchisement of Art*, New York: Columbia University Press.

Dash, P. (2005) 'Cultural demarcation, the African diaspora and art education', in D. Atkinson and P. Dash (eds), *Social and Critical Practices in Art Education*, Stoke-on-Trent UK and Sterling USA: Trentham Books.

Davey, N. (1994) 'Aesthetics as the foundation of human experience', *Journal of Art and Design Education* 13, 1.

Davies, T. (2002) 'Drawing on the past: reflecting on the future', *International Journal of Art and Design Education*, 21, 3: 284–291.

Davison, J. and Dowson, J. (eds) (1998) *Learning to Teach English in the Secondary School*, London: Routledge.

Dawe Lane L. 'Using contemporary art', in L. Dawtrey *et al.* (eds) (1996) *Critical Studies and Modern Art*, New Haven, CT and Yale University Press in Association with The Open University, the Arts Council of England and the Tate Gallery London.

Dawtrey, L., Jackson, T., Masterson, M., Meecham, P., and Wood, P. (eds) (1996) *Critical Studies and Modern Art*, New Haven, CT: Yale University Press.

Day, C. (1993) 'Reflection: a necessary but not sufficient condition for professional development', *BERA*, 19, 1: 83–93.

Dear, J. (2001) 'Motivation and meaning in contemporary art: from Tate Modern to the primary school classroom', *Journal of Art and Design Education*, 20, 3.

Dearing Report (1997) *Higher Education in the Learning Society*, London: HMSO.

Deepwell, K. (ed.) (1995) *New Feminist Art Criticism*, Manchester: Manchester University Press.

Denvir, B. (1986) *The Late Victorians: Art, Design and Society 1852–1910*, London: Longman.

DES (1988) *National Curriculum Design & Technology Working Group, Interim Report*, London: HMSO.

—— (1991a) *Art for Ages 5 to 14*, London: HMSO.

—— (1991b) *NC Art Working Group, Interim Report*, London: HMSO.

—— (1992) *Art in the National Curriculum*, London: HMSO.

—— (1995) *NC Design and Technology Order*, London: HMSO.

Design Council (1994) *Design Focus in Schools, Consultation Document*, London: Design Council.

Designs on Britain. Online. Available HTTP: http://www.artsinform.com/dob/teacher-spack.html (accessed 30 October 2006).

Dewdney, A. (1996) 'Art photography: continuing historical narratives', *Journal of Art and Design Education* 15, 1.

Dewey, J. (1934) (edition 1979) *Art as Experience*, New York: Paragon Books.

Dexter, E. (2005) *Vitamin D: New Perspectives in Drawing*, London, New York: Phaidon.

DFE (1994) *Education Act 1993: Sex Education in Schools*, Circular number 5/94, London: HMSO.

—— (1995) *Art in the National Curriculum*, London: HMSO.

DfEE (1995) *A Guide to Safe Practice in Art and Design*, London: HMSO.

—— (1997) *Teaching: High Status, High Standards*, London: HMSO.

—— (1998) *Consultation Document on the New Induction Arrangements*, London: HMSO.

—— (1999) *The National Curriculum for England: Art and Design*, London: Stationery Office and QCA.

DfES (2000) *Working with Teaching Assistants: A Good Practice Guide*, Nottingham: DfES.

—— (2001a) *Key Stage 3 National Strategy: Literacy across the Curriculum*, London: DfES.

—— (2001b) *The Revised Special Educational Needs Code of Practice*, London: DfES.

—— (2003) *SEN: Training Materials for the Foundation Subjects*, London: DfES.

—— (2004a) *Supporting ITT Providers: The Effective Use of Plenaries*, London: DfES.

—— (2004b) *Removing Barriers to Achievement: The Government's Strategy for SEN*, London: HMSO.

—— (2005) *Youth Matters Green Paper*, London: DfES.

Dickson, M. (1995) *Art with People*, AN Publication.

(*The*) *Disability Discrimination Act 1995* London: HMSO.

Dissanayake, E. (1992) *Homoaestheticus: Where Art Comes from and Why*, New York: Macmillan.

Dormer, P. (1991a) 'Beyond the dovetail: crafts, skills and imagination', in C. Frayling (ed.) *Beyond the Dovetail: Crafts, Skills and Imagination*, London: Crafts Council.

—— (1991b) *The Meanings of Modern Design*, London: Thames and Hudson.

—— (1994) *The Art of the Maker*, London: Thames and Hudson.

—— (1997) *The Culture of Craft*, London: Crafts Council.

Doy, G. (2000) *Black Visual Culture: Modernity and Postmodernity*, London and New York: I.B. Tauris.

Downing, D. and Watson, R. (2004) *School Art: What's in It?: Exploring Visual Arts in Secondary Schools*, Berkshire: NFER.

(The) Drumcroon Gallery Website (2006) online. Available HTTP: http://www.drumcroon.org.uk/Sketchbooks/sketchbooks.html (accessed 11 April 2006).

Dryden, G. and Voss, J. (2001) *The Learning Revolution: To Change the Way the World Learns*, Stafford: Network Educational Press in association with Learning Web.

Duffin, D. (1995) 'Exhibiting strategies', in K. Deepwell (ed.) *New Feminist Art Criticism*, Manchester: Manchester University Press.

Duncan, C. (1995) *Civilising Rituals: Inside Public Art Museums*, London: Routledge.

Dyer, R. (1997) *White*, London and New York: Routledge.

Dyson, A. (1989) 'Art history in schools: a comprehensive strategy', in D. Thistlewood (ed.) *Critical and Contextual Studies in Art and Design Education*, Burnt Mill, Essex: Longman.

Earle, K. and Curry, G. (2005) *Meeting SEN in the Curriculum: Art*, London: David Fulton.

Easthope, A. and McGowan, K. (eds) (1992) *A Critical and Cultural Theory Reader*, Buckingham: Open University.

EDEXCEL (2006) *Advanced Subsidiary GCE in Art and Design (8030–8036) and Advanced GCE in Art and Design (9030–9036): For Assessment from Summer 2007*, London: EDEXCEL. Online. Available HTTP: http://www.edexcel.org.uk/VirtualContent/47923/Spec_Issue_5.pdf (accessed 14 December 2006).

Education Reform Act (ERA) (1988) London: HMSO.

Edwards, A. and Furlong, V. (1978) *The Language of Teaching: Meaning in Classroom Interaction*, London: Heinemann Education.

Efland, A. (1970) 'Conceptualising the curriculum problem', *National Society for Education in Art and Design Bulletin*, April.

—— (1976) 'School art's style of functional analysis', *Studies in Art Education* 17, 2.

Efland, A., Freedman, K., and Stuhr, P. (eds) (1996) *Postmodern Art Education: An Approach to Curriculum*, Virginia: The National Art Education Association.

Eggleston, J. (1996) (2nd edition) *Teaching Design & Technology*, Milton Keynes: Open University Press.

—— (ed.) (1998) *Learning through Making: A National Enquiry into the Value of Creative Practical Education in Britain*, London: Crafts Council.

Ehrenzweig, A. (1967) (edition 1993) *The Hidden Order of Art*, London: Weidenfeld.

Eisner, E. (1972) *Educating Artistic Vision*, New York: Macmillan.

—— (1982) *Cognition and Curriculum: A Bases for Deciding what to Teach*, Harlow, Essex: Longman.

—— (1985) *The Educational Imagination*, New York, Macmillan.

—— (1998) 'Does experience in the arts boost academic achievement?', *Journal of Art and Design Education* 17, 1.

Elliot, J. (1991) 'A model of professionalism and its implications for teacher education', *British Educational Research Journal* 17, 4: 309–18.

Elkins, J. (2003) *Visual Studies: A Sceptical Introduction*, New York; London: Routledge.

Emery, L. (2002) *Teaching Art in a Postmodern World*, Altona Victoria: Common Ground.

Engage (2006) *En-quire: Inspiring Learning in Galleries*, London: Engage.

English Heritage. Online. Available HTTP: http://www.english-heritage.org.uk (accessed 30 October 2006).

Every Child Matters (2005) online. Available HTTP: http://www.everychildmatters.gov.uk/ (accessed 30 October 2006).

Feldman, E. (1987) *Varieties of Visual Experience*, New York: Harry N. Abrams.

Ferguson, B. (1995) 'Race, gender and a touch of class', in R. Prentice (ed.) *Teaching Art and Design: Addressing Issues and Identifying Directions*, London: Cassell.

Fernie, E. (1995) *Art History and Its Methods*, London: Phaidon.

Field, R. (1970) *Change in Art Education*, New York: Routledge and Kegan Paul.

Fisher, M. (1994) 'Labour plans to get architecture taught in schools', *Building Design Magazine* 1202.

Fiske, J. (1982) *Introduction to Communication Studies*, London: Methuen.

—— (1989) *Understanding Popular Culture*, London: Routledge.

Foster, H. (1992) 'The primitive unconscious', in F. Frascina and J. Harris (eds) (1992) *Art in Modern Culture*, London: Phaidon.

—— (1996) *The Return of the Real*, London and Cambridge, MA: MIT Press.

Foucault, M. (1976) *Discipline and Punish: The Birth of the Prison*, Trans. A. Sheridan, New York: Pantheon Books.

—— (1988) 'Technologies of the Self', in L. Martin, H. Gutman and P. Hutton (eds) *Technologies of the Self: A Seminar with Michel Foucault*, London: Tavistock.

Franks, A. (2003) 'The role of language within a multimodal curriculum', in N. Addison and L. Burgess (eds) *Issues in Art and Design Teaching*, London: RoutledgeFalmer.

Frascina, F. and Harris, J. (eds) (1992) *Art in Modern Culture*, London: Phaidon.

Frascina, F. and Harrison, C. (eds) (1982) *Modern Art and Modernism: A Critical Anthology*, London: Open University Press.

Frayling, C. (1982) *Crafts Conference For Teachers*, London: Crafts Council.

—— (1990) 'Some perspectives on the crafts revival in the twentieth century', in D. Thistlewood (ed.) *Issues in Design Education*, Harlow, Essex: Longman and NSEAD.

Frederickson, N. and Cline, T. (2002) *SEN, Inclusion and Diversity: A Textbook*, Buckingham: OUP.

Freedman, K. (1994) 'Interpreting gender and visual culture in art classrooms', *Studies in Art Education*, 35, 3.

—— (1997) 'Visual art/virtual art: teaching technology for meaning', *Art Education* 50, 4; Freedman refers to N. Brown (1989) 'The Myth of Visual Literacy', *Australian Art Education* 13, 2: 28–32.

Freire, P. (1985) *The Politics of Education: Culture, Power and Liberation*, South Hadley, MA: Bergin and Garvey.

Fuller, P. (1983) 'The necessity of art education', in *The Naked Artist*, London: Writers and Readers.

—— (1993) *Modern Painters; Reflections on British Art*, London: Methuen.

Gablik, S. (1991) *The Reenchantment of Art*, London: Thames and Hudson.

—— (1995) *Conversations Before the End of Time*, New York: Thames and Hudson.

Gardner, H. (1993) *The Unschooled Mind*, London: Fontana.

—— (1999) *Intelligence Reframed. Multiple intelligences for the 21st century*, New York: Basic Books.

Garner, S. W. (1990) 'Drawing and designing: the case for reappraisal', *Journal of Art and Design Education* 9, 1: 39–55.

Geertz, C. (2000 edition) *The Interpretation of Cultures*, New York: Basic Books.

Gentle, K. (1988) *Children and Art Teaching*, London: Routledge.

—— (1990) 'The significance of art making for individuals', *Journal of Art and Design Education* 9, 3.

Gere, C. (2001) 'The canon and the network society', *Engage 8*: 57–60, London: Engage.

Gill, A. (1998) 'Deconstruction fashion: the making of unfinished, decomposing and re-assembled clothes', *Fashion Theory*, 2, 1.

Gillborn, D. and Mirza, H. (2000) *Educational Inequality: Mapping Race, Class and Gender: A Synthesis of Research Evidence*, London: Office for Standards in Education.

Gipps, C. (1994) *Beyond Testing: Towards a Theory of Educational Assessment*, London: Falmer Press.

Gipps, C. and Stobart, G. (1993) *Assessment: A Teacher's Guide to the Issues*, London: Hodder and Stoughton.

Giroux, H. (1992) *Border Crossings*, New York; London: Routledge.

—— (1994) *Disturbing Pleasures*, New York: Routledge.

Giroux, H. and McLaren, P. (eds) (1994), *Between Borders*, London: Routledge.

Glenister, S. (1968) *The Technology of Craft Teaching*, London: Harrap.

Glimcher, A. and M. (eds) (1986) *Je Suis le Cahier: The Sketchbooks of Picasso*, New York: The Pace Gallery.

Global Leap website. Online. Available HTTP: http://www.global-leap.com (accessed 29 September 2006).

Golby, J. M. (1986) *Culture and Society in Britain 1850–1890*, Oxford: Oxford University Press in association with The Open University.

Goldwater, R. (1986) *Primitivism in Modern Art*, Cambridge, MA: Belknap Press.

Goleman, D. (1985) *Vital Lies, Simple Truths: The Psychology of Self-deception*, New York: Simon and Schuster.

Golomb, C. (1989) 'Sculpture: the development of representational concepts in three-dimensional medium', in D. J. Hargreaves (ed.) *Children and the Arts*, Milton Keynes: Open University Press.

Gombrich, E. (1992) 'Absolute standards', *Crafts*, April.

—— (1950) *The Story of Art*, London: Phaidon.

Goodman, N. (1976) *Languages of Art: An Approach to a Theory of Symbols*, Indianapolis: Hackett Publishing Company.

Greenberg, C. (1965) 'Modernist painting', in F. Frascina and C. Harrison (eds) (1982) *Modern Art and Modernism*, London: The Open University.

Greenblatt, S. (1991) 'Resonance and wonder', in S.D. Lavine and I. Karp (eds) *Exhibiting Cultures: The Poetics and Politics of Museum Display*, Washington and London: Smithsonian Institution Press.

Greenfield, S. (20 April 2006) '*Education: Science and Technology*', Online. Available HTTP: http://www.publications.parliament.uk/pa/ld199900/ldhansrd/pdvn/lds/text/60420-18.htm (accessed 29 September 2006).

Greenhalgh, M (1997) 'The history of craft', in P. Dormer (ed.) *The Culture of Craft*, Manchester: Manchester University Press.

Gregory, R. (1977) *Eye and Brain: The Psychology of Seeing*, London: Weidenfeld and Nicolson.

Gretton, T. (2003) 'Loaded canons', in N. Addison and L. Burgess (eds) *Issues in Art and Design Teaching*, London: RoutledgeFalmer.

Griffiths, M. and Tann, S. (1991) 'Ripples in the reflection', in P. Lomax (ed.) *BERA Dialogues, No 5*: 82–101.

Grigsby, J. (1977) *Art and Ethnics*, Dubuque, IA: Wm. C. Brown Publishers.

Grossman, P. L., Shulman, L., and Wilson, S. (1989) 'Teachers of substance: subject matter knowledge for teaching', in M. Reynolds *The Knowledge Base for the Beginning Teacher*, Oxford: Pergamon Press.

Grosz, E. (1995) *Space, Time and Perversion*, London: Routledge.

(GTC) General Teaching Council (2004) 'Internal and External Assessment: What is the Right Balance for the Future? Advice to the Secretary of State for Education and Others'. Online. Available HTTP: http://arg.educ.cam.ac.uk/CIE3.pdf (accessed 31 August 2006).

(The) Guardian Newspaper (1998) Schools Education Syllabuses, 1 September.

Guerilla Girls (1995) *Confessions of Guerilla Girls*, London: Pandora.

Hall, D. and Fifer, S.J. (1990) *Illumination Video*, New York: Aperture/BAVC.

Hall, J. (2004) 'Art education and spirituality', in R. Hickman (ed.) (2nd Edition) *Art Education 11–18: Meaning, Purpose and Direction*, London, New York: Continuum.

Hall, S. (1990) 'Cultural identity and diaspora', in J. Rutherford (ed.) *Identity, Community, Culture, Difference*, London: Lawrence and Wishart.

—— (1997) *Representation: Cultural Representations and Signifying Practices*, London: Sage in association with the Open University.

—— (2005) 'Assembling the 1980: The Deluge – And After', in D. A. Bailey, I. Baucom, and S. Boyce (eds) *Shades of Black*, Durham, NC: Duke University Press in association with inIVA and Aavaa.

Hargreaves, D. H. (1983) 'The teaching of art and the art of teaching: towards an alternative view of aesthetic learning', in M. Hammersly and A. Hargreaves (eds) *Curriculum Practice: Some Sociological Case Studies*, London: Falmer.

Hargreaves, D. J. (1989) *Children and the Arts*, Milton Keynes: Open University.

Harland, J., Haynes, J., Kinder, K., and Schagen, I. (1998) *The Effects and Effectiveness of Arts Education in Schools, interim report on research organised by the National Foundation for Educational Research (NFER) and the Royal Society for the encouragement of the Arts, Manufacturers and Commerce (RSA)*, Slough: NFER.

Harries, D. (ed.) (2002) *The New Media Book*, London: British Film Institute Publishing.

Harris, J. (1996) 'Grounding Postmodernism', in S. West (ed.) *Guide to Art*, London: Bloomsbury.

—— (2001) *The New Art History: A Critical Introduction*, London: Routledge.

Harrison, C. and Wood, P. (eds) (1992) *Art in Theory: 1900–1990*, Oxford: Blackwell.

Harrod, T. (ed.) (1997) *Obscure Objects of Desire*, London: Crafts Council.

Hatt, M. and Klonk, C. (2006) *Art History: A Critical Introduction to its Methods*, Manchester, Manchester University Press.

Hauser, A. (1951) *The Social History of Art*, J. Harris (ed.) (1999 edition), London: Routledge.

HC Education and Skills Committee (2006) *Special Educational Needs: Third Report of Session 2005–06*, London: The Stationery Office.

Hebdidge, D. (1988) *Hiding in the Light*, London: Routledge.

Heidegger, M. (1954) *What is Called Thinking*, trans. J. Gray (edition 1968), London, New York: Harper and Row.

Hermon, A. and Prentice, R. (2003) 'Positively different: art and design in Special Education', *International Journal of Art and Design Education*, 22, 3.

Hickman, R. (1996) *Finding Your First Job – A Guide for Prospective Art and Design Teachers*, Corsham: NSEAD.

—— (ed.) (2004) (2nd Edition) *Art Education 11–18: Meaning, Purpose and Direction*, London and New York: Continuum.

—— (ed.) (2005) *Critical Studies in Art & Design Education*, Bristol: Intellect Books.

Hiett, C. and Orbach, S. (1996) *Venus Re-Defined*, Resource Pack, Liverpool: Tate, Liverpool.

Higgins, C. (2006) *The Guardian*, 11 July 2006.

Hiller, S. (ed.) (1991) *The Myth of Primitivism*, London: Routledge.

—— (1996) *Thinking About Art: Conversations with Susan Hiller*, Manchester: Manchester University Press.

HMSO (2001) *The Special Needs and Disability Act (SENDA)*, London: HMSO.

—— (2003) *Every Child Matters*, London: HMSO.

Hodgson, A. and Spours, K. (eds) (1997) *Dearing and Beyond: 14–19 Qualifications, Frameworks and Systems*, London: Kogan Page.

Holt, J. (1996) 'Art for Earth's sake', in Dawtrey *et al.* (eds) *Critical Studies and Modern Art*, Milton Keynes: Open University.

Honour, H. and Fleming, J. (1982) *A World History of Art*, London: Macmillan.

hooks, b. (1992) 'Representing Whiteness on the Black imagination', in M. L. Grossberg *et al.* (eds) *Cultural Studies*, New York, London: Penguin.

—— (1994) *Teaching to Transgress*, New York; London: Routledge.

Hoopes, J. (ed.) (1991) *Peirce on Signs; Writings on Semiotics*, Chapel Hill and London: University of North Carolina Press.

Hughes, A. (1989) 'The copy, the parody and the pastiche: observations on practical approaches to critical studies', in D. Thistlewood (ed.) *Critical Studies in Art and Design Education*, Burnt Mill, Essex: Longman.

—— (1998a) 'A further reconceptualisation of the art curriculum', in *Broadside 1* UCE: ARTicle Press.

—— (1998b) 'Reconceptualising the art curriculum', *Journal of Art and Design Education* 17, 1.

—— (1999) 'Art and intention in schools: towards a new paradigm' *Journal of Art and Design Education* 18, 1.

Hughes, R. (1993) *Culture of Complaint*, New York and Oxford: New York Public Library and Oxford University Press.

Hulks, D. (2003) 'Measuring artistic performance: the assessment debate and art education', in N. Addison and L. Burgess (eds) *Issues in Art and Design Teaching*, London: RoutledgeFalmer.

Hyde, S. (1997) *Exhibiting Gender*, Manchester: Manchester University Press.

inIVA (Institute of International Visual Arts). Online. Available HTTP: http://www.iniva.org/ (accessed 30 October 2006).

International Journal of Education Through Art, Bristol: Intellect Publishers.

Isherwood, S. and Stanley, N. (eds) (1994) *Creating Vision*, London: ACE.

Jefferies, J. (1995) 'Weaving across borderlines', in K. Deepwell (ed.) *New Feminist Art Criticism*, Manchester: Manchester University Press.

Jefferies, J. (ed.) (2001) *Reinventing Textiles: Gender and Identity*, Winchester: Telos.

International Journal of Art and Design Education (iJADE) three editions yearly, is the journal of the NSEAD (National Society for Education in Art and Design)

Journal of Art and Design Education (1983) 1.

Johnson, P. (1995) 'Naming of parts', *Crafts* 34: 42–45.

—— (1997a) *Crafts Council Conference*, London: Crafts Council.

—— (1997b) 'Crafts', *Crafts Magazine* 146, June.

—— (1998a) 'Material culture', *Blueprint* 147, February: 24–26.

—— (ed.) (1998b) *Ideas in the Making*, London: Crafts Council.

Jones, A. (1996) *Sexual Politics*, Los Angeles: University of California Press.

Jones, P. (1998) 'Art', in OFSTED, *The Arts Inspected*, Oxford: OFSTED and Heinemann.

Jung, C. (1964) *Man and His Symbols*, London: Aldus Books Ltd.

Kaelin, E. F. (1989) *An Aesthetics for Art Educators*, New York: Teachers College Press.

Kelly, G. (1955) *The Psychology of Personal Constructs*, New York: Norton.

Kemp, M. (ed.) (2000) *The Oxford History of Western Art*, Oxford: Oxford University Press.

Kennedy, H. (1997) *Learning Works: Widening Participation in FE Education*, FEFC.

Kennedy, M. (1995a) 'Issue-Based Work at KS4: Crofton School – A Case Study', *Journal of Art and Design Education* 14, 1.

—— (1995b) 'Approaching assessment', in R. Prentice (ed.) *Teaching Art and Design: Addressing Issues and Identifying Directions*, London: Cassell.

Ker, M. (1997) 'Grounds for design: improvement of school and college grounds', *Journal of Art and Design Education* 16, 1.

Kester, G. H (2004) *Conversation Pieces: Community + Communication in Modern Art*, Berkeley and LA: University of California Press.

Kimbell, R., Stable, K., and Green, R. (1996) *Understanding Practice in Design and Technology*, Milton Keynes: Open University Press.

Klages, M. (2004) Online. Available HTTP: http://www.colorado.edu/English/courses/ENGL2012Klages/1derrida.html (accessed 29 October 2006).

Klee, P. (1968) *The Pedagogic Sketchbook*, London: Faber and Faber.

Kovats, T. (ed.) (2005) *The Drawing Book. A Survey of Drawing: the Primary Means of Expression*, London: Black Dog.

Krauss, R. (1978) 'Sculpture in the expanded field', in R. Krauss (1986) *The Originality of the Avant-Garde and Other Modernist Myths*, Cambridge, MA.: MIT Press.

—— (1986) *The Originality of the Avant-Garde and Other Modernist Myths*, Cambridge, MA.: MIT Press.

Kress, G., Jewitt, C., Ogborn, J. and Tsatsarelis, C. (2001) *Multimodal Teaching and Learning: The Rhetorics of the Science Classroom*, London: Continuum.

Kress, G. and Leeuwen, T. (2006) 2nd edition *Reading Images: The Grammar of Visual Design*, London: Routledge.

Kristeva, J. (1982) *Powers of Horror: An Essay on Abjection*, New York: Columbia University Press.

Lacy, S. (1995) *Mapping the Terrain: New Genre Public Art*, Seattle: Bay Press.

Ladson-Billings, G. and Gillborn, D. (eds) (2004) *The RoutledgeFalmer Reader in Multicultural Education*, London: RoutledgeFalmer.

La Trobe Bateman, R. (1997) Frontispiece to Introduction in P. Dormer (ed.) *The Culture of Craft*, Manchester: Manchester University Press.

Lave, J. and Wenger, E. (1991) *Situated Learning: Legitimate Peripheral Participation*, Cambridge: Cambridge University Press.

Levi-Strauss, C. (1963) 'Do dual organisations exist?', in *Structural Anthropology*, Middlesex: Penguin.

Lippard, L. (1995) 'Looking around: where we are, where we could be', in S. Lacy (ed.) *Mapping the Terrain*, Seattle: Bay Press.

Lloyd, F. (1999) *Contemporary Arab Women's Art: Dialogues of the Present*, London: WAL.

Loeb, H. (1984) *Changing Traditions*, Birmingham: UCE.

Loeb, H., Slight, P. and Stanley, N. (1993) *Designs We Live By*, Corsham: NSEAD.

Lovink, G. (2002) *Dark Fiber: Tracking Critical Internet Culture*, Cambridge, MA: The MIT Press.

Lowenfeld, V. and Brittain, W. (1966) *Creative and Mental Growth*, London: Macmillan.

—— (1987) *Creative and Mental Growth*, London: Macmillan.

Lyons, J. (ed.) (1985) *Artists' Books: A Critical Anthology and Sourcebook*, New York: Visual Studies Workshop Press.

Mac an Ghaill, M. (1995) *The Making of Men: Masculinities, Sexualities and Schooling*, Buckingham: Open University Press.

—— (1996) 'The making of men', in P. Woods, *Contemporary Issues in Teaching and Learning*, London: Routledge.

MacDonald, S. (1970) *The History and Philosophy of Art Education*, London: London University Press.

MacPherson, W. (1999) *The Stephen Lawrence Inquiry: Report of an Inquiry*, London: HMSO.

McFee, J. K. (1998) *Cultural Diversity and the Structures and Practice of Art Education*, National Art Education Association, Oregon.

McRobbie, A, (1994) *Postmodernism and Popular Culture*, London: Routledge.

—— (1998) 'But is it Art?', *Marxism Today*, November/December: 55.

Maharaj, S. (2000) 'Interculturalism' (unpublished lecture), *Art Histories in Action*, Conference, London: Tate Britain: in association with the AAH.

Make is the quarterly journal of the Women's Art Library London.

Malik, R. (1996) *Portraiture: Education Pack*, London: InIVA.

Manser, S. and Wilmot, H. (1995) *Artists in Residence: A Teachers' Handbook*, London: London Arts Board.

Mason, R. (1995) (2nd Edition) *Art Education and Multiculturalism*, Corsham: NSEAD.

—— (1996) *Beyond Tokenism: Towards Criteria for Evaluating the Quality of Multicultural Curricula*, in Dawtrey *et al.* (eds) *Critical Studies and Modern Art*, New Haven, CT: Yale University Press in Association with The Open University, the Arts Council of England and the Tate Gallery London.

Mason, R. and Steers, J. (2006) 'The impact of formal assessment procedures on teaching and learning in art & design in secondary schools', *International Journal of Art and Design Education*, 25, 2: 119–133.

Matthews, J. (1999) *The Art of Childhood and Adolescence: The Construction of Meaning*, London: Falmer Press.

—— (2003) (2nd Edition) *Drawing and Painting: Children and Visual Representation*, London: Paul Chapman Publishing.

Mayer, M. (1998) 'Can philosophical change take hold in the American art museum?', *Art Education*, Virginia, USA: The Journal of the National Art Education Association.

Mealing, S. (2002) (2nd edition) *Computers and Art*, Bristol: Intellect.

Meecham, P. (1996) 'What's in a National Curriculum?', in L. Dawtrey *et al.* (eds) *Critical Studies in Modern Art*, Milton Keynes: Open University Press.

—— (1999) 'Of Webs and Nets and Lily Ponds', *Journal of Art and Design Education* 18, 1.

Meecham, P. and Sheldon, J. (2004) (2nd Edition) *Modern Art: A Critical Introduction*, London: Routledge.

Meskimmon, M. (1996) *The Art of Reflection: Women Artists' Self-portraiture in the 20th Century*, London: Scarlet Press.

—— (2003) *Women Making Art*, London: Routledge.

Miliband, D. (2004) 'Delivering for All Students', Speech given at the *Diversity and Leadership in the 21st Century, Next Steps Conference*, London: 22 June 2004. Online. Available HTTP: http://www.davidmiliband.info/sarchive/speech04_02.htm (accessed 8 October 2006).

Millar, S. (1998) 'Artists' holes start mother of all rattles in krazy Golf War', *The Guardian*, 1 August, Home News.

Mirza, H. (1992) *Young, Female and Black*, London: Routledge.

Mirzoeff, N. (2002) (2nd edition) *An Introduction to Visual Culture*, London: Routledge.

Mitchell, S. (2005) (Goldsmiths PGCE student) 'Assignment 2: Carnival Banner Project', unpublished course paper.

de Monchaux, C. (1997) interviewed by L. Buck in *Tate: The Art Magazine*, Summer, in M. Parkin (1998) *The Turner Prize 1998*, London: Tate Gallery.

Moon, B. and Mayes, A. S. (eds) (1994) *Teaching and Learning in the Secondary School*, London: Open University and Routledge.

Moore, H. (1952) Notes on sculpture from M. Evans, 'The Painters Object', in *The Creative Process: A Symposium*, London: Mentor Books, The New English Library Ltd.

—— (1972) *Henry Moore's Sheep Sketchbook*, London: Thames and Hudson.

Morgan, S. and Morris, F. (eds) (1995) *Rites of Passage: Art for the End of the Century*, London: Tate Gallery.

Morley, D. (1992) *Television Audiences and Cultural Studies*, London: Routledge.

Muf (2001) *This is What We Do: A Muf Manual*, London: Ellipsis Press.

Mugerauer, R. (2001) 'Openings to each other in the technological age' in N. Alsayyad (ed.) *Consuming Tradition, Manufacturing Heritage: Global Norms and Urban Forms in the Age of Tourism*, London: Routledge .

Myerson, J. (1993) *Design Renaissance*, London: Open Eye.

National Union of Teachers (annually) *Your First Teaching Post*, London, NUT.

NC (1992) *Art Non-Statutory Guidance Booklet*, York: National Curriculum Council.

NCET (1998) *Fusion: Art and IT in Practice Supporting National Curriculum Art at Key Stage 3 and Beyond*, Coventry: NCET (now BECTA).

Neperud, R. (ed.) (1995) *Context, Content and Community in Art Education: Beyond Postmodernism*, New York: Teachers' College Press.

(*The*) *New Shorter Oxford Dictionary* (1993) Oxford: Clarendon Press.

Nochlin, L. (ed.) (1991) *Women, Art and Power and other Essays*, London: Thames and Hudson.

—— (1999) *Representing Women*, London: Thames and Hudson.

Norwich, B. (1996) *Special Needs Education, Inclusive Education or Just Education for All?* London: Institute of Education.

NSEAD (2001) *A'N'D The Newsletter of the National Society for Education in Art & Design*, 2, Autumn Issue.

NSEAD and DATA (2006) *Design Education: Now You See it Now You Don't*, Seminar held at the Royal College of Art, London.

Objects of Desire (Exhibition Catalogue) (1997) London: Hayward Gallery.

Oddie, D. and Allen, G. (1998) *Artists in Schools: A Review*, London: OFSTED.

OFSTED (1995a) *The Framework for the Inspection of Schools*, London: HMSO.

—— (1995b) *Art: A Review of Inspection Findings 1993–94*, London: HMSO.

—— (1997) *Secondary Education 1993–1997: A Review of Secondary Education Schools in England* online. Available HTTP: http://www.archive.official-documents.co.uk/document/ofsted/seced/chap-7d.htm (accessed 11 December 2006).

—— (1998) *The Arts Inspected*, Oxford: Heinemann.

—— (2000) *OFSTED Subject Reports, 1999–2000: Secondary Art and Design*, London: HMSO.

—— (2001) *Secondary Subjects Reports 2000/01: Art and Design*, London: HMSO.

—— (2004a) *Special Educational Needs and disability: Towards Inclusive Schools*, London: OFSTED.

—— (2004b) *OFSTED Subject Reports 2002/03: Art and Design in Secondary Schools*, London: OFSTED.

—— (2005) *The Annual Report of her Majesty's Inspector of Schools 2004/5: Art and Design in Secondary Schools*, online. Available HTTP: http://www.ofsted.gov.uk/publications/annualreport0405/4.2.1.html (accessed 8 October 2006).

Oguibe, O. (1993) 'In the "Heart of Darkness" ', in E. Fernie (ed.) (1995) *Art History and its Methods*, London: Phaidon.

Oguibe, O. and Enwezor, O. (1999) *Reading the Contemporary: African Art from Theory to the Marketplace*, London: inIVA.

Osbourn, R. (1991) 'Speaking, listening and critical studies in art', *Journal of Art and Design Education* 10, 1.

Overy, P. (1991) *De Stijl: Art, Architecture, Design*, London: Thames and Hudson.

Paley, N. (1995) *Finding Art's Place: Experiments between Contemporary Art and Education*, New York; London: Routledge.

Palmer, F. (1989) *Visual Elements of Art and Design*, Burnt Mill, Harlow: Longman.

Panting, L. (1999) 'Debating Gender', *MAKE*, 33.

Papanek, V. (1971) *Design for the Real World*, London: Thames and Hudson.

—— (1995) *The Green Imperative*, London: Thames and Hudson.

Parker, R. (1984) *The Subversive Stitch: Embroidery and the Making of the Feminine*, London: Women's Press.

Parker, R. and Pollock, G. (1981) *Old Mistresses: Women, Art and Ideology*, London: Pandora.

Parsons, M. (1987) *How We Understand Art: A Cognitive Developmental Account of Aesthetic Experience*, Cambridge: Cambridge University Press.

Parsons, M. J. and Blocker, G. (1993) *Aesthetics and Education*, Champaign, IL: University of Illinois Press.

Pateman, T. (1991) *Key Concepts: A Guide to Aesthetics, Criticism and the Arts in Education*, London: Falmer Press.

Perry, L. (1998) In conversation with the authors at the Institute of Education.

Petersen, K. and Wilson, J. (1976) *Women Artists: recognition and reappraisal from the early middle ages to the 20th century*, London: The Women's Press.

Phillips, T. (1980) *A Humument*, London: Thames and Hudson.

Piaget, J. (1962) *Judgement and Reasoning in the Child*, London: Routledge and Kegan Paul.

Picasso, P. (1974) 'La Tete d'obsidienne', A. Malraux, in E. Cowling and J. Golding (eds) (1994) *Picasso Sculpture/Painter*, London: Tate Gallery.

Pietersie, J. (1992) *White on Black: Images of Africa and Blacks in Western Popular Culture*, London: Yale University Press.

Pinder, K.N. (ed.) (2002) *Race-ing Art History: Critical Readings in Race and Art History*, New York; London: Routledge.

Piper, K. (1997) *Relocating the Remains*, a monograph and CD-ROM, London: InIVA.

Placecheck. Online. Available HTTP: http://www.placecheck.info/ (accessed 30 October 2006).

Plato (c. 380 BC) 'The Republic', in S. Buchanan (ed.) *The Portable Plato*, Harmondsworth: Penguin.

Pliny the Elder (1st century AD) *Natural History: A Selection*, trans. J. F. Healy (edition 1991), Harmondsworth: Penguin.

(*The*) *Pocket Oxford Dictionary of Current English*, (1992) Oxford: Clarendon Press.

Pointon, M. (1990) *Naked Authority*, Cambridge: Cambridge University Press.

Polanyi, M. (1964) *Personal Knowledge*, New York: Harper and Row.

Pollock, G. (1982) 'Vision, voice and power', in *BLOCK* 6.

—— (1988) *Vision and Difference: Femininity, Feminism and the Histories of Art*, London and New York: Routledge.

—— (1996a) 'Art, art school, culture', in J. Bird *et al. The BLOCK Reader in Visual Culture*, London: Routledge.

—— (1996b) (ed.) *Generations and Geographies in the Visual Arts: Feminist Readings*, London: Routledge.

—— (1996c) 'Art, artists and women', in Dawtrey *et al.* (eds) *Critical Studies and Modern Art*, Milton Keynes: Open University.

Poupeye, V. (1998) *Caribbean Art*, London: Thames and Hudson.

Powell, R. (1997) *Black Art and Culture in the 20th Century*, London: Thames and Hudson.

Prentice, R. (1995) 'Learning to teach: a conversational exchange', in R. Prentice (ed.) *Teaching Art and Design: Addressing Issues and Identifying Directions*, London: Cassell.

—— (2003) 'Changing places?', in N. Addison and L. Burgess (eds) *Issues in Art and Design Teaching*, London: RoutledgeFalmer.

Press, M. (1996) 'Recycling', in *Crafts Council Catalogue*, Birmingham: Craftspace Touring.

—— (1997) 'A new vision in the making', *Crafts* 147, August: 42–45.

Preziosi, D. (1998) *The Art of Art History: A Critical Anthology*, Oxford: Oxford University Press.

—— (2006) *In the Aftermath of Art: Ethics, Aesthetics, Politics*, London and New York: Routledge.

Price, D. (1998) *The New Neurotic Realists*, London: The Saatchi Gallery.

Price, E. (1982) 'National criteria for a single system of examining at 16+ with reference to Art and Design', *Journal of Art & Design Education*, 1, 3: 397–407, Abingdon: Carfax Publishing.

Price, M. (1989) 'Art history and critical studies in schools: an inclusive approach', in D. Thistlewood (ed.) *Critical Studies in Art and Design Education*, Burnt Mill, Essex: Longman.

Pringle, E. (2002) 'We did stir things up: the role of artists in sites for learning'. Online. Available HTTP: http://www.artscouncil.org.uk/documents/publications/phpShHUNy.pdf (accessed 23 October 2006).

—— (2006) *Learning in the Gallery: Context, Process, Outcomes*, London: Engage.

Proctor, N. (1996) 'Is Women's Art Homeless?', *Make* September, 71.

Pye, J. (1988) *Invisible Children: Who are the Real Losers at School?*, Oxford: Oxford University Press.

QCA (1997) *A Survey of Teachers' Reference to the Work of Artists, Craftspeople and Designers in Teaching Art*, London: QCA.

—— (1998a) 'A Survey of artists, craftspeople and designers used in teaching art research' in *Analysis of Educational Resources in 1997/98*, research coordinated by R. Clements (1996–97) CENSAPE: University of Plymouth.

—— (1998b) *Education for Citizenship and the Teaching of Democracy in Schools*, London: QCA on behalf of the DfEE.

—— (1999) *QCA's Work in Progress to Develop the School Curriculum: Pack B*, London: QCA.

—— (2001) *Planning, Teaching and Assessing the Curriculum for Pupils with Learning Difficulties: Art and Design*, London: QCA.

—— (2002) *National Curriculum Inclusion Statement*. Online. Available HTTP: http://www.nc.uk.net/inclusion.html (accessed 8 December 2006).

—— (2004) *Key Skills*, Online. Available HTTP: http://www.qca.org.uk/603.html (accessed 30 October 2006).

—— (2006) *The Annual Report of Her Majesty's Chief Inspector of Schools 2004/5*. Online. Available HTTP: http://www.qca.org.uk/downloads/qca-05-2163-and-report.pdf (accessed 23 October 2006).

—— *Gifted and Talented Guidance*. Online. Available HTTP: http:// www.qca.org.uk/8774.html (accessed 8 December 2006).

(The) Race Relations Act 1976, as amended 2000/2003, London: HMSO.

Rampton Report (1981) 'West Indian children in our schools: Interim Report of the Committee of Inquiry into the Education of Children of Ethnic Minority Groups', London: DES, HMSO.

Read, H. (1943, 1950 and 1958) *Education through Art*, London: Faber and Faber.

Recycling: Focus for the Next Century – Austerity for Posterity (1996) Crafts Council Exhibition Catalogue.

(The) Rehabilitation of Offenders Act 1974 (Exceptions) (Amendment) Order 1986, London: HMSO.

Reid, L. A. (1969) *Meaning in the Arts*, London: Allen and Unwin.

—— (1986) *Ways of Understanding and Education*, London: Heinemann.

Renaissance in the Regions (2006). Online. Available HTTP: http://www.mla.gov.uk (accessed 1 November 2006).

Rhodes, C. (1994) *Primitivism and Modern Art*, London: Thames and Hudson.

RIBA (Royal Institute of British Architects) Online. Available HTTP: http:// www.architecture.com (accessed 30 October 2006).

Richardson, M., Lane, K. and Flanagan, J. (eds) (1995) *School Empowerment*, Switzerland: Technomic Publications.

Riding, R.J. and Rayner, S. (1998) *Learning Styles and Strategies*, London: David Fulton.

Robins, C. (2003) 'In and out of place: cleansing rites in art education', in N. Addison and L. Burgess (eds) *Issues in Art and Design Teaching*, London: RoutledgeFalmer.

Robinson, C. (1992) Unpublished paper given at the NSEAD Conference in Brighton.

—— (1995) 'National Curriculum for Art: translating it into practice', in R. Prentice (ed.) *Teaching Art and Design: Addressing Issues and Identifying Directions*, London: Cassell.

Robinson, G. (1993) 'Tuition or intuition? making and using sketchbooks with a group of ten-year old children', *Journal of Art and Design Education* 12, 1: 73–84.

—— (1995) *Sketchbooks: Explore and Store*, London: Hodder and Stoughton.

Robinson, K. (ed.) (1982) *The Arts in Schools: Principles, Practices and Provision*, London: Calouste Gulbenkian Foundation.

—— (1998) *All Our Futures: Creativity, Culture & Education*, London: National Advisory Committee on Creative and Cultural Education.

Rogers, C. (1969) (1983 edition) *Freedom to Learn*, New York: Macmillan.

Rogers, R. and Bacon, S. (eds) (2002) *Space for Art: A Handbook for Creative Learning Environments*, London: Clore Duffield Foundation.

Rogers, R., Edwards, S. and Steers, J. (2001) *Survey of Art and Design Resources in Schools*, Artworks/NSEAD. Available online HTTP: http://www.artworks.org.uk (accessed 28 October 2006).

Rogoff, (1998) 'Studying visual culture', in N. Mirzoeff (ed.) *Visual Culture Reader*, London: Routledge.

Rose, C. (1991) *Design After Dark the Story of Dancefloor Style*, London: Thames and Hudson.

Rose, G. (2001) *Visual Methodologies: An Introduction to the Interpretation of Visual Materials*, London: Sage.

Rosenberg, H. (1952) 'The American Action Painters', in D. Shapiro (ed.) (1990) *Abstract Expressionism: A Critical Record*, Cambridge: Cambridge University Press.

—— (1967) 'Art – where to begin', *Encounter*, March.

Ross, M. (1993) *Assessing Achievement in the Arts*, Milton Keynes: Open University Press.

Rothenstein, J. (ed.) (1986) *Michael Rothenstein: Drawings and Paintings Aged 4–9, 1912–1917*, London: Redstone Press.

Rowley, A. (1996) 'On viewing three paintings by Jenny Saville: rethinking the feminist practice of painting', in G. Pollock (ed.) *Generations and Geographies in the Visual Arts: Feminist Readings*, London: Routledge.

Rowley, S. (ed.) (1999) *Reinventing Textiles: Traditions and Innovation*, Winchester: Telos.

RSA (1998) *The Effects and Effectiveness of Arts Education in Schools*, Slough: NFER.

Rubin, W. (1969) 'Some reflections prompted by the recent work of Louise Bourgeois', *Art International*, 13, 4: April 20.

—— (ed.) (1984) *Primitivism in Twentieth Century Art*, New York, MOMA.

Rycroft, C. (1968) *A Critical Dictionary of Psychoanalysis*, Harmondsworth: Penguin.

Said, E. (1980) *Orientalism*, Harmondsworth: Penguin.

—— (1993) *Culture and Imperialism*, London: Chatto and Windus.

—— in R. Hughes (1993) *The Culture of Complaint*, New York and Oxford: Oxford University Press.

Salmon, P. (1988) *Psychology for Teachers: An Alternative View*, London: Hutchinson.

—— (1995) 'Experiential learning', in R. Prentice (ed.) *Teaching Art and Design: Addressing Issues and Identifying Directions*, London: Cassell.

Saussure, F. (1974) *Course in General Linguistics*, trans. W. Baskin, New York: Fontana, Collins.

SCAA (1996) *Consistency in Teachers' Assessment – Exemplification of Standards KS3*, London: SCAA.

—— (1997) *Survey and Analysis of Published Resources for Art (5–19)*, London: SCAA.

Schilling, C. (2005) *The Body in Culture, Technology and Society*, London: Sage Publications.

Schneider, N. (1990) *The Art of Still Life*, Koln: Taschen.

Schodt, F. (1993) *GManga, Manga, the World of Japanese Comics*, New York: Kodansha International.

Schofield, K. (1995) 'Objects of desire', in R. Prentice (ed.) *Teaching Art and Design: Addressing Issues and Identifying Directions*, London: Cassell.

Schon, D. (1987) *Educating the Reflective Practitioner*, San Francisco: Jossey Bass.

Searle, A. (ed.) (1993) *Talking Art 1*, London: ICA.

Secondary Examinations Council (SEC) (1986a) *CDT GCSE Guide for Teachers*, London: HMSO.

—— (1986b) *Policy and Practice in School-based Assessment (Working Paper 3)*, London: Secondary Examinations Council.

—— (1986c) *Report of the Working Party: Art and Design Draft Grade Criteria*, London: Secondary Examinations Council.

Sefton-Green, J. (ed.) (1998) *Digital Diversions: Youth Culture in the Age of Multimedia*, London: University College Press.

Selwood, S., Clive, S. and Irving, D. (1994) *Cabinets of Curiosity*, London: Art & Society for ACE.

Sensations Exhibition Catalogue (1997), London: Royal Academy of Arts.

(*The*) *Sex Discrimination Act 1975*, updated in April 2007 to include the Gender Equality Duty (2007). Online. Available HTTP: http://www.eoc.org.uk (accessed 30 October 2006).

(*The*) *Sexual Offences Act 2003*, London: HMSO.

Shahn, B. (1967) *The Shape of Content*, Cambridge, MA: Harvard University Press.

Sharp, C. and Dust, K. (1998) *Artists in Schools*, Slough: NFER.

Sheets (2002) *Kara Walker's Cutout Silhouettes of Antebellum Racial Stereotypes are Lewd, Provocative-and Beautiful*. Online. Available HTTP: http://artnews.com/issues/article.asp?art_id=1097 (accessed 11 December 2006).

Shohat, E. and Stam, R. (1994) *Unthinking Eurocentrism: Multiculturalism and the Media*, London and New York: Routledge.

—— (1995) 'The politics of multiculturalism in the postmodern age', in *Art & Cultural Difference: Hybrids & Clusters*, Art & Design Magazine, 43.

—— (1998) 'Narrativizing visual culture: towards a polycentric aesthetics', in N. Mirzoeff, *The Visual Culture Reader*, London: Routledge.

Shreeve A. (1998) 'Tacit knowledge in textile crafts', in P. Johnson (ed.) *Ideas in the Making*, London: Crafts Council.

Shulman, L. (1986) 'Those who understand: knowledge growth in teaching', *Educational Researcher*, February, 4–14.

Simpson, A. (1987) 'What is art education doing?', *Journal of Art and Design Education* 6, 3.

Skeggs, B. (2004) *Class, Self, Culture*, London: Routledge.

Smith, A. (2001) (ed.) *Exposed: The Victorian Nude*, London: Tate Publishing.

Smith, M. (1995) 'Joseph Beuys: life as drawing', in D. Thistlewood (ed.) *Joseph Beuys: Diverging Critiques*, Liverpool: Liverpool University Press and Tate Gallery Liverpool.

Smolik, N. (2006) 'Resurrection and cultural renewal', in H. Fischer and S. Rainbird (eds) *Kandinsky: The Path to Abstraction*. London: Tate.

Soganci, I. (2005) 'Mom, why isn't there a picture of our prophet?', *International Journal of Education Through Art*, 1, 1: 21–27.

Sontag, S. (1977) *On Photography*, London: Penguin.

Spare, C. (2006) *Rationale for BEd Year 3 Art and Design Assignment*, Goldsmiths, University of London.

Sparke, P. (1995) *As Long as its Pink: The Sexual Politics of Taste*, London: Pandora.

Sparrow, F. (ed.) (1996) *Bill Viola: The Messenger*, Durham: The Chaplaincy to the Arts and Recreation in North East England.

Spencer, R. (1989) *Whistler: A Retrospective*, New York: Hugh Lanter Levin Associates.

Spender, D. (1989) *Invisible Women: The Schooling Scandal*, London: Women's Press.

—— (1995) *Nattering on the Net: Women, Power and Cyberspace*, Melbourne: Spinifox.

Stallabrass, J. (1999) *High Art Lite*, London: Verso.

Stanley, N. (1986) 'Anthropology and the visual arts', *Journal of Art and Design Education* 5, 1 and 2.

—— (1996) 'Photography and the politics of engagement', *Journal of Art and Design Education*, 15, 1.

—— (1998) *Being Ourselves for You: The Global Display of Cultures*, London: Middlesex University Press.

—— (2007) ' "Anything you can do": a proposal for lesbian and gay art education', Special Edition: Lesbian and Gay Issues in Art & Design Education, *International Journal of Art and Design Education*, 26, 1.

Stanworth, M. (1987) 'Girls on the margins: a study of gender divisions in the classroom', in M. Arnot and G. Weiner (eds) *Gender & the Politics of Schooling*, London: Hutchinson.

Steers, J. (1994) 'Technology in the National Curriculum: the 1994 Draft Order', paper given at *NSEAD Conference*, Westminster, 2 July.

—— (1998) Lecture given to PGCE Students at the Institute of Education, University of London.

—— (2003) 'Art and design in the UK: the theory gap', in N. Addison and L. Burgess (eds) *Issues in Art and Design Teaching*, London: RoutledgeFalmer.

Stibbs, A. (1998) 'Language in art and art in language', *Journal of Art and Design Education* 17, 1.

Stungo, N. (1998) 'Smart dressing', in *Blueprint* February.

Sturrock, J. (1993) *Structuralism*, London: Fontana Press.

Swann, *The Swann Report* (1985) *Education for All: The Report of the Committee of Inquiry into the Education of Children from Ethnic Minority Groups*, London: HMSO.

Swift, J. (1992) 'Marion Richardson's contribution to art teaching', in D. Thistlewood (ed.) *Histories of Art and Design Education*, Burnt Mill, Essex: Longman.

—— (1995) 'Controlling the masses: the reinvention of a "national" curriculum', *Journal of Art and Design Education* 14, 2.

Swift, J. and Steers, J. (1999) 'A manifesto for art in schools', *Directions: Journal of Art and Design Education* 18, 1: 7–14.

Tate, N. (1997) 'Keynote address', in *The Arts in the Curriculum*, London: SCAA.

Taylor, B. (1989) 'Art history in the classroom: a plea for caution', in D. Thistlewood (ed.) *Critical Studies in Art and Design Education*, Burnt Mill, Essex: Longman.

—— (1995) *The Art of Today*, London: Everyman Art Library, Weidenfeld and Nicolson.

Taylor, R. (1986) *Educating for Art*, London: Longman.

—— (1989) 'Critical studies in art and design education', in D. Thistlewood (ed.) *Critical Studies in Art and Design Education*, Burnt Mill, Essex: Longman.

—— (1991) *Artists in Wigan Schools*, London: Calouste Gulbenkian Foundation.

—— (1992) *Visual Arts in Education*, London: Falmer Press.

—— (1999) *Understanding and Investigating Art*, London: National Gallery.

Taylor, R. and Taylor, D. (1990) *Approaching Art and Design: A Guide for Students*, London: Longman.

TDA (2006a) *The Revised Standards for the Recommendation of Qualified Teacher Status (QTS)*, London: TDA. Online. Available HTTP: http://www.tda.gov.uk/upload/resources/doc/draft_qts_standards_17nov2006.doc (accessed 11 December 2006).

TDA (2006b) *Career Entry and Development Profile*. Online. Available HTTP: http://www.tda.gov.uk/upload/resources/pdf/c/cedpdividers.pdf (accessed 11 December 2006).

Teachernet website (2006) Online. Available HTTP: http://www.teachernet.gov.uk (accessed 11 April 2006).

Teaching as a Career Unit (TASC) (annually), *Getting a Teaching Job*, University of Greenwich.

(The) Third Text Reader on Art, Culture and Theory (2002) edited by R. Araeen, S. Cubitt and Z. Sardar, London: Continuum.

Third Text is a quarterly journal presenting perspectives on contemporary art from the developing world.

Thistlewood, D. (ed.) (1989) *Critical and Contextual Studies in Art and Design Education*, Burnt Mill, Essex: Longman.

Thistlewood, D. (ed.) (1990) 'Editorial: the essential disciplines of design education', *Journal of Art and Design Education* 9, 1: 6.

—— (ed.) (1992) *Histories of Art and Design Education*, Burnt Mill, Essex: Longman.

—— (1996) 'Critical development in critical studies', in Dawtrey *et al.* (eds) *Critical Studies and Modern Art*, Milton Keynes: Open University.

Thomas, G., Walker, D. and Webb, J. (1998) *The Making of the Inclusive School*, London: Routledge.

Thurber, F. (1997) 'A site to behold: creating curricula about local urban environmental art', *Art Education* November.

Times Educational Supplement (annually, around the middle of January) First Appointments Supplement, TES.

Todd, S. (1997) 'Psychoanalytical questions', in H. Giroux (ed.) with P. Shannon, *Education and Cultural Studies*, New York: Routledge.

(The) Tomlinson Report (1996) *Inclusive Learning, Report of the Learning Difficulties and/or Disabilities Committee*, London: HMSO.

—— (2004) *14–19 Curriculum and Qualifications Reform: Final Report of the Working Group on 14–19 Reform*. Online. Available HTTP: http://www.dfes.gov.uk/14–19/documents/Final%20Report.pdf (accessed 11 December 2006).

Torres Pereira de Eça, M. (2005) 'Using portfolios for external assessment: an experiment in Portugal', *International Journal of Art & Design Education*, 24, 2: 209–218.

Troyna, B. (1992) 'Can you see the join? An historical analysis of multicultural and antiracist education policies', in D. Gill *et al.* (eds) *Racism and Education Structures and Strategies*, London, Newbury Park and Delhi: Sage.

Troyna, B. and Hatcher, R. (1993) *Racism in Children's Lives*, London: Routledge.

Tucker, W. (1974) 'What Sculpture is?' *Studio International*, in A. Gouk, 'Carbon, Diamond', in *Have you seen Sculpture from the Body?*, London: Tate Gallery.

Turner, J. M. W. (1987) *The 'Ideas of Folkestone' Sketchbook, 1845*, London: The Tate Gallery.

Turner, S., Tyson, I., and Courtney, C. (eds.) (1993) *Facing the Page: British Artists' Books, A Survey, 1983–1993*, London: Estamp Editions.

Urry, J. (2002) 2nd edition, *The Tourist Gaze*, London: Sage.

Usher, R. and Edwards, R. (1994) *Postmodernism and Education*, London: Routledge.

Usherwood, P. (1998) 'Park Life', *Art Monthly* 219, September.

Van Der Volk, J. (1987) *The Seven Sketchbooks of Vincent Van-Gogh*, London: Thames and Hudson.

van Manen, M. (1991) *The Tact of Teaching: The Meaning of Pedagogical Thoughtfulness*, Albany, NY: SUNY.

Victoria and Albert Museum, London (1985) *John Constable: Sketchbooks 1813–14*, London: Victoria and Albert Museum.

Victoria and Albert Museum's *Architecture for All*. Online. Available HTTP: http://www.vam.ac.uk/collections/architecture/index.html (accessed 30 October 2006).

Vygotsky, L. (1986) *Thought and Language*, trans A. Kozulin (revised edition), Cambridge, MA: The MIT Press.

Walker, J. (1983) *Art in the Age of Mass Media*, London: Pluto Press.

—— (1992) *A History of Design and Design History*, London: Pluto Press.

—— (1998) 'Marcus Harvey's sick disgusting painting of Myra Hindley: a semiotic analysis', *The Art Magazine* 14: Spring, London: Tate Gallery.

Warnock, M. (1978) *Special Educational Needs: The Report of the Committee of Enquiry into the Education of Handicapped Children and Young People*, London: DES, HMSO.

—— (2004) cited in R. Garner, 'Mainstreaming fails "fragile" disabled pupils', *The Independent*, 12 October, p.5.

—— (2005) *SEN: A New Look*, London: Philosophy of Education Society of Great Britain.

Watkins, C. (2005) *Classrooms as Learning Communities: What's in it for Schools?* London: Routledge.

Watkins, C., Carnell, E., Wagner, P. and Whalley, C. (2003) *Effective Learning in NSIN: Research Matters, No. 17, Summer 2002*, London: Institute of Education.

Wells, L. (ed.) (1997) *Photography: A Critical Introduction*, London: Routledge.

Wertsch, J. V. (1998) *Mind as Action*, Oxford: Oxford University Press.

White, J. (1998) *Do Howard Gardner's Multiple Intelligences Add Up?* London: Institute of education, University of London.

White, M. (1998) 'The pink's run out', in N. Yelland (ed.) *Gender and Early Childhood*, London: Routledge.

Willats, J. (1997) *Art and Representation*, Princeton, NJ: Princeton University Press.

Williams, G. (1994) Inaugural lecture, Royal College, in G. Norman, 'Work fuel debate on meaning of "sculpture" ', in *The Independent*, 1 May.

Williams, R. (1971) *The Long Revolution*, Harmondsworth: Pelican.

—— (1976) *Keywords*, London: Fontana.

—— (1979) *Television: Technology and Cultural Form*, Glasgow: Fontana.

Willis, P. (1990a) *Moving Culture*, London: Calouste Gulbenkian Foundation.

—— (1990b) *Common Culture*, Milton Keynes: Open University Press.

Wilson, S. M., Shulman, L. and Richert, A. (1987) '150 ways of knowing: representation of knowledge in teaching', in J. Calderhead (ed.) *Exploring Teacher Thinking*, London: Cassell.

Witkin, R. (1974) *The Intelligence of Feeling*, London: Heinemann.

Wolff, J. (1990) *Feminine Sentences: Essays on Women and Culture*, Cambridge: Polity Press. Reprint 1995 in association with London: Blackwell.

Wolheim, R. (1983) *Art and its Objects*, Cambridge: Cambridge University Press.

—— (1987) *Painting as an Art*, London: Thames and Hudson.

Woods, P. (1996) *Contemporary Issues in Teaching & Learning*, London: Routledge.

Xanthoudaki, M., Tickle, L. and Sekules, V. (eds) (2003) *Researching Visual Arts Education in Museums and Galleries: An International Reader*, Dordrecht; London: Kluwer Academic Publishers.

Yeomans, R. (1993) 'Islam: the abstract expressionism of spiritual values', in D. Starkings (ed.) *Religion and the Arts*, Sevenoaks: Hodder and Stoughton.

Youdell, D. (2004) 'Identity traps or how black students fail', in G. Ladson-Billings and D. Gillborn (eds) *Reader in Multicultural Education*, London: RoutledgeFalmer.

Young, M., Young, D. and Yew, T.H. (2004) *Introduction to Japanese Architecture (Periplus Asian Architecture)*, London: Periplus UK Ltd.

Youth Matters (2005) Online. Available HTTP: http://www.dfes.gov.uk/publications/youth/ (accessed 30 October 2006).

Zisek, S. (1991) *Looking Awry: An introduction to Jacques Lacan through Popular Culture*, Cambridge, MA and London: MIT Press.

Name and Author Index

Subject Index